Sustainable Fibres for Fashion and Textile Manufacturing

The Textile Institute Book Series

Incorporated by Royal Charter in 1925, The Textile Institute was established as the professional body for the textile industry to provide support to businesses, practitioners, and academics involved with textiles and to provide routes to professional qualifications through which Institute Members can demonstrate their professional competence. The Institute's aim is to encourage learning, recognize achievement, reward excellence, and disseminate information about the textiles, clothing and footwear industries and the associated science, design and technology; it has a global reach with individual and corporate members in over 80 countries.

The Textile Institute Book Series supersedes the former "Woodhead Publishing Series in Textiles" and represents a collaboration between The Textile Institute and Elsevier aimed at ensuring that Institute Members and the textile industry continue to have access to high caliber titles on textile science and technology.

Books published in The Textile Institute Book Series are offered on the Elsevier website at: store.elsevier.com and are available to Textile Institute Members at a substantial discount. Textile Institute books still in print are also available directly from the Institute's website at: www.textileinstitute.org

To place an order, or if you are interested in writing a book for this series, please contact Matthew Deans, Senior Publisher: m.deans@elsevier.com

Recently Published and Upcoming Titles in the Textile Institute Book Series:

Handbook of Natural Fibres: Volume 1: Types, Properties and Factors Affecting Breeding and Cultivation, 2nd Edition, Ryszard Kozlowski Maria Mackiewicz-Talarczyk, 978-0-12-818398-4
Handbook of Natural Fibres: Volume 2: Processing and Applications, 2nd Edition, Ryszard Kozlowski Maria Mackiewicz-Talarczyk, 978-0-12-818782-1
Advances in Textile Biotechnology, Artur Cavaco-Paulo, 978-0-08-102632-8
Woven Textiles: Principles, Technologies and Applications, 2nd Edition, Kim Gandhi, 978-0-08-102497-3
Auxetic Textiles, Hong Hu, 978-0-08-102211-5
Carbon Nanotube Fibres and Yarns: Production, Properties and Applications in Smart Textiles, Menghe Miao, 978-0-08-102722-6
Sustainable Technologies for Fashion and Textiles, Rajkishore Nayak, 978-0-08-102867-4
Structure and Mechanics of Textile Fibre Assemblies, Peter Schwartz, 978-0-08-102619-9
Silk: Materials, Processes, and Applications, Narendra Reddy, 978-0-12-818495-0
Anthropometry, Apparel Sizing and Design, 2nd Edition, Norsaadah Zakaria, 978-0-08-102604-5
Engineering Textiles: Integrating the Design and Manufacture of Textile Products, 2nd Edition, Yehia Elmogahzy, 978-0-08-102488-1
New Trends in Natural Dyes for Textiles, Padma Vankar Dhara Shukla, 978-0-08-102686-1
Smart Textile Coatings and Laminates, 2nd Edition, William C. Smith, 978-0-08-102428-7
Advanced Textiles for Wound Care, 2nd Edition, S. Rajendran, 978-0-08-102192-7
Manikins for Textile Evaluation, Rajkishore Nayak Rajiv Padhye, 978-0-08-100909-3
Automation in Garment Manufacturing, Rajkishore Nayak and Rajiv Padhye, 978-0-08-101211-6
Sustainable Fibres and Textiles, Subramanian Senthilkannan Muthu, 978-0-08-102041-8
Sustainability in Denim, Subramanian Senthilkannan Muthu, 978-0-08-102043-2
Circular Economy in Textiles and Apparel, Subramanian Senthilkannan Muthu, 978-0-08-102630-4
Nanofinishing of Textile Materials, Majid Montazer Tina Harifi, 978-0-08-101214-7
Nanotechnology in Textiles, Rajesh Mishra Jiri Militky, 978-0-08-102609-0
Inorganic and Composite Fibers, Boris Mahltig Yordan Kyosev, 978-0-08-102228-3
Smart Textiles for In Situ Monitoring of Composites, Vladan Koncar, 978-0-08-102308-2
Handbook of Properties of Textile and Technical Fibres, 2nd Edition, A. R. Bunsell, 978-0-08-101272-7
Silk, 2nd Edition, K. Murugesh Babu, 978-0-08-102540-6

The Textile Institute Book Series

Sustainable Fibres for Fashion and Textile Manufacturing

Edited by

Rajkishore Nayak
School of Communication and Design (Fashion Enterprise),
RMIT University Vietnam,
Ho Chi Minh, Vietnam

Woodhead Publishing is an imprint of Elsevier
50 Hampshire Street, 5th Floor, Cambridge, MA 02139, United States
The Boulevard, Langford Lane, Kidlington, OX5 1GB, United Kingdom

Copyright © 2023 Elsevier Inc. All rights reserved.

No part of this publication may be reproduced or transmitted in any form or by any means, electronic or mechanical, including photocopying, recording, or any information storage and retrieval system, without permission in writing from the publisher. Details on how to seek permission, further information about the Publisher's permissions policies and our arrangements with organizations such as the Copyright Clearance Center and the Copyright Licensing Agency, can be found at our website: www.elsevier.com/permissions.

This book and the individual contributions contained in it are protected under copyright by the Publisher (other than as may be noted herein).

Notices
Knowledge and best practice in this field are constantly changing. As new research and experience broaden our understanding, changes in research methods, professional practices, or medical treatment may become necessary.

Practitioners and researchers must always rely on their own experience and knowledge in evaluating and using any information, methods, compounds, or experiments described herein. In using such information or methods they should be mindful of their own safety and the safety of others, including parties for whom they have a professional responsibility.

To the fullest extent of the law, neither the Publisher nor the authors, contributors, or editors, assume any liability for any injury and/or damage to persons or property as a matter of products liability, negligence or otherwise, or from any use or operation of any methods, products, instructions, or ideas contained in the material herein.

ISBN: 978-0-12-824052-6

For information on all Woodhead Publishing publications visit our website at https://www.elsevier.com/books-and-journals

Publisher: Matthew Deans
Acquisitions Editor: Sophie Harrison
Editorial Project Manager: Sara Greco
Production Project Manager: Anitha Sivaraj
Cover Designer: Christian J. Bilbow

Typeset by TNQ Technologies

Contents

Contributors	**xi**
Part One Introduction to sustainable fibres	**1**

1   Traditional fibres for fashion and textiles: Associated problems and future sustainable fibres   **3**
Rajkishore Nayak, Lalit Jajpura and Asimanda Khandual

1.1	Introduction	**3**
1.2	Textile fibres	**4**
1.3	Environmental impacts of textile fibre production	**9**
1.4	Future directions	**16**
1.5	Conclusions	**21**
	References	**22**

2   Introduction to sustainable fibres for fashion and textiles   **27**
Lebo Maduna and Asis Patnaik

2.1	Introduction	**27**
2.2	Textile fibres-environmental impacts and sustainability	**28**
2.3	Consumer behaviour and sustainability	**42**
2.4	Sustainable designing	**43**
2.5	Summary and future directions	**44**
	Acknowledgements	**45**
	References	**45**

Part Two Sustainable natural fibres	**49**

3   Organic cotton and BCI-certified cotton fibres   **51**
Ashvani Goyal and Mayank Parashar

3.1	Introduction	**51**
3.2	Cotton fibre	**52**
3.3	Organic cotton	**54**
3.4	BCI (Better Cotton Initiative)	**66**
3.5	Bt cotton	**68**
3.6	Application of organic cotton	**69**
3.7	Way ahead	**70**
3.8	Conclusions	**70**
	References	**71**

4 Hemp, flax and other plant fibres   75

Ryszard Kozlowski and Malgorzata Muzyczek

4.1 Introduction—natural fibres, yarns, fabrics and knitting for fashion   75
4.2 The sustainability aspects of natural fabrics and knitting from flax, hemp, ramie, curaua, bamboo, pineapple fibres. Example of apparels   78
4.3 Recycling of natural textiles as a sustainable solution   89
4.4 Future trends and further information and advice   90
4.5 Conclusions   91
References   91

5 Lotus fibre drawing and characterization   95

Ritu Pandey, Amarish Dubey and Mukesh Kumar Sinha

5.1 Introduction   95
5.2 Lotus cultivation   96
5.3 Lotus fibre drawing   96
5.4 Fibre physical properties   99
5.5 Chemical analysis of lotus fibre   101
5.6 Comparison of lotus fibre with cotton fibre   102
5.7 Application of lotus fibre for commercial product   103
5.8 Lotus inspired design culture   105
5.9 Conclusion   106
References   106

6 Macrophyte and wetland plant fibres   109

Ritu Pandey, Mukesh Kumar Sinha and Amarish Dubey

6.1 Introduction   109
6.2 Classification of macrophyte and wetland plants   110
6.3 Fibre morphology   121
6.4 Physicomechanical properties   121
6.5 Chemical composition   121
6.6 Application of macrophytes in effluent treatment   123
6.7 Conclusion   124
References   124

7 Mushroom and corn fibre—the green alternatives to unsustainable raw materials   129

Yamini Jhanji

7.1 Detrimental impact of textile and fashion supply on environment   129
7.2 Eco leather/environmentally preferred leather   133
7.3 Mycelium and mushroom leather   134
7.4 Introduction to corn fibre   148
7.5 Conclusions   156
References   157

Contents   vii

8   Wool and silk fibres from sustainable standpoint   **159**
Vinod Kadam and N. Shanmugam
   8.1   Introduction   **159**
   8.2   Wool   **160**
   8.3   Silk   **168**
   8.4   Concluding remarks   **174**
       References   **176**

9   Sustainable protein fibres   **181**
Asim Kumar Roy Choudhury
   9.1   Introduction   **181**
   9.2   Animal protein fibres   **182**
   9.3   Vegetable protein fibres   **203**
   9.4   Green composites   **219**
   9.5   Conclusion   **221**
       References   **222**
       Further reading   **226**

Part Three   Sustainable synthetic fibres   **227**

10   Regenerated synthetic fibres: bamboo and lyocell   **229**
C. Prakash and S. Kubera Sampath Kumar
   10.1   Bamboo fibre   **229**
   10.2   Research on bamboo fibre   **232**
   10.3   Lyocell   **234**
   10.4   Conclusions   **243**
       References   **243**

11   Sustainable polyester and caprolactam fibres   **247**
Sanat Kumar Sahoo and Ashwini Kumar Dash
   11.1   Introduction   **247**
   11.2   Polyester fibre   **248**
   11.3   Caprolactam or nylon fibre   **257**
   11.4   Conclusions   **265**
   11.5   Sources for further information   **265**
       References   **266**

Part Four   Fibres derived from waste   **271**

12   Orange fibre   **273**
Subhankar Maity, Pranjul Vajpeyee, Pintu Pandit and Kunal Singha
   12.1   Introduction   **273**
   12.2   The orange fruit   **274**
   12.3   Orange peel waste as a textile raw material   **274**

12.4	Structure and chemical composition of the orange peel	**276**
12.5	Fibre extraction method	**277**
12.6	Preparation of film from orange peel extracts	**278**
12.7	Fibre morphology and properties	**279**
12.8	Chemical composition of orange fibre	**279**
12.9	Burning behaviour of orange fibre	**280**
12.10	Solubility behaviour of orange fibre	**280**
12.11	Moisture absorbency behaviour of orange fibre	**280**
12.12	FTIR spectroscopy	**281**
12.13	Thermal characterization of orange fibre	**282**
12.14	Anti-microbial efficacy of orange fibre	**282**
12.15	Benefits of textiles made of orange peel extracts	**283**
	References	**284**

13   Coffee fibres from coffee waste   **287**

Ajit Kumar Pattanayak

13.1	Introduction	**287**
13.2	Coffee botanicas	**288**
13.3	Recycled PET (rPET)	**291**
13.4	Coffee fibres	**293**
13.5	Sustainable products from coffee waste	**297**
13.6	Sustainability of coffee fabric manufacturing	**299**
13.7	Conclusions and futuristic trends	**303**
	References	**303**

14   Recycled fibres from polyester and nylon waste   **309**

Sanat Kumar Sahoo and Ashwini Kumar Dash

14.1	Introduction	**309**
14.2	Textile recycling	**310**
14.3	Polyester	**310**
14.4	Nylon	**321**
14.5	Conclusion	**329**
14.6	Sources of further information	**329**
	References	**330**

15   Composites derived from biodegradable Textile wastes:
A pathway to the future   **333**

Saniyat Islam

15.1	Introduction	**333**
15.2	Textile waste	**333**
15.3	Material thinking	**335**
15.4	Designing out waste with a material circularity approach	**335**
15.5	What are textile composites?	**336**
15.6	What are biocomposites?	**336**
15.7	Aspects of biodegradability of natural cellulose-based fibres	**337**

	15.8	Natural fibres as reinforcement for composites materials	**342**
	15.9	Opportunities and challenges around natural fibres reinforced polymers	**342**
	15.10	Design innovations	**343**
	15.11	Streamlining waste	**344**
	15.12	Way forward	**345**
	15.13	Conclusion	**346**
		References	**346**

Part Five   Organizations, standards and challenges   **353**

16   Organizations and certifications relating to sustainable fibres   355
Kunal Singha, Subhankar Maity and Pintu Pandit

	16.1	Introduction	**355**
	16.2	Key sustainability organizations and certifications	**355**
	16.3	Fair labour schemes and initiatives	**371**
	16.4	Examples of sustainable textile fibres and fabric materials	**378**
	16.5	Conclusion	**382**
		References	**383**

17   Challenges and future directions in sustainable textile materials   385
Lebo Maduna and Asis Patnaik

	17.1	Introduction	**385**
	17.2	Clothing production	**386**
	17.3	Consumer behaviour	**386**
	17.4	Sustainability approach	**388**
	17.5	Recycling	**391**
	17.6	Second-hand clothing	**391**
	17.7	Fibres	**392**
	17.8	Dyeing	**393**
	17.9	Recycling methods	**394**
	17.10	Ecolabel	**397**
	17.11	Future direction	**397**
	17.12	Conclusion	**398**
		Acknowledgements	**399**
		References	**399**

18   Life cycle analysis of textiles and associated carbon emissions   403
Yamini Jhanji

	18.1	Introduction to life cycle assessment (LCA)	**403**
	18.2	Environmental impact, carbon emissions & the ardent need of LCA	**405**

18.3	Carbon footprint, classification & related parameters	**413**
18.4	LCA framework methodology	**419**
18.5	Conclusion	**428**
	References	**429**

Index **433**

Contributors

Ashwini Kumar Dash   Department of Textile Engineering, Odisha University of Technology and Research, Bhubaneswar, Odisha, India

Amarish Dubey   Rajiv Gandhi Institute of Petroleum Technology, Jais, Uttar Pradesh, India

Ashvani Goyal   Department of Textile Technology, The Technological Institute of Textile & Sciences, Bhiwani, Haryana, India

Saniyat Islam   School of Fashion and Textiles, RMIT University, Melbourne, VIC, Australia

Lalit Jajpura   Department of Textile Technology, Dr BR Ambedkar National Institute of Technology, Jalandhar, Punjab, India

Yamini Jhanji   Fashion and Apparel Engineering Department, The Technological Institute of Textile and Sciences, Bhiwani, Haryana, India

Vinod Kadam   Textile Manufacturing and Textile Chemistry Division, ICAR-Central Sheep and Wool Research Institute, Avikanagar, Rajasthan, India

Asimanda Khandual   Department of Textile Engineering, Odisha University of Technology and Research, Bhubaneswar, Odisha, India

Ryszard Kozlowski   FAO/ESCORENA European Cooperative Research Network on Flax and other Bast Plants, Institute of Natural Fibres, Poznan, Poland

S. Kubera Sampath Kumar   Department of Chemical Engineering (Textile Technology), Vignan's Foundation for Science, Technology & Research, Vadlamudi, Guntur, Andhra Pradesh, India

Lebo Maduna   Technology Station: Clothing and Textiles, Faculty of Engineering and the Built Environment, Cape Peninsula University of Technology, Cape Town, Western Cape, South Africa

Subhankar Maity   Department of Textile Technology, Uttar Pradesh Textile Technology Institute, Kanpur, Uttar Pradesh, India

Malgorzata Muzyczek   Institute of Natural Fibres and Medicinal Plants, FAO/ESCORENA European Cooperative Research Network on Flax and other Bast Plants, Poznan, Poland

Rajkishore Nayak School of Communication and Design (Fashion Enterprise), RMIT University Vietnam, Ho Chi Minh, Vietnam

Ritu Pandey Department of Textiles & Clothing, Chandra Shekhar Azad University of Agriculture & Technology, Kanpur, Uttar Pradesh, India

Pintu Pandit Department of Textile Design, National Institute of Fashion Technology, Patna, Bihar, India

Mayank Parashar Department of Textile Technology, The Technological Institute of Textile & Sciences, Bhiwani, Haryana, India

Asis Patnaik Department of Clothing and Textile Technology, Faculty of Engineering and the Built Environment, Cape Peninsula University of Technology, Cape Town, Western Cape, South Africa

Ajit Kumar Pattanayak The Technological Institute of Textile & Sciences, Bhiwani, Haryana, India

C. Prakash Department of Handloom and Textile Technology, Indian Institute of Handloom Technology, Fulia Colony, Shantipur, Nadia, West Bengal

Asim Kumar Roy Choudhury KPS Institute of Polytechnic, Belmuri, West Bengal, India; Govt. College of Engineering and Textile Technology, Hooghly, West Bengal, India

Sanat Kumar Sahoo Department of Textile Engineering, Odisha University of Technology and Research, Bhubaneswar, Odisha, India

N. Shanmugam Textile Manufacturing and Textile Chemistry Division, ICAR-Central Sheep and Wool Research Institute, Avikanagar, Rajasthan, India

Kunal Singha Department of Textile Design, National Institute of Fashion Technology, Patna, Bihar, India

Mukesh Kumar Sinha Technical Textile Division, Ministry of Textiles, New Delhi, New Delhi, India

Pranjul Vajpeyee Department of Textile Technology, Uttar Pradesh Textile Technology Institute, Kanpur, Uttar Pradesh, India

Part One

Introduction to sustainable fibres

Traditional fibres for fashion and textiles: Associated problems and future sustainable fibres

1

Rajkishore Nayak[1], Lalit Jajpura[2] and Asimanda Khandual[3]
[1]School of Communication and Design (Fashion Enterprise), RMIT University Vietnam, Ho Chi Minh, Vietnam; [2]Department of Textile Technology, Dr BR Ambedkar National Institute of Technology, Jalandhar, Punjab, India; [3]Department of Textile Engineering, Odisha University of Technology and Research, Bhubaneswar, Odisha, India

1.1 Introduction

Textile fibres are the building blocks for manufacturing of fashion and textiles. Natural textile fibres such as cotton, flax, wool and jute have been used to make clothes for the last 5000 years (Kozłowski and Mackiewicz-Talarczyk, 2012). The other widely used natural fibres are animal hair fibres (such as camel, alpaca, angora and mohair), hemp, sisal and silk (Nayak et al., 2012). Synthetic fibres started wider acceptance in the fashion and textiles after the invention of rayon in 1930s. Although nylon and polyester were invented almost at the same time at the DuPont lab (in the late 1930s) (Britannica, 2019), the scientist W. H. Caruthers set his experimental work on nylon first. Later a group of British scientists worked on W. H. Caruthers' incomplete work to produce the first commercial polyester fibre called Terylene, in 1941.

Natural textile fibres are grown almost all over the world ranging from Northern to Southern Arctic Circle and are well known for recycling carbon dioxide (CO_2) (Kozłowski and Mackiewicz-Talarczyk, 2012). With the increased importance of the use of sustainable fibres, the natural fibres are gaining increased attention from the textile manufacturers as well as agricultural fibre growers (Duarte et al., 2019). The measure attentions have been to resolve: (i) environmental issues such as water and land pollution, (ii) excessive chemical and freshwater usage, (iii) intensive use of toxic chemicals, and biodiversity loss; (iv) excessive energy usage and (v) social issues such as unhealthy working conditions, use of forced and child labour, and animal cruelty (Duarte et al., 2019; Nayak, 2020).

The first step for fashion and textile manufacturing is the selection of raw material (i.e. fibre), which is finalized in the conceptualization stage by designers and product developers (Nayak and Padhye, 2015b). The conceptualization stage plays a vital role in the environmental impact of garment production (Nayak and Padhye, 2017; Nayak, 2019). During the conceptualization stage, raw materials (i.e., type of fibre) and trims are finalized in addition to the style and design. The fibre selection process greatly influences the environmental impact of fashion and textile manufacturing (Nayak and

---

Sustainable Fibres for Fashion and Textile Manufacturing. https://doi.org/10.1016/B978-0-12-824052-6.00013-5
Copyright © 2023 Elsevier Ltd. All rights reserved.

Patnaik, 2021). This is because the process routes to manufacture a specific garment are selected based on the type of raw materials used in the garment. Hence, in order to avoid the most environmentally damaging processes especially in chemical processing, raw materials should be carefully selected (Nayak and Padhye, 2015a).

In this chapter, the environmental impacts of manufacturing various textile fibres have been discussed. In the beginning, a brief introduction has been given to the traditional textile fibres based on their source. The information on the recent trends in the textile fibre usage is being highlighted. The major textile fibres such as cotton, wool, silk, polyester, nylon and rayon had been discussed in terms of various environmental, social and economic impacts. The chapter has been prepared by collecting the existing data from a variety of resources, mainly secondary data. The challenges relating to sustainability, which covers air, water and land pollution are covered. The impact of fibre manufacturing on human health and animal cruelty is also covered. The use of new fibre types such as polylactic acid (PLA), regenerated protein fibres (soy fibre and milk fibre); and other new fibres are also discussed, which is the major discussion point in this book. Furthermore, the recycling concepts to deal with the hard polyester and nylon waste are also discussed in brief. The end-of-life (EOL) clothes and the hard wastes of various fibres are also highlighted.

1.2 Textile fibres

1.2.1 Fibre types

Textile fibres can be broadly divided into two categories such as: (1) natural and (2) synthetic fibres (Eichhorn et al., 2009). Natural fibres are derived mainly from plants and animals; whereas the synthetic fibres are either manufactured from petroleum resources (e.g., polyester and nylon) or natural resources (e.g., viscose and lyocell). Among these, the use of natural fibres such as cotton, flax, sisal, hemp, jute and pineapple has certain advantages over synthetic fibres which include renewability, biodegradability, better specific properties, non-abrasiveness and natural availability (Nayak et al., 2019; Thomas et al., 2011). Furthermore, the use of plant-based natural fibres such as flax and hemp is more ecofriendly than the synthetic fibres as the plants absorb carbon dioxide (CO_2) from the air and release oxygen (O_2). A detailed list of natural and synthetic textile fibres has been given in Table 1.1.

Natural fibres such as cotton need a lot of resources (e.g., fresh water) and toxic chemicals for harvesting. However, the energy consumption to produce natural fibres is lower than the synthetic counterparts. On the other hand, synthetic fibres consume significantly lower (about 1/100) amount of water and free from the use of toxic chemicals compared to cotton. For example, acrylic fibre needs only 0.3—15 L/kg of water (Cupit, 1996), and polyester fibre needs 17.2 L/kg of water (Table 1.2). However, the production of synthetic fibres consumes a substantial higher amount of energy. For example, the polymerization, spinning and finishing of synthetic fibres such as polyester consume 369—432 MJ/kg of fibre production, whereas the production of cotton requires much lower energy, in the range of 38—46 MJ/kg of cotton (Cupit, 1996).

Table 1.1 List of natural and synthetic textile fibres.

Natural fibres			Synthetic fibres		
From plants	**From animals**	**Inorganic or mineral**	**From natural polymer (vegetable and animal)**	**From synthetic polymer**	**Inorganic or mineral**
Seed fibres • Cotton • Coir • Kapok • Soy • Poplar **Leaf fibres** • Pineapple • Sisal • Abaca • Curaua • Palm • Jukka **Bast fibres** • Flax • Hemp • Jute • Ramie • Kenaf	**Fruit fibres** • Coir • Luffa **Grass fibres** • Bagasse • Bamboo • Canary • Corn • Sabai **Wood fibres** • Hard wood • Soft wood **Stalk fibres** • Barley • Maize • Oat • Rice • Wheat	**Wool and animal hair** • Sheep • Alpaca • Angora • Camel • Cashmere • Rabbit • Yak **Silk fibres** • Bombyx • Eri • Muga • Mulberry • Tussar • Spider silk	• Asbestos • Basalt • Carbon • Glass • Mineral wool	**Vegetable (Cellulosic)** • Acetate • Bamboo • Cuprammonium • Lyocell • Modal • Viscose **Regenerated protein** • Casein • Soybean **Elastomeric** • Rubber **Biodegradable polyester** • Polylactic acid (PLA)	• Acrylic • Aramid • Chloro fibre • Modacrylic • Ployamide • Polyester • Polyethylene • Polypropylene	• Carbon fibres • Glass fibres • Boron fibres • Ceramic fibre

Table 1.2 Environmental factors involved in the production of cotton and polyester fibres.

Parameters	Cotton	Polyester
Energy consumption	38—46 MJ/kg	369—432 MJ/kg
Water	20,000 L	17.2 L
CO_2 emission	3.0 kg	2.3 kg
Oil or gas	—	1.5 kg
Fertilisers	457 g	—
Pesticides	16 g	—
Cost (approx.) (as of 2017)	US $ 1—1.6/kg	US $ 0.85—1.0/kg
Organic cotton	US $ 1.2—2.3/kg	—

Hence, the use of traditional synthetic fibres does not guarantee a sustainable approach to the garment production. Some of the environmental factors involved in the production of cotton (natural fibre) and polyester (synthetic fibre) have been shown in Table 1.2.

The natural fibres are produced from renewable resources unlike the synthetic fibres produced from limited petroleum resources, which leads to the depletion of resources. Furthermore, the use of natural fibres avoids the problem of a large amount of waste generation at the EOL, as they are biodegradable (Muthu et al., 2012). For example, cellulose is the major component of plant-based fibres, which can be converted to glucose by enzymatic hydrolysis of microorganisms at the EOL (Muthu, 2014). Due to the above reasons, some natural fibres are comparatively more sustainable than synthetic fibres due to less harmful impacts on the environment. However, the production of natural fibres is not completely sustainable. For example, the use of various toxic chemicals, large amount of fresh water and energy for turning cellulose of plants into fibres; use of pesticides and hormones in wool farming, killing of silkworms and other animals, are not favourable for sustainable garment production.

The use of synthetic fibres is the least choice for designing sustainable garments when the environmental impact is considered. However, some of the regenerated fibres such as Lyocell and Tencel are considered to be more ecofriendly than synthetic fibres such as polyester, nylon and acrylic, due to their biodegradability and renewable source of raw material (Muthu, 2014). In the last 2 decades, the technological developments in polymer science have resulted in the commercial production of fibres such as soybean protein, bamboo, Lenpur, SeaCell and polylactic acid (PLA). These fibres are prepared from renewable sources and they are biodegrade after disposal at the EOL (Mohanty et al.,2000). Hence, these fibres provide a solution to the recycling problems involved with synthetic fibres (Gross and Kalra, 2002).

Natural fibres such as cotton and flax provide excellent comfort properties (relating to moisture absorbency, moisture management properties and breathability) (Nayak et al., 2009; Bliss, 2005), which is the major reason for cotton and flax being exclusively selected in the apparel sector. However, these fibres are prone to wrinkle, take long time to dry and low in durability. On the other hand, synthetic fibres are such as polyester and nylon are stronger, dry readily and provide good wrinkle recovery properties. Therefore, the blends of natural and synthetic fibres are prepared to derive the benefits of each fibre and compensate the drawbacks (Nayak and Padhye, 2015c).

1.2.2 Fibre usage

The global textile fibre production has been substantially increased (almost five times) over the last 5 decades (Fig. 1.1). For example, in 1975, the total textile fibre production was approximately 30 million metric tonnes, which surpassed 108 million metric tonnes in 2020. The synthetic fibres increased from 14 million metric tonnes to 81 million metric tonnes in the same period. However, the natural fibres such as cotton or wool had a production volume of 27.4 million metric tonnes in 2020 from 15 million metric tonnes, which is approximately double compared to 1975. Chemical fibres include synthetic fibres such as polyesters or polyamides, and manmade cellulosic fibres like viscose or rayon.

Fig. 1.2 shows the global textile fibre production per capita (in kgs) from 1980s to 2020. It can be observed that in 1980, the global per capita textile fibres production was totalled to be about 8.4 kgs; which reached 14 kgs in 2020. The per capita fibre production is expected to reach approximately 15.2 and 17.1 kgs, in 2025 and 2030, respectively.

The total global textile fibre consumption reached 108 million metric tonnes in 2020, in spite of COVID-19 pandemic, which is projected to reach 127.3 million metirc tonnes by 2027 with a compound annual growth rate (CAGR) of 2.2%. The CAGR rate of the manmade or synthetic fibres will be 2.5%, which will reach 84.3 million metric tonnes, whereas cotton will grow at a CAGR of 0.7% in the mentioned period from 2022 to 2027. The regenerated fibre market sector will reach 6.5 million metirc tonnes in 2027 from the market size of five million metirc tonnes in 2020.

In the global fibre market, China will be the fastest-growing country, which will reach at 25.1 million metirc tonnes by the year 2027 with a CAGR of 4.2%

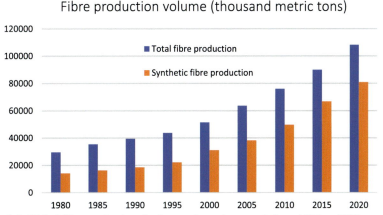

Figure 1.1 Global fibre production (in thousand metric tonnes) from 1980 to 2020.

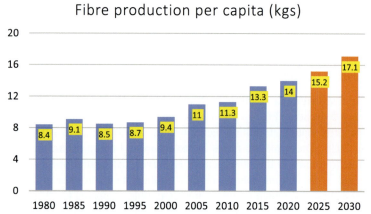
Figure 1.2 Global textile fibre production per capita (in kgs) from 1980s to 2020, forecasted values for 2025 and 2030.

(PRNNews, 2021). Other leading Asia—Pacific countries such as India, Australia, and South Korea, will reach at a total volume of 17.2 million metirc tonnes by 2027. Other countries such as Japan, Germany, and Canada, are projected to grow at a CAGR of 0.4%, 0.9% and 1.6%, respectively in fibre manufacturing.

Fig. 1.3 shows the global share of various types of textile fibres in fashion and textile sector during 2020 (Statista, 2021). It can be observed that the predominant fibre was polyester (52%) followed by cotton (24.2%) in the global fibre market. Regenerated cellulosic fibres and other plant fibres each contributed 5.9%. Polyamide and other synthetic fibres contributed 5% and 5.4%, respectively. The contribution of animal fibres was a mere 1.6%.

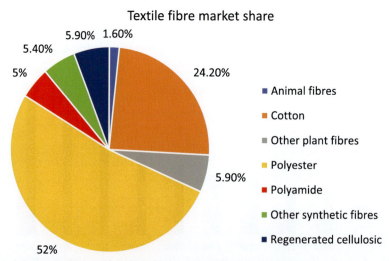
Figure 1.3 Textile fibre market share in the year 2020 (Statista, 2021).

Traditional fibres for fashion and textiles: Associated problems and future sustainable fibres 9

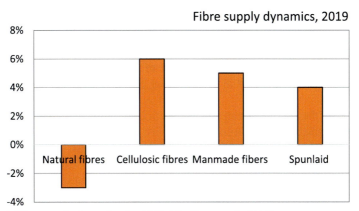

Figure 1.4 Fibre supply dynamics for 2019 (Textilenetwork, 2020).

The fibre supply dynamics for 2019 has been sown in Fig. 1.4 (Textilenetwork, 2020). It can be observed that the natural fibre supply was decreased in 2019. This is due to the cultivation of natural fibres was reduced by 3% to 32 million metirc tonnes, with cotton fibre reduced to 26 million metirc tonnes, wool reduced by 1%, and bast fibres increased by 1%, while other natural fibres were relatively stable. However, the overall cellulosic fibres (including regenerated fibres), manmade fibres and spunlaid fibres showed an increasing trend.

A brief idea of various textile fibres and their trends relating to usage have been given in the above section. The following section describes various fibres and their impacts on the planet earth during fibre manufacturing process.

1.3 Environmental impacts of textile fibre production

1.3.1 Cotton

Cotton is the most widely used natural fibre derived from the Gossypium plant. The clothing made from cotton fibre is preferred by many consumers due to its superior properties such as moisture absorbency, breathability and softness (Ravandi and Valizadeh, 2011). Cotton is a seed-based, single-celled, cellulosic fibre derived from renewable resources and it is biodegradable. It can be used for a range of clothing types for men, women and children, especially for hot climatic conditions (Khanzada et al.,2020). Cotton fibre is often blended with the synthetic fibre, such as polyester to improve some of its properties such as strength, durability, creasing properties and ease of drying.

In spite of several advantages, the cotton fibre production and fabric manufacturing has created environmental and social impacts. For example, the cultivation of cotton fibre requires a substantial amount of freshwater, which can be in the range of 8000–22,000 L/kg of cotton fibre production (Roy and Kumar, 2013; McKinion et al., 2001). As the cotton plats are highly vulnerable to varieties of pests and insects, a substantial number of pesticides and insecticides are applied to the plants during the fibre cultivation. Although cotton is cultivated in about 3% of the world's farmland, approximately 11% of the pesticides and 26% of the insecticides of world's total consumption have has been used in conventional cotton production (Roy and Kumar, 2013).

In addition, the seeds are treated with insecticides, herbicides are applied to control weed growth, and synthetic fertilizers are applied to the soil (Li et al., 2017). The synthetic fertilizer usage can be approximately of about 33% of the weight of cotton fibre production. Many synthetic chemicals contribute to acute environmental toxicity, which can deteriorate soil quality, make it infertile and potentially drift into neighbouring ecosystem (Fletcher, 2013). Some of these chemicals can lead to acute health problems such as cancer, birth defects, tumours and mutations in humans (Mancini et al., 2005). The people working in the farming and harvesting of cotton fibre may face these problems due to the inadequate supply of the safety devices. Further, these chemicals also cause harm to the land and aquatic animals.

Before the cotton fibre is harvested, the leaves are removed by application of chemical defoliation (Wang et al., 2019). The defoliant chemicals are applied when the balls are almost open about 60% or higher. Similarly, the use of chemical desiccant is also in practice to make the plants dry and ready for harvesting. During the picking of fibres, the use of automatic spindle pickers leads to the compactness of soil and removes air from it. This influences the soil quality for the future crop plantation. Intensive farming of cotton fibre can lead to reduced soil fertility and soil salinization (Dai and Dong, 2014). Due to the application of excessive fertilizers and chemicals, the natural soil fertility reduces substantially. Excessive irrigation can lead to the salinization of the soil due to the presence of some salts in the water. In some instances, the soil is no more usable for 2−3 years due to fertilizers and salinization. For example, as much as 50% of the farming land in Uzbekistan are not usable due to cotton farming in these lands (Shaumarov et al., 2012).

The chemical processing of fabrics made of cotton fibre also causes water pollution. For example, scouring and mercerization of cotton fibre use large amount of caustic soda or sodium hydroxide (NaOH) solution (Madhav et al., 2018). Once, these processes are finished, the unused water containing chemicals are discharged into the water. The durable press finish applied to the cotton fabrics using formaldehyde creates acute toxicity when released into the water courses (Dehabadi et al., 2013). As formaldehyde is carcinogenic, it creates health hazards to the workers.

In spite of the above-mentioned problems associated with the cultivation of the traditional cotton fibre, still it is the most widely used natural fibres. In some countries, there have been reports of cotton growers suiciding due to inadequate demand for their cotton fibre (Rao, 2004). The use of child labour and forced labour has been reported in some countries involved in cotton fibre farming (Bhat, 2013). Hence, traditional cotton fibre is at odds when sustainability is considered. Although the use of other types of cotton fibres such as organic cotton, low chemical cotton and BCI (better cotton initiative) certified cotton fibre is there, the market share of these fibres is too low in the global fashion and textile supply chain.

1.3.2 Wool

Wool is one of the most widely used animal-based natural fibres derived from sheep. Wool fibre is derived from renewable resources and it is biodegradable. Wool fibre is preferred for a range of garment types for winter and summer. The clothing made from wool fibre is preferred by the consumers due to its durability, high moisture absorbency, resiliency, resistance to fire and warmth in cold climate (Simpson and Crawshaw, 2002). Further, the EOL clothing of wool can be recycled to new products that avoid the problem of waste management ending in the landfill. In addition, apparels and textiles made from wool fibre can be used in many thermal and acoustic insulation purposes in industries (Padhye and Nayak, 2016) or in pads to soak up oil spills (Loh et al., 2021).

Although wool fibre farming does not need large quantities of fertilizers and pesticides, production of wool is not free from negative environmental challenges or animal cruelty. Shearing of wool fibres is done carelessly on some farms, which can cause injury and pain to the sheep (Ross, 2021). In some instances, the skin, ear and tail are being ripped off during the sheep shearing. Further, in order to keep sheep free from parasites, chemical pesticides and insecticides are often applied to sheep (Fletcher, 2013). These chemicals are present in the wool fibre, which is removed during the wool scouring and discharged to the water sources. The chemical pesticides used in 'Sheep dip' causes contamination of land and water. For example, the use of toxic organophosphate in the 1990s resulted in the debilitating health conditions of many sheep farmers (Peta, 2021).

Furthermore, a large amount of the farmland needs to be reserved for the sheep growth. For this, a massive amount of forest needs to be cleared by cutting the trees, which leads to deforestation, soil erosion and decreased biodiversity (Henry et al., 2015). For example, about 20% of the global pastureland has been degraded because of overgrazing, compaction, and erosion due to sheep farming (Peta, 2021). A large-scale sheep farming can result in soil depletion and loss of biodiversity. This can be a threat to the soil fertility and there is a need for fodder crops growing, which may take up to 50% of the total global farmland.

For commercial-scale fibre production, large herds of sheep are raised, which lead to the emission of greenhouse gases such as methane (CH_4) due to the natural life process of the sheep (Howden et al., 1996). As a sheep needs a large amount of vegetable matter, they generate a large amount of CH_4 gas that goes into the atmosphere. One sheep can generate about 30 L of CH_4 every day (Peta, 2021). The severity of this can be understood as in New Zealand, the sheep farming contributes 90% of the CH_4 gases generated in the country (Saggar et al., 2004). Further, the sheep manure pollutes the land, and water systems due to the run-off waste, resulting in eutrophication, leading to excessive growth of aquatic plant life (McDowell et al., 2009). This in turn may lead to the depleting level of oxygen dissolved in the water, which is harmful to the other aquatic life. After shearing, the wool fibres are cleaned from greases and other impurities by using soap and alkaline solutions, which cause water pollution. Chemicals used to prevent the felting shrinkage of wool fibres are also toxic and cause water pollution.

Considering the above facts, wool fibre can't be considered as a source of sustainable fibre. It's high time for the designers and fashion brands to switch to alternative sustainable raw materials from wool considering the negative environmental impacts, impacts to the human health and the cruelty that sheep are subjected to during their lifetime.

1.3.3 Silk

Silk is one of the natural fibres derived from silkworms. There has been a wide use of silk-made fabrics from the past. The silk clothes are well known for their shine, luster, high strength and durability among the natural fibre-based textiles (Grishanov, 2011). There has been a wide demand for specific silk fabrics such as chiffon, georgette, silk satin and organza for various clothing. Silk fibres and fabrics can be used for a range of clothing such as suits, shirts, nightgowns, evening wear, blouses, sport coats, pullovers and dresses (Franck, 2001). In spite of several developments in the spinning and weaving technologies, the silk manufacturing process to make yarns and fabrics still follows the traditional approach as it was done in the ancient world.

Although the environmental impact of silk sericulture is less, it is not absolutely free from animal cruelty. Generally, silk fibre is extracted from the cocoons by gruesome killing of the silkworm in hot boiling water (Gupta et al., 2014). There is an alternative route of allowing the silk moth to fly away from the cocoon, and then collect the silk fibre. Some activists working for silk rights have protested the killing of silkworm and suggested to extract silk fibre following the alternate route. However, some industries follow the first route to get silk in continuous form, where thousands of moths are being killed.

Majority of silk sericulture has been done in the hot climates in Asia such as India, Chain, Korea, Vietnam and Thailand (Wicker, 2020). Due to hot weather, the temperature and humidity need to be precisely controlled for a good growth of the silkworm, which consumes a lot of energy. After harvesting, the silk farms use hot air or steam to dry the cocoons, which is also energy-intensive. The energy can be obtained by burning mulberry wood, or from the electric grid.

Some silk farms apply disinfectants such as lime, formaldehyde, or chlorine on the cocoons to protect silkworm from disease (Barcelos et al., 2021). Chemical-free options such as hot-air sterilization, steaming and sun drying are also available for this. In some instances, the silk farms apply growth hormones for better production of silk. There has been the use of metallic salts for weighted silk or to make silk heavier and more lustrous. This in turn could increase the generation of toxic wastewater and intoxicate the silk fibre. The silk dyeing and printing industries also use a lot of chemicals, which pollute the water courses.

In addition to these problems, there have has been reports of the use of child labor (Wicker, 2020). The leading global producers of silk are China, India, Uzbekistan, Brazil, Iran and Thailand. Out of these countries, the use of child labour has been reported countries such as India, and Uzbekistan. However, silk cultivation provides employment to many rural people who use it as a craft to produce silk fabrics and clothing. In many countries, some traditional fabrics are manufactured by artisans from silk.

1.3.4 Polyester

Polyester is the predominant synthetic fibre derived from petroleum resources. Approximately 52% of the total global fibre consumption is covered by polyester (Fig. 1.3) (Statista, 2021). The clothing made from polyester fibre is durable, light weight, wrinkle resistant and dries easily. Due to these properties, polyester fabrics are easier to maintain compared to their cotton counterpart. However, the thermal comfort properties are not as good as cotton due to poor moisture absorbency and breathability. Hence, it is blended with cotton or viscose to improve thermal comfort properties.

Polyester fabrics are suitable for a range of applications such as suiting, shirting, dress, active wear, sportswear and industrial applications. Now the use of recycled polyester from plastic bottles is gaining significant attention due to environmental impacts (Muthu, 2019). The manufacturing of polyester fibre results in less environmental impact when the water usage and the use of toxic chemicals' impact is considered in the land, water and air. However, the energy usage is more compared to the natural fibres (approximately double compared to cotton), which may be the cause of global warming.

Polyester, technically known as polyethylene terephthalate (PET) is a thermoplastic fibre manufactured from ethylene glycol (alcohol) and terephthalic acid (acid) (Deopura et al., 2008). The resources used for polyester fibre are nonrenewable. Further, at the EOL, the clothing made from polyester fibres is non-biodegradable, which may take up to 200 years to biodegrade, hence, lead to plastic pollution. The clothing made from polyester fibres shreds microfibres into the water courses (e.g., ocean). These microfibres enter the food chain through the aquatic food and can cause harm to the human digestive system.

Polyester has some disadvantages as well as some advantages when compared with cotton while selecting as a textile raw material for manufacturing fashion and textiles. The market segment of using recycled polyester is increasing, which will reduce the usage of virgin polyester. Further, research is underway to prepare renewable polyester fibres from plant resources, which will be renewable and biodegradable.

1.3.5 Nylon

Nylon is the first synthetic fibre that was liked by women, due to the longevity of stockings compared to silk, when it was introduced in 1938. The clothing made from nylon are well known for strength, durability and elasticity. Nylon fibre (i.e., nylon 6,6) is produced by the condensation polymerization of a dicarboxylic acid and a diamine, which are obtained from petroleum resources (Deopura et al., 2008). Nylon fibres are used for producing a range of fabrics including active wear, swimwear, stockings, hosiery and industrial textiles (carpets, parachutes, and packaging materials). Like polyester, the fabrics made from nylon fibre are strong, durable, lightweight and smooth. Further, the clothing of nylon fabrics dries quickly, and resistant to wrinkle.

The manufacturing of nylon fibre is energy-intensive that can lead to global warming (three times energy consumed compared to cotton). The manufacturing process releases nitrous oxide, which is a greenhouse gas and more harmful than carbon dioxide in depleting the ozone layer. The washing of nylon clothing also generates microfibres, which can lead to the marine pollution. Similar to polyester, the microfibres enter the food chain through the aquatic food and can be harmful to the human digestive system.

1.3.6 Rayon

Rayon or viscose fibre is a synthetic or regenerated fibre made from wood pulp (of trees such as eucalyptus and pine). It can be produced with similar properties to silk and cotton, with a cheaper price (Chen, 2015). A range of garments can be produced from viscose fibres both for men and women. Although the raw material is derived from plant, the manufacturing process uses a large amount of energy, and chemicals that can lead to negative environmental impact, impacts on workers, and society.

During the manufacturing of rayon fibre the wood is cut into chips, then the wood pulp is prepared with treatment of chemicals such as carbon disulphide (CS_2) (You-xin and De-zhen, 1985). Once, the fibres are formed, the unused carbon disulphide is released into the water courses leading to water pollution. The use of carbon disulphide can lead to coronary heart disease (CHD) in workers and employees coming in direct contact. Further, it can lead to birth defects, skin allergies and cancer, in the people who live near the viscose manufacturing plant (Chang et al., 2007).

In addition to the carbon disulphide, a large amount of sodium hydroxide (NaOH) is also used during the xanthation process. Sodium hydroxide is a strong acid and corrosive, which can lead to skin burns, eye damage and ingestion. Furthermore, there have been instances of deforestation of old trees to make the raw materials for viscose fibre. The ideal approach should be growing trees, especially for viscose fibres and replacing them once they are cut.

1.3.7 Acrylic

Acrylic fibre is used to replace wool as it provides warmth like wool. The clothes made from acrylic fibres are resistant to acids and they are not attacked by insects or moths like wool. Acrylic fibres can be used to make a range of clothing for men, women and children, especially for cold climates; blankets, carpets, fleece and pile fabrics. The clothing made from acrylic is cheaper than the woolen clothing due to cheaper price of the raw materials. In many countries, a range of acrylic products is manufactured for winter clothing, jumpers, and blankets.

Acrylic fibres are derived from petroleum resources like the polyester and nylon fibre using chemical processes. Acrylic fibres are produced by dry spinning of acrylonitrile copolymers with at least 85% acrylonitrile monomer and some other chemicals (Gupta and Kothari, 1997). For dry spinning, the polymer is dissolved in dimethyl formamide (DMF), the solution is then heated and passed through the spinneret to make continuous filament. Subsequently the filaments are washed, dried and cut into staple fibres.

The manufacturing process used for acrylic fibre is energy-intensive as heat and pressure are used during the polymerization process. Manufacturing of 1 kg of acrylic fibre generates about 5.5 kg of CO_2 equivalent, which is the major greenhouse gas; and 0.013 kg of sulfur dioxide (SO_2), which can lead to acid rain. In addition, the use of arsenic (As), chromium (Cr), cadmium (Cd) and zinc (Zn) can lead to carcinogenic problems with direct exposure or released into the water and land. Acrylic fibre manufacturing also involves non-hazardous waste production such as empty cans, papers, plastics, and filter cloth (Yacout et al., 2016).

1.3.8 Summary of impacts caused by textile fibre production

From the above discussions, it can be concluded that the production of textile fibres involves various environmental and social impacts. The production of the most widely used textile fibre, cotton, involves several environmental and social impacts. The environmental impacts include land, air, water pollution, production of greenhouse gases and depletion of ozone layer. The social impacts include human health hazards, animal cruelty and use of child and forced labour. On the other hand, the production of synthetic fibres is associated with the use of a large amount of energy, release of microplastics and creations of plastic wastes at the EOL of the fibres. The environmental and social impacts of textile fibres are shown in Fig. 1.5. Various environmental impacts of

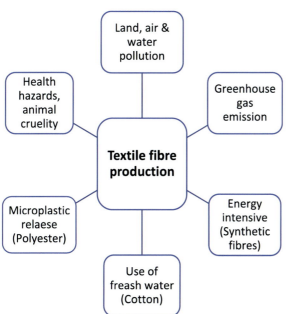

Figure 1.5 Environmental and social impacts of textile fibres.

Table 1.3 Various environmental impacts of textile fibres in decreasing order (NorthFjord, 2017).

	Energy use	Water use	Greenhouse gases	Waste water	Direct land use
Decreasing environmental impact →	Acrylic	Cotton	Nylon	Wool	Wool
	Nylon	Silk	*Synthetic*	*Regen.*	Ramie
	Polyester/PTT	Nylon	Polyester	*cellulosic*	Cotton
	Regen. cellulosic	*Regen.*	Lyocell	*Natural*	Flax
	(viscose, Modal)	*cellulosic*	PLA	*bast fibres*	Hemp
	PLA/Cotton/Lyocel	Acrylic	Viscose	Nylon	Viscose and
	Wool	Hemp	Modal	Polyester	Modal
	Natural bast fibres	Wool	Cotton		Jute
	[1](nettle,	*Natural*	*Natural*		PLA
	hemp,flax)	*bast fibres*	*bast fibres*		Lyocell
		Polyester	Wool		*(Synthetic)*

textile fibres in decreasing order have been given in Table 1.3 and the summary of environmental impacts of major textile fibres has been shown in Table 1.4.

1.4 Future directions

Various problems associated with the production of natural and synthetic fibres are addressed in the previous section. Hence, in order to eliminate the negative impacts, new fibres are paving their way into the fashion and textile world. The major polluting natural fibre cotton being replaced with other alternative fibres such as organic cotton, low chemical cotton, low water or rainfed cotton and BCI cotton (Fletcher, 2013). Other natural fibres include hemp, flax, lotus, natural polyester (PLA), mechanically extracted bamboo, banana, milk protein and soy protein fibres. Similarly, the process of extracting fibres from the wastes such as orange peels or coffee grounds are also have been established and clothes have been manufactured.

There has been tremendous growth in the use of recycled synthetic fibres, which reduces the use of virgin materials, energy and chemical consumption. The recycling of plastics such as polyester and nylon drink bottles to make different types of clothes are also increasing significantly, which can reduce the dependency on the freshly derived raw materials. The mechanical process route involves grinding, melting and extrusion to make the new fibres, whereas the chemical process route involves chemical breakdown of the polymer, which then is repolymerized to make fibres with consistent quality. Both the mechanical and chemical process routes being widely used in many countries that solve the problem of plastic pollution, where a vast amount of plastics are discarded to landfill (Horton, 2022).

The recycling of used clothing of natural fibres, synthetic fibres and their blends are also getting commercially success (Nayak et al., 2020). Fibres are extracted from the

Table 1.4 Summary of environmental impacts of major textile fibres.

Fibre name	Derived from renewable resources	Biodegradability at EOL	Environmental and social impacts	Water and resource usage	Energy usage	Health and other hazards
Cotton	Yes	Yes	Soil, water and air pollution due to: • Use of synthetic fertilisers, pesticides, herbicides and defoliants. • Use of caustic soda and other chemicals in dyeing and printing. • Use of formaldehyde in permanent press finishing. • Use of forced labour in some countries. • Use of child labour in the supply chain.	Large amount of water (8000– 22,000 L/,kg)	Low	Cancer, birth defects, tumours, and mutations in humans
Wool	Yes	Yes	Soil, water and air pollution due to: • Chemical pesticides and insecticides applied on sheep to protect from parasites. • Use of toxic organophosphate for 'Sheep dip'. • Soap and alkali used in scouring. • Generation of methane gas by sheep.	Use of large amount of farmland.	Low to medium	Sometimes the skin, ear and tail are ripped off during sheep shearing.

Continued

Table 1.4 Summary of environmental impacts of major textile fibres.—cont'd

Fibre name	Derived from renewable resources	Biodegradability at EOL	Environmental and social impacts	Water and resource usage	Energy usage	Health and other hazards
Silk	Yes	Yes	Soil, water, and air pollution due to: • Disinfectants such as lime, formaldehyde, or chlorine are used which can pollute water, air, and land. • Metallic salts used to make weighted silk, which could increase the generation of toxic wastewater. • Silk dyeing and printing industries use a lot of chemicals, which pollute the water courses. • There has been reports of the use of child labour in silk production.	Use of farmland for growing silk.	• The temperature and humidity control consumes a lot of energy. • Use of hot air or steam to dry the cocoons, is also energy intensive.	Gruesome killing of the silkworm in hot boiling water.

Polyester	No	No	Environmental pollution produced due to: • Polyester fibres are non-biodegradable and may take up to 200 years to biodegrade. • Cause the problem of plastic pollution. • Can generate greenhouse gases in the landfill. • Can generate microplastics, which can enter into the food value chain.	Consumes monomers of dialcohols, diacids.	Energy usage is more compared to cotton (approx. double) which may be the cause of global warming.	Polyester clothing shred microfibres into river and ocean, which enter food chain through the aquatic food and can harm to human digestive system.
Nylon	No	No	Environmental pollution produced due to: • Manufacturing process releases nitrous oxide, which is a greenhouse gas and more harmful than carbon dioxide in depleting the ozone layer. • The washing of nylon clothing also generates microfibres, which can lead to the marine pollution. • Polyester fibres are non-biodegradable and may take up to several years to biodegrade. • Cause the problem of plastic pollution.	Consumes monomers for polymerization of the fibres	Energy intensive manufacturing process that can lead to global warming (three times energy usage compared to cotton).	Microfibres entering the food chain through the aquatic food and can cause harm to the human digestive system

Continued

Table 1.4 Summary of environmental impacts of major textile fibres.—cont'd

Fibre name	Derived from renewable resources	Biodegradability at EOL	Environmental and social impacts	Water and resource usage	Energy usage	Health and other hazards
Rayon	Yes	Yes	Environmental concerns are caused by: • Land, water and air pollution due to carbon disulphide, and sodium hydroxide use. • Deforestation of old trees to make the raw materials for viscose fibre.	Use of large amount of farmland and involves deforestation.	Energy intensive manufacturing process.	Uses carbon disulphide, and sodium hydroxide. Sodium hydroxide can lead to skin burns, eye damage and ingestion.
Acrylic	No	No	Environmental concerns are caused by: • Generation of CO_2 and SO_2, which are GHGs and can lead to acid rain. • Non-hazardous wastes such as empty cans, papers, plastics and filter cloth are produced.	Uses 10 kg of steam for producing 1 kg of fibre	Energy intensive manufacturing process.	The use of arsenic (As), chromium (Cr), cadmium (Cd) and zinc (Zn) can lead to carcinogenic problems.

Table 1.5 Textile raw materials (fibre groups) for environmental sustainability.

Fibre group	List of fibres
Organic	Organic cotton, organic wool, and wild silk.
Other natural fibres	Low chemical cotton, low water cotton, hemp, mechanically extracted bamboo, flax and other bast fibres.
Regenerated fibres (from wood)	Bamboo, PLA (corn), lyocell, soy protein, casein, lenpur, SeaCell, chitin and chitosan.
Other fibres	Lotus, natural polyester (PLA), mechanically extracted bamboo, banana, milk protein and soy protein fibres.
Recycled natural fibres	Cotton, wool, hemp and flax.
Recycled synthetic fibres	Polyester, nylon and polypropylene recycled by mechanical or chemical routes.
Recycled blends	Blends of synthetic and natural fibres.

used clothing by mechanical methods, then these short fibres are re-spun into yarn, which then are converted to fabrics and garments. The mechanical recycling consumes lower energy, saves the use of virgin materials and chemicals. Selection of appropriate colour combinations and processing them to make fabrics can avoid dyeing process (Fletcher, 2013). Fibre-based composites are also being widely researched, where the fibres are being extracted from the waste clothing and composites are developed for automotive, acoustic and laminate purposes (Echeverria et al., 2019; Nayak and Patnaik, 2021). Although several alternatives are available to solve the problems of sustainable fibres, many of them are still in research stage and/or not received wider commercial applications.

This chapter has discussed the problems associated with manufacturing of the major traditional textile fibres and introduced some sustainable fibres. The list of all the sustainable fibres and some important aspects are briefly covered in Chapter 2 and discussed in detail in other chapters in this book. The list of all the fibres that are considered to be environmentally sustainable has been included in Table 1.5.

1.5 Conclusions

This chapter has been prepared to give an idea of the environmental as well as social impacts of manufacturing textile fibres for fashion and textiles. Different problems associated with harvesting the traditional natural textile fibres and manufacturing of synthetic fibres, which has very high detrimental impact on the environment has been discussed in this chapter. The associated problems relating to environmental pollution with various fibre manufacturing such as cotton, wool, silk, polyester, nylon, rayon and acrylic have been discussed. Various environmental problems such as water and air pollution; use of a large amount of fresh water, use of hazardous toxic chemicals, generation of greenhouse gases and generation of large amount of wastewater are some of the problems associated with the production of textile fibres, which are highlighted in this chapter. This chapter also gives an update on the current usage of textile fibres and their future directions. Some sustainable fibres (from plant and animal resources) that will be widely used in the future in the sustainable fashion and textile design have also been discussed briefly in this chapter.

References

Barcelos, S.M.B.D., Salvador, R., Guedes, G., Pinheiro, E., Moro Piekarski, C., de Francisco, A.C., 2021. Socioeconomic and environmental aspects of the production of silk cocoons in the Brazilian sericulture. In: Sustainable Fashion and Textiles in Latin America. Springer.

Bhat, B.A., 2013. Forced labor of children in Uzbekistan's cotton industry. International Journal on World Peace 30, 61—85.

Bliss, R.M., 2005. Flax fiber offers cotton cool comfort. Agricultural Research 53, 12—13.

Britannica, 2019. Learn How Synthetic Materials in Astronauts' Suits Help Them Survive the Hostile Environment of Space. https://www.britannica.com/science/nylon. (Accessed 10 November 2021).

Chang, S.-J., Chen, C.-J., Shih, T.-S., Chou, T.-C., Sung, F.-C., 2007. Risk for hypertension in workers exposed to carbon disulfide in the viscose rayon industry. American Journal of Industrial Medicine 50, 22—27.

Chen, J., 2015. Synthetic textile fibers: regenerated cellulose fibers. In: Textiles and Fashion-Materials, Design and Technology. Elsevier.

Cupit, M.J., 1996. Opportunities and Barriers to Textile Recycling. AEA Technology, Recycling Advisory Unit.

Dai, J., Dong, H., 2014. Intensive cotton farming technologies in China: achievements, challenges and countermeasures. Field Crops Research 155, 99—110.

Dehabadi, V.A., Buschmann, H.-J., Gutmann, J.S., 2013. Durable press finishing of cotton fabrics: an overview. Textile Research Journal 83, 1974—1995.

Deopura, B.L., Alagirusamy, R., Joshi, M., Gupta, B., 2008. Polyesters and Polyamides. Elsevier.

Duarte, L.O., Kohan, L., Pinheiro, L., Fonseca Filho, H., Baruque-Ramos, J., 2019. Textile natural fibers production regarding the agroforestry approach. SN Applied Sciences 1, 1—10.

Echeverria, C.A., Handoko, W., Pahlevani, F., Sahajwalla, V., 2019. Cascading use of textile waste for the advancement of fibre reinforced composites for building applications. Journal of Cleaner Production 208, 1524—1536.

Eichhorn, S., Hearle, J.W.S., Jaffe, M., Kikutani, T., 2009. Handbook of Textile Fibre Structure: Natural, Regenerated, Inorganic and Specialist Fibres. Elsevier.

Fletcher, K., 2013. Sustainable Fashion and Textiles: Design Journeys. Routledge.

Franck, R.R., 2001. Silk, Mohair, Cashmere and Other Luxury Fibres. Elsevier.

Grishanov, S., 2011. Structure and properties of textile materials. In: Handbook of Textile and Industrial Dyeing. Elsevier.

Gross, R.A., Kalra, B., 2002. Biodegradable polymers for the environment. Science 297, 803—807.

Gupta, D., Agrawal, A., Rangi, A., 2014. Extraction and characterization of silk sericin. Indian Journal of Fibre and Textile Research 39, 364−372.

Gupta, V.B., Kothari, V.K., 1997. Manufactured Fibre Technology. Springer Science & Business Media.

Henry, B.K., Butler, D., Wiedemann, S.G., 2015. Quantifying carbon sequestration on sheep grazing land in Australia for life cycle assessment studies. The Rangeland Journal 37, 379−388.

Horton, A.A., 2022. Plastic pollution: when do we know enough? Journal of Hazardous Materials 422, 1−5.

Howden, S.M., White, D.H., Bowman, P.J., 1996. Managing sheep grazing systems in southern Australia to minimise greenhouse gas emissions: adaptation of an existing simulation model. Ecological Modelling 86, 201−206.

Khanzada, H., Khan, M.Q., Kayani, S., 2020. Cotton based clothing. In: Cotton Science and Processing Technology. Springer.

Kozłowski, R., Mackiewicz-Talarczyk, M., 2012. Handbook of Natural Fibres. Elsevier.

Li, R., Tao, R., Ning, L., Chu, G., 2017. Chemical, organic and bio-fertilizer management practices effect on soil physicochemical property and antagonistic bacteria abundance of a cotton field: implications for soil biological quality. Soil and Tillage Research 167, 30−38.

Loh, J.W., Goh, X.Y., Nguyen, P.T.T., Thai, Q.B., Ong, Z.Y., Duong, H.M., 2021. Advanced aerogels from wool waste fibers for oil spill cleaning applications. Journal of Polymers and the Environment 30, 681−694.

Madhav, S., Ahamad, A., Singh, P., Mishra, P.K., 2018. A review of textile industry: wet processing, environmental impacts, and effluent treatment methods. Environmental Quality Management 27, 31−41.

Mancini, F., Van Bruggen, A.H.C., Jiggins, J.L.S., Ambatipudi, A.C., Murphy, H., 2005. Acute pesticide poisoning among female and male cotton growers in India. International Journal of Occupational and Environmental Health 11, 221−232.

McDowell, R.W., Larned, S.T., Houlbrooke, D.J., 2009. Nitrogen and phosphorus in New Zealand streams and rivers: control and impact of eutrophication and the influence of land management. New Zealand Journal of Marine and Freshwater Research 43, 985−995.

McKinion, J.M., Jenkins, J.N., Akins, D., Turner, S.B., Willers, J.L., Jallas, E., Whisler, F.D., 2001. Analysis of a precision agriculture approach to cotton production. Computers and Electronics in Agriculture 32, 213−228.

Mohanty, A.K., Misra, M., Hinrichsen, G., 2000. Biofibres, biodegradable polymers and biocomposites: an overview. Macromolecular Materials and Engineering 276, 1−24.

Muthu, S.S., 2014. Roadmap to Sustainable Textiles and Clothing: Eco-Friendly Raw Materials, Technologies, and Processing Methods. Springer.

Muthu, S.S., 2019. Recycled Polyester: Manufacturing, Properties, Test Methods, and Identification. Springer Nature.

Muthu, S.S., Li, Y., Hu, J.Y., Mok, P.Y., 2012. Quantification of environmental impact and ecological sustainability for textile fibres. Ecological Indicators 13, 66−74.

Nayak, R., Padhye, R., 2017. Automation in Garment Manufacturing. Elsevier.

Nayak, R., Padhye, R., 2015a. Introduction: the apparel industry. In: Nayak, R., Padhye, R. (Eds.), Garment Manufacturing Technology. Woodhead Publishing.

Nayak, R., 2019. Sustainable Technologies for Fashion and Textiles. Woodhead Publishing.

Nayak, R., 2020. Supply Chain Management and Logistics in the Global Fashion Sector: The Sustainability Challenge. Routledge.

Nayak, R., Houshyar, S., Patnaik, A., Nguyen, L.T.V., Shanks, R.A., Padhye, R., Fegusson, Mac, 2020. Sustainable reuse of fashion waste as flame-retardant mattress filing with ecofriendly chemicals. Journal of Cleaner Production 251, 1−9.

Nayak, R., Padhye, R., 2015b. Garment Manufacturing Technology. Elsevier.

Nayak, R., Patnaik, A., 2021. Waste Management in the Fashion and Textile Industries. Elsevier.

Nayak, R., Singh, A., Panwar, T., Padhye, R., 2019. A review of recent trends in sustainable fashion and textile production. Current Trends in Fashion Technology and Textile Engineering 13, 102−118.

Nayak, R.K., Padhye, R., 2015c. The care of apparel products. In: Sinclair, R. (Ed.), Textiles and Fashion: Materials, Design and Technology. Elsevier, London, UK.

Nayak, R.K., Padhye, R., Fergusson, S., 2012. Identification of natural textile fibres. In: Handbook of Natural Fibres. Elsevier.

Nayak, R.K., Punj, S.K., Chatterjee, K.N., Behera, B.K., 2009. Comfort properties of suiting fabrics. Indian Journal of Fibre and Textile Research 34, 122−125.

NorthFjord, 2017. The Most Sustainable Fabrics for Garment Production. https://www.northfjordshop.com/post/the-most-sustainable-fabrics-for-garment-production. (Accessed 10 October 2021).

Padhye, R., Nayak, R., 2016. Acoustic Textiles. Springer.

Peta, 2021. Sheep Farming and the Wool Industry's Damaging Environmental Impact. PETA. https://www.peta.org.au/issues/clothing/cruelty-wool/environmental-hazards-wool-production/. (Accessed 21 October 2021).

PRNNews, 2021. Global textile fibers market report 2021−2027: Asia-Pacific—the dominant consumer. PRNNews. https://www.prnewswire.com/news-releases/global-textile-fibres-market-report-2021-2027-asia-pacific—the-dominant-consumer-301350339.html. (Accessed 10 October 2021).

Rao, R.M.M., 2004. Suicides Among Farmers: A Study of Cotton Grower. Concept Publishing Company.

Ravandi, S.A.H., Valizadeh, M., 2011. 'Properties of fibers and fabrics that contribute to human comfort. In: Improving Comfort in Clothing. Elsevier.

Ross, C.B., 2021. Environmental and Ethical Issues in the Production of Natural Fabrics and Fibres. PETA. https://www.the-sustainable-fashion-collective.com/2017/02/27/environmental-ethical-issues-production-natural-fabrics-fibres. (Accessed 21 October 2021).

Roy, C., Kumar, A., 2013. Green chemistry and the textile industry. Textile Progress 45, 3−143.

Saggar, S., Bolan, N.S., Bhandral, R., Hedley, C.B., Luo, J., 2004. A review of emissions of methane, ammonia, and nitrous oxide from animal excreta deposition and farm effluent application in grazed pastures. New Zealand Journal of Agricultural Research 47, 513−544.

Shaumarov, M., Toderich, K.N., Shuyskaya, E.V., Ismail, S., Radjabov, T.F., Kozan, O., 2012. 'Participatory management of desert rangelands to improve food security and sustain the natural resource base in Uzbekistan. In: Rangeland Stewardship in Central Asia. Springer.

Simpson, W.S., Crawshaw, G., 2002. Wool: Science and Technology. Elsevier.

Statista, 2021. Distribution of Textile Fibers Production Worldwide in 2020, by Type. https://www.statista.com/statistics/1250812/global-fiber-production-share-type/. (Accessed 10 November 2021).

Textilenetwork, 2020. The Fiber Year 2020: Characterized by the Coronavirus. https://textile-network.com/en/Technical-Textiles/Fasern-Garne/The-Fiber-Year-2020-characterized-by-the-coronavirus. (Accessed 10 November 2020).

Thomas, S., Paul, S.A., Pothan, L.A., Deepa, B., 2011. Natural fibres: structure, properties and applications. In: Cellulose Fibers: Bio-And Nano-Polymer Composites. Springer.

Wang, F.-yong, Han, H.-yong, Hai, L., Bing, C., Kong, X.-hui, Ning, X.-zhu, Wang, X.-wen, Jing-de, Y.L.Y., 2019. Effects of planting patterns on yield, quality, and defoliation in machine-harvested cotton. Journal of Integrative Agriculture 18.

Wicker, A., 2020. Why does silk have such a bad environmental rap? Ecocult. https://ecocult.com/why-does-silk-have-such-a-bad-environmental-rap/. (Accessed 20 October 2020).

Yacout, D.M.M., Abd El-Kawi, M.A., Hassouna, M.S., 2016. 'Cradle to gate environmental impact assessment of acrylic fiber manufacturing. International Journal of Life Cycle Assessment 21, 326−336.

You-xin, L., De-zhen, Q., 1985. Cost-benefit analysis of the recovery of carbon disulfide in the manufacturing of viscose rayon. Scandinavian Journal of Work, Environment and Health 4, 60−63.

Introduction to sustainable fibres for fashion and textiles

2

Lebo Maduna[1] and Asis Patnaik[2]
[1]Technology Station: Clothing and Textiles, Faculty of Engineering and the Built Environment, Cape Peninsula University of Technology, Cape Town, Western Cape, South Africa; [2]Department of Clothing and Textile Technology, Faculty of Engineering and the Built Environment, Cape Peninsula University of Technology, Cape Town, Western Cape, South Africa

2.1 Introduction

The fashion and textile industry are very diverse with various interest groups such as suppliers of raw materials, manufacturers of fashion and textile products, retailers, regulators, non-governmental organizations and consumers. They all have different viewpoints when it comes to the sustainability of the fashion and textile industry (Greco and De Cock, 2021). The fashion and textile products are made from natural fibres (wool, cotton, flax etc.) and man-made fibres (polyester, nylon, polypropylene etc.) and because of their origin the fibres are at the centre of sustainability. Natural fibres are from sources that are renewables and biodegradable (Shafie et al., 2021; Stenton et al., 2021; Tiza et al., 2021) whereas man-made fibres from non-renewable fossil fuels and pose risk to the environment as they are not easily degraded (Tshifularo and Patnaik, 2020). The biggest challenge in the use of natural fibres is the lack of land for planting plants and breeding animals. In order to increase production, farmers tend to use more chemicals which contaminate the fibres, soil and water. Contaminated fibres end up in finished products and consumers come into contact with the finished products when they buy them. When consumers dispose of products, the soil and water are contaminated and the cycle continues (Sumner, 2015).

It is well known that the current consumption of fashion and textile products is not sustainable, and the waste generated has a huge impact on the environment. It is very difficult to completely eliminate waste and what can be done is to reduce processes and behaviours that generate it and one of the ways that is being suggested is by recycling and reuse. Recycling and reuse waste that ends up in landfills and incineration plants. A circular economy concept is also suggested which recommends recycle, reduce, reuse and up cycle waste fibre materials and products which are circulated in a close loop to avoid waste disposal. Waste can be treated using heat, chemicals and enzymes recycling techniques to recover the fibres. Heat and chemical techniques damage the fibres especially when used to recover fibres from blended products. The fibre which is susceptible to either heat or chemical will be damaged. Biological enzymes cause

---

Sustainable Fibres for Fashion and Textile Manufacturing. https://doi.org/10.1016/B978-0-12-824052-6.00016-0
Copyright © 2023 Elsevier Ltd. All rights reserved.

very less damage but they have a very low yield (Claxton and Kent, 2020; Navone et al., 2020; Nayak et al., 2020; Wagner and Heinzel, 2020).

Consumer behaviour plays an important role in reducing waste that ends up in landfills and incineration plants. Excessive consumption and disposal of products are not good for the environment and the consumers must think about their role in this. When they buy products which are from sustainable materials they are protecting the environment (Nayak et al., 2020). Designers also play an important role in achieving sustainability because when they select and use less fibre materials in their products, they are protecting the environment (Claxton and Kent, 2020; Dissanayake and Weerasinghe, 2021).

This chapter highlights cotton and polyester fibres which are the two most dominant fibres that are widely used to make fashion and textile products. Other fibres like wool and silk are also highlighted. Cotton and polyester fibres dominate due to their fibre properties and their low price. The rapid consumption of the fibres to meet increased demand for fashion and textile products have raised questions about the sustainability aspect as some of the fibres are from non-renewable sources. Renewable fibres have their own limitations as the renewable sources are limited by land and water. Recycling, reduce and reuse have been suggested as some of the solutions in addressing environmental problems and achievement of sustainability goals, however there are some challenges when it comes to the recovery of blended fibres as currently there is no technology that can separate blended fibres without causing damage. The influence of designers and consumers is also highlighted. Designers influence sustainability and environmental impact along the value chain by selecting fibre and production processes that are sustainable and have low impact on the environment. Consumers also play their role when they buy products that are made from sustainable fibre sources. However the fact that a consumer is environmentally conscious does not always translate into them buying products that are made from sustainable sources and cause minimal damage to the environment.

2.2 Textile fibres-environmental impacts and sustainability

2.2.1 Textile fibre sources

Fibres can be classified into natural and man-made fibres. Generally, natural fibres are considered as sustainable raw materials as they are from renewable sources while man-made fibres are not as they are from non-renewable fossil fuels. Some of the examples of natural fibres are cotton, flax and wool and whereas for man-made fibres are polyester, nylon, polypropylene and regenerated cellulosic fibres. Regenerated fibres are man-made fibres from cellulosic plant sources. Natural fibres used to make clothing products can be sourced from animals and plants. Animal fibres include silk, mohair and wool. Plant fibres include cotton, hemp, flax and kenaf.

Natural fibres have been used for many years to make fashion and textile products. Their advantage over man-made fibres is that they are biodegradable and biocompatible; hence, they are used in medical applications where these characteristics are required. They are the preferred choice for sustainable raw materials due to their low negative impact on the environment and human health. They provide better body protection against radiation and have good thermal resistivity. Wool is a good thermal insulator. Cotton is soft, lightweight and absorbs moisture easily making it one of the most widely used natural fibres to make products where these properties are required. Hemp, flax and kenaf are not only used to make clothing products as they have found application in composite products as reinforcement materials due to their good mechanical properties.

Plant and animal fibres are from renewable sources and can be used in such a way that minimises deforestation and overgrazing. Plants or animals can be planted or breed again. The disadvantage of natural fibres is that they are susceptible to fungal and insect attack. Some of the variation in properties of natural fibres is attributed to the region where they were sourced. Man-made and natural fibres can be blended with each other to make a product which incorporates the properties of each fibre (Das and Chaudhary, 2021; Navone et al., 2020; Tiza et al., 2021; Tonk, 2021). Flax fibres are long fibres which are extracted from a bark of a flax plant called Linum usitatissimum. The fibres have been used for a very long time to make clothing and textile products. A fabric made from the flax fibres is known as linen which has a soft feel. The fibres have good moisture absorption, mechanical and feel properties. Another plant fibre which can be used is Hemp which is from the Cannabis sativa plant. Although the fibres can be used to make clothing, the fibres have poor spinnability and as such they are blended with other fibres such as cotton (Morton and Hearle, 2008).

Wool fibres are popular animal fibres which are widely used to make clothing products. Fibres are harvested from the fibrous materials covering a sheep. Wool has natural lustre, flame retardancy, moisture absorption, insulation and feel which makes it a good raw material for clothing products. Wool fibres can be recovered from recycled wool products to make new products such as insulation for the construction industry (Patnaik et al., 2015). It is not uncommon for recycled wool to be used to make wool rags and this improves the sustainability of wool and protection of the environment.

Conventional petrochemical fibres used to make rags are not renewables resulting in significant damage to the environment (Morton and Hearle, 2008; Tiza et al., 2021). Silk is the only natural filament fibre produced by a silkworm called *Bombyx mori*. Silk fibres are flexible, elastic and smooth which make the fabric flexible, lustrous and soft. The fibres can be used to make women's dresses, lingerie and sportswear. The fibres are expensive and as such cheaper thermoplastic fibres which have the same properties as silk fibres have replaced them. An example is nylon stockings which have replaced silk stockings. This economic benefit is not sustainable as nylon fibres are from non-renewable fossil fuels and furthermore nylon fibres are not easily degraded. Silk fibres like other natural fibres face the same dilemma as others as they require land for silkworm cultivation and water which are limited resources (Franck, 2001).

Regenerated cellulosic fibres such as viscose are produced from plant materials containing cellulose. Even though the source of the raw materials is from natural renewables, the fact that it uses chemicals and energy is of a serious concern on the environment. Wood is the most abundant source of cellulose, and it must be managed sustainably in order to avoid deforestation. Trees require land, water and chemicals. Huge track of land is cleared when trees are planted and in some places where there are limited water resources, water is diverted to tree plantations and deny the communities and animals the little water that is available. The use of chemicals to control diseases contaminates the soil and water resources and as a result the ecosystem is destroyed. In order to protect the environment, new advanced production technologies which have low environmental impact should be used (Islam, 2020; Stenton et al., 2021).

Research on alternative synthetic organic fibres is still in its early stages. These are said to have low environmental impact compared to conventional synthetic fibres. Few fibres that have been synthesized are not scalable for commercial operations due to the low yields and high cost. A fibre called Qmilk was synthesized using milk waste with no waste generated from the process. Another fibre called Orange was synthesized from citrus juice by-products. The use of cellulose fibres with polylactic acid and petroleum resins is also being explored as an alternative artificial leather (Stenton et al., 2021).

2.2.2 Global market of textile fibres

The global market for the textile fibres is dominated by the petrochemical fibres and plant cotton fibres accounting for 64% and 24%, respectively. Synthetic cellulosic fibres account for 6% whereas animal wool fibres accounts for 1%. Other natural fibres account for the remaining 5% (Navone et al., 2020). Cotton is the most dominant fibre out of all the natural fibres whereas polyester is the dominant man-made fibre accounting for just more than half. The dominance of the polyester fibres is due to their low production cost and good fibre properties as raw materials for making fabrics.

In 2017 the annual production of polyester, polyamide, regenerated cellulose and other synthetic fibres (polypropylene) was 53.7, 5.7, 6.7 and six million tonnes respectively and for cotton, wool, other plant fibres (flax, hemp) and other natural fibres it was 25.8, 1.2, 5.6 and 0.4 million tonnes, respectively (Table 2.1). A natural fabric consumes more energy when it is washed whereas for a synthetic fabric more energy is consumed in the production of fibres (Clancy et al., 2015). Laundering causes a fabric to release micro-fibres. The micro-fibres released by a synthetic fabric are called micro-plastics which cause serious environmental and health problems as they end up in soil, rivers and oceans. Micro-plastics have been found in marine animals (Kebler et al., 2021; Stenton et al., 2021; Wagner and Heinzel, 2020).

2.2.3 Environmental impacts

2.2.3.1 Impacts of cotton farming

For many rural communities, land is their only resource that they can use to earn a living. Cotton farmers grow cottons that they sell to provide for their families. In

Introduction to sustainable fibres for fashion and textiles

Table 2.1 Global fibre production in 2017 (Stenton et al., 2021).

Fibres		Million tonnes
Synthetic	Polyester	53.7
	Polyamide	5.7
	Regenerated cellulose	6.7
	Other synthetic fibres (polypropylene)	6
Natural	Cotton	25.8
	Wool	1.2
	Other plant fibres (flax, hemp)	5.6
	Other natural fibres	0.4

Pakistan cotton accounts for 1% of the GDP. China cotton farmers have one of the highest yields of about 1438 kg/ha which is higher than the global average of 769 kg/ha (Dai and Dong, 2014; Imran et al., 2019). Cotton cultivation is limited by land, water and pests. Farmers use pesticides to protect their crops and the chemicals have a negative effect on the environment, biodiversity and health of other plants, animals and humans. Cotton cultivation accounts for 10% of the chemical pesticides. Since cotton competes with food crops for cultivation land, it is not a priority when the country's food security is taken into consideration and the first priority will be given to crops when land is allocated.

Shortages of labour during the planting and harvesting seasons are barriers in improving yields and quality of cotton. In places where the yields are low, farmers respond by using more water, pesticides and fertilizers to improve yields and profits and these cause damage to the soil and depletion of water resources. Pesticides prevent insect attack and excessive use can cause pests to develop resistance and to overcome the resistance, farmers will respond by using even more pesticides. Increased use of fertilizers to improve soil causes the accumulation of nitrogen in the soil which leads to increased nitrous oxide emissions to the environment.

An alternative to this conventional farming is organic farming and genetically engineered cotton which have low environmental impact. Unlike conventional farming, organic farming promotes the use of natural resources to conserve water, improve soil quality and control pests in order to improve yields and make the farming sustainable which has limited damage on the environment and health. Organic waste is used as a fertilizer or as a raw material for bioenergy. One of the reasons why organic farming is not widely used by farmers is due to extensive labour required and the natural resources which are limited. Genetically engineered cotton has been developed which produces a toxin that kills the bollworms that attacks it (Bwana et al., 2021; Imran et al., 2019; Mancini et al., 2008; Nibouche et al., 2007; Patil et al., 2014).

2.2.3.2 Other environmental impacts

Emissions from fashion and textile products are not only from the production processes, it is also from the extraction and transportation of raw materials because even when products have reached their end of life, they still have their share of contribution (Shen et al., 2020). Even though natural fibres are from sustainable sources they too have their own share in carbon footprint emissions. In the UK, the production of wool fibres contributes 30% of the fashion industry carbon footprint (Stenton et al., 2021). When it comes to the carbon footprint, a cotton shirt produces 2.1 kg of CO_2 compared to 5.5 kg of CO_2 of a polyester shirt which is more than double. Cotton fabric consumes more water than polyester and as such the cotton fabric water footprint is much higher. Viscose, a synthetic cellulosic fibre occupies the second place as shown in Fig. 2.1. For a cotton to compensate for its high water footprint, it must be used 20,000 times as it can take up to 10,000–20,000 L of water to produce 1 kg of a cotton fabric (Wu and Li, 2019).

The demand for plastics is expected to increase greenhouse emissions from 84 to 309 million tons by the year 2030–2050. When polymers are burned, they release CO_2 and other toxic chemicals they contain. Products made from petrochemical fibres are not easily decomposed and the microplastics they release not only contaminate the environment but also enter the food chain and pose health risks to animals and human beings. They have been detected in intestinal tracts of animals. All the fibres reach a stage where the fibres have been damaged to such an extent that it is no longer practicable to recycle and reuse them as a source of new raw materials to make the fibres. When such a stage is reached the fibres are either disposed of in landfills or incineration facilities.

Traditional way of managing plastic fibres is to dispose of them in landfills and incineration facilities and with increasing production and utilization of fibres there is

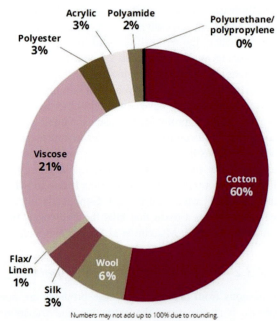

Figure 2.1 Fibre water consumption (Wu and Li, 2019).

Introduction to sustainable fibres for fashion and textiles

a huge environmental risk that comes with this method as treatment sites are accepting and treating more waste. Historically raw materials and products were designed for single use and as such there is a need to minimize waste and make efforts to achieve fibre sustainability by recycling and reuse. Recycling is a process of recovering an item without damaging its molecular structure. It reduces waste that would have ended up in landfills and incineration facilities. Furthermore, recycling reduces the need for original raw materials.

New original raw materials consume more energy to produce. High energy consumption affects the environment. Blended fabrics pose challenges in sorting and recycling as different fibres must be recovered which is not always possible. This can discourage recycling and reuse. Furthermore, collection and sorting facilities which are far away from industries that beneficiate the products can discourage recycling and reuse as the transportation cost can be too prohibitive. It is estimated that recycling about 3.17 million tons of plastics can save about 3.2 million tons of CO_2 equivalent.

Incineration, one of the methods used to treat waste, uses heat to reduce a large quantity of waste into a small ash. Incineration is not eco-friendly as it uses energy and releases toxic gasses and in order to offset its impact the heat that is generated when waste is burned can be used as a source of new energy to power the facility. Alternative sources of energy for incineration which are eco-friendly are solar and wind power. One ton of plastic fibres that is incinerated is estimated to release about 2.9 ton of CO_2. Landfills receive the majority of fashion and textile waste and leaching of toxic chemicals from the waste will contaminate soil and ground and surface water. The contribution of landfills to the greenhouse emissions is still lacking (Shen et al., 2020; Sun et al., 2021).

2.2.4 Sustainability by recycling

Demand for fibres used to make new garments is increasing, and since most of the fast fashion items are not reused or recycled, it creates an enormous waste that predominantly ends up in landfills or incineration facilities. In the UK, it is estimated that less than 1% of the clothing products are recycled into new clothing products meaning 99% end up in landfills or incineration facilities. In order to reduce the negative impact on the environment, the circular economy concept is suggested. It aims to minimize the use of materials to make products and when the materials reach their end of their life they must be collected, recycled and reused. It promotes the collection, recycling, reuse and repair in a circular manner where waste or products are circulated in a close loop. It reduces the need for new materials and products and the impact on the environment is minimal.

Recycle, reuse and repair are called the 3R and the addition of redesign and reimagine to the 3R makes it the 5R. Fibre sustainability and protection of the environment is very complex as shown in Fig. 2.2 due to the different factors that are intertwined. It is not a linear approach as shown that even cotton fibres can pose a risk to workers along its value chain. Major barrier to the reuse or recycling of postconsumer waste is the composition of the waste as most of the products are blended with different types of fibres. It causes difficulties in the separation according to the

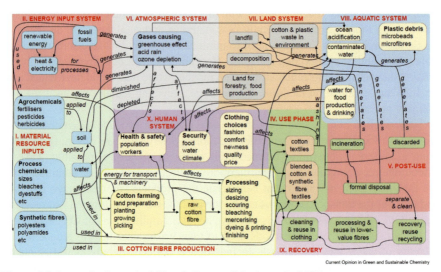

Figure 2.2 Intertwined sustainability challenges (Martin et al., 2020).

different types of fibres as different fibres have different thermal, mechanical and chemical properties.

For a blended fabric like poly-cotton, heat can be used to melt the polyester fibres; however, this will cause the destruction of cotton fibres. When chemicals are used to recover the fibres from blended fabrics, the mechanical and chemical properties of fibres can be affected. Damaged fibres affect the quality of the raw materials if they are used as a source of raw materials. In order to avoid this, biological enzymatic treatment can be used to recover the fibres as the damaging effect is minimal on the fibres. It should be noted that technological advancement in the separation and recycling of blended fabrics is still not advanced and it is a major barrier for industrial application (Claxton and Kent, 2020; Kebler et al., 2021; Martin et al., 2020; Navone et al., 2020; Nayak et al., 2020; Wagner and Heinzel, 2020). Lack of machines that can sort and separate waste according to the fibre type contributes to low recycling (Sun et al., 2021).

Fabrics made of blended fibres are difficult to sort and separate which makes recycling and reuse to be lower. Recovering from fibres blended polyester and cotton or polyester and wool fabric is not easy as polyester is a thermoplastic fibre. Recovered thermoplastic fibres can be melted using heat to make new fibres whereas the natural fibres cannot be melted as they are degraded by heat. Chemicals used to recover natural fibres from the blended products cause damage to the fibres. Using chemicals can cause health and environmental problems. The use of ionic liquids to separate polyester/cotton and polyester/wool blends is being investigated. Cotton and wool fibres are dissolved, whereas the polyester fibres are not. The solid polyester can be recovered and melted to produce fibres. The dissolved cellulose and wool keratin is regenerated using the wet spinning. Used ionic liquid is then recovered via the distillation process to be reused again and minimal chemical is disposed of to the environment (Sun et al., 2021).

Recovered fibres can be used to make musical instruments, packaging (bags, boxes), sport goods, automotive (door panels, boot liner), military (helmet), household (ceiling panels, sun blinds) (Das and Chaudhary, 2021).

2.2.4.1 Cellulosic fibres recycling

Chemicals can be used to recover recycled cellulosic fibres. Chemicals such as NaOH, phosphoric acid and methylmorpholine-N-oxide hydrolyse cellulosic fibres. Cellulosic fibres that are not treated with alkalis tend to show some resistance against enzymatic hydrolysis. Cellulosic waste can be used as a feedstock for the bacteria to produce bacterial cellulose. Conventional feedstock such as sucrose, glucose, fructose and ethanol are expensive which limits their use for large industrial production given that the yield of bacterial cellulose is not too high. Acetic acid bacteria produces pure cellulose which is not contaminated with lignin and hemicellulose and given that it is biocompatible and has high water absorption it is a good material for the biomedical industry. The use of textile waste containing cellulosic fibres reduces the need for new conventional materials and production costs associated with them. This protects the environment (Hong et al., 2012; Wu and Liu, 2013).

2.2.4.2 Polyester recycling

The global consumption for thermoplastic fibres is expected to grow as the demand for fashion and textile products grows. The polyester fibres dominate the thermoplastic fibres. In 2020, the market for polyester fibres was about 52% and their dominance is expected to grow in coming years (Textile Exchange, 2021). The fibres are inexpensive to produce and possess good mechanical strength and show some resistance when exposed to chemicals and heat (Gholamzad et al., 2014; Guo et al., 2021; Park and Kim, 2014; Provin et al., 2021).

There are efforts to recover and recycle the products made from polyester fibres in order to protect the environment. It is estimated that it can take up to 200 years to degrade polyester fibres. Clothing with good quality can be made from recycled polyester bottles (Wagner and Heinzel, 2020). Polyester waste can be recycled using various methods. Thermal recycling method involves the melting of fibre. Received waste first washed to remove contaminants and dry and then fibres are melted to produce fibres. Chemical method uses chemicals to degrade the fibres. It involves the depolymerisation to ethylene glycol (EG), terephthalic acid (TPA), mono(2-hydroxyethyl) terephthalic acid (MHET) and bis(2-hydroxyethyl) terephthalate (BHET). The cost of operating the chemical method is too high and often there are safety concerns and strict measures must be in place to avoid accidents.

Biological method uses biological components like enzymes to degrade the waste and the method is considered to be environmentally friendly. Enzymes that can degrade polyester are lipases and cutinases. The MHET can act as an inhibitor of the enzymes, especially the cutinases. Regenerated polyester fibres that have less than 3dtex are reported to be susceptible to breakage during the spinning process and in order to minimize breakage the fibre titre is increase to above 3.5dtex (Guo et al., 2021; Ion et al., 2021).

2.2.5 Wool reuse and sustainability

Wool fibres are treated with chemicals such as dichloroisocyanuric acid, chlorine and sodium hypochlorite solution for surface treatment in order to reduce felting and shrinkage properties. Since wool fabric suffers from shrinkage and felting, changes in the dimensions and appearance of the fabrics make them unusable. After a first wash of a non-treated fabric, it can shrink by up to 55% of its original dimensions. Chemical treatment is cheaper and easier to apply. Chemicals pose risk to the environment and as such alternative enzymatic treatments are being explored, but there is a risk of destroying the fibres as enzymes like protease degrade the fibres. In order to avoid this, the proteolytic enzymes can be genetically modified. Enzymatic treatment is a slow process which produces a residue.

Other environmentally friendly methods which can be used to treat wool fibres are plasma and radiation treatments. Radiation treatment uses radiation to change the fabric characteristics by exciting the fibre atoms. Plasma treatment uses plasma to change the individual fibre scales to change the fibre surface. Radiation and plasma treatments are processes that can be used to treat fabrics on a running production line without stopping the production. However, they can cause damage to the fibres. Radiation can also cause yellowing of the fibres. Waste wool fibres can be used as a construction material for a sustainable future given that conventional are not renewables and cause significant damage on the environment (Jiang et al., 2021; Hassan and Carr, 2019; Tiza et al., 2021; Patnaik et al., 2015).

2.2.6 Sustainable natural fibres

This section will briefly discuss the sustainable natural fibres, which are currently being used or in the research stage. Important aspects of some sustainable fibres are discussed in other chapters of this book. This chapter will provide a basic understanding of the sustainable textile fibres used for fashion and textile manufacturing.

2.2.6.1 Organic cotton

Organic cotton fibres are grown from seeds that are not genetically modified and without the use of any chemicals such as pesticides, herbicides, fertilizers and defoliants or growth regulators (Myers and Stolton, 1999). Organic cotton reduces soil and water pollution in addition to its superior properties such as free from allergic reactions and soft feel as compared to traditional cotton. Cultivation of organic cotton reduces the toxicity to almost zero and the overall product toxicity is reduced by 93% compared to traditional cotton. Organic cotton heals the planet, supports a true economy and protects our health. In organic fibre production not only, the synthetic fertilizers are replaced with organic ones, but also the systematic approach of diversified

Introduction to sustainable fibres for fashion and textiles 37

farming activities is followed. For switching from traditional to organic cotton, the farms are prepared 2 years prior to the cultivation; mechanical and manual methods are adopted for weed control, which are more environmentally sustainable. The organic cotton type is carefully selected depending on the soil, climate and resistant to diseases and pests.

Organic cotton cultivation needs to be certified by third-party agencies who certify that the producers follow the standard procedure, depending on the needs of the buyers. Organic cotton certifiers around the world are accredited by the International Federation of Organic Agricultural Movements (IFOAM) (Luttikholt, 2007). In the United States, the organic cotton needs to be certified by the inspecting agencies registered with the US Department of Agriculture (USDA), irrespective of the place the cotton was cultivated (Fletcher, 2013). The certifying agencies in other counties in Europe and Asia (i.e., Japan and India); Australia and New Zealand use their own organic standards to certify organic cotton. Many other countries use standard that are formulated by different organization that covers organic production and processing of cotton into garments.

The major problem associated with large-scale usage of organic cotton is the limited cultivation. Due to the time involved with the land preparation; vulnerability to attacks by pests and insecticides; and low productivity (about half the yield of traditional cotton), the cotton farmers are not adopting the farming of organic cotton. Furthermore, due to limited availability, the problem of making an intimate blend for good quality yarn is a challenge as organic cotton from different sources varies in their properties (Muthu, 2014). Although the contribution of organic cotton is still less than 1% of the total cotton consumption in 2015 (which was 26 million metric tons), the demand for it is ever increasing. However, the low productivity creates questions of cultivating organic cotton due to the need for high amount of arable land. Low productivity and limited availability make the organic cotton price higher than the traditional cotton (Table 2.2).

2.2.6.2 Low water and low chemical cotton

Low water cotton uses lower amount of water compared to the traditional cotton (Fletcher, 2013). The water usage can be minimized by the proper utilization of rainwater. When needed, the cotton can be irrigated by efficient methods such as microirrigation or sprinkler irrigation rather than the surface irrigation, which can save up

Table 2.2 Average production of various natural fibres per hectare of land.

Fibre name	Average fibre production (kg/ha)
Hemp	1600
Flax	1000
Cotton	800
Organic cotton	500
Wool	65

to 50% of water. However, the quality of low water cotton is slightly inferior to the traditional cotton due to an inadequate water supply. The other type of cotton known as low chemical cotton (LCC) uses fewer chemical than the traditional cotton. Genetically modified crop and the integrated pest control (IPC) techniques are used for the growth of the LCC, which ensures the minimal impact to the environment and human health. The IPC focuses on natural, mechanical, and biological pest control mechanisms; critically monitors plant's growth and allows acceptable pest levels when needed.

2.2.6.3 Organic wool

The production of organic wool involves the practices where the sheep are not exposed to any chemicals and hormones and are grown in humane and good farm conditions. Although there is not much difference between the fibre properties, organic wool production ensures animal ethics and free from toxic chemicals and hormones compared to conventional wool. Organic wool is considered to be more sustainable as the environmental impact is almost negligible. The global demand for organic wool is also growing. Similar to cotton, wool needs to be certified by third-party certifying agencies if the following terms are satisfied: (a) the feed and forage used by the sheep must be organic, and the grazing land is not treated with synthetic pyrethroids or pesticides; (b) there is no use of genetic engineering and synthetic hormones for the growth of sheep; (c) the sheep should only be injected with permitted chemicals to prevent sheep-scab; (d) the sheep is not treated with any form of pesticides that is harmful to the sheep, people and the environment (Fletcher, 2013); and (e) the finished woollen yarn is coloured with organic colorants.

2.2.6.4 Wild silk

Wild silk is cultivated in open forest, which is an easy source of food for silkworms and there is no use of hazardous chemicals (Fletcher, 2013). The production of wild silk is considered to be ecofriendly and sustainable, due to the benefits such as (a) the wild silk cultivation can promote the forest preservation and generate income for the tribes; (b) the wild silk cocoons are collected after the moth has emerged as butterfly compared to killing of the moth while in situ for the cultivated silk, hence, wild silk is considered as vegetarian silk; (c) as the wild silk cocoons are broken, the continuous silk filament cannot be extracted. Hence wild silk is inferior in quality compared to the cultivated silk; and (d) the wild silk is spun as staple fibre similar to cotton using staple spinning systems. The wild silk is difficult to bleach or dye but have naturally attractive gold sheen.

2.2.6.5 Hemp

Hemp was first used about 10,000 years ago and one of the fastest growing natural fibres. Hemp has texture similar to flax and very good drape, which can be used to produce textiles and clothing. Hemp can be used as single fibre or blended with fibres such as cotton, polyester and spandex. Among the natural fibres, the durability of hemp fibre

Introduction to sustainable fibres for fashion and textiles

is the highest. The other advantages of hemp are: higher strength than cotton (about 300%), high breathability and absorbency, good abrasion resistance, resistant to mould and mildew, ultraviolet (UV) resistant and antimicrobial properties (Muthu, 2014). Hemp fabrics are dyed easily, soften after washing and can be washed or dry-cleaned.

Hemp production is considered to be environmental friendly due to no herbicides and pesticides or genetically modified (GMO) seeds are needed for its production and requires little water to grow (Muthu et al., 2012). Hemp fibre has the highest yield among all the plant fibres as shown in Table 2.2. The cultivation of hemp replenishes the essential nutrients in soil and prevents soil erosion. Hemp can be cultivated by using only organic ingredients such as natural manure, compost and rainfall. The cultivation of hemp is considered to be illegal in many countries due to hemp plant's resemblance with the marijuana plant. Hemp fibre production can reduce the ecological footprint by 50% compared to organic cotton (Fletcher, 2013). The only environmental problem associated with hemp fibre production is the retting used for fibre extraction, which impacts the water. This can be avoided by adopting processes such as dew retting, steam explosion and enzymatic retting.

2.2.6.6 Flax

The production of flax fibre is more sustainable than cotton as the growth of flax needs no pesticides or herbicides and consumes less fresh water (Muthu, 2014; Avinc and Khoddami, 2010). Furthermore, the flax seeds need less deep ploughing, hence, less energy consumption and less soil erosion on the sloping fields. The environmental benefit of flax includes biodegradability, easy waste-disposal, reduced water usage, preservation of non-renewable resources, and reduced carbon dioxide emission as flax plant absorb high amount of CO_2 during its lifetime. Flax fibre is stronger than cotton, has good resistance to abrasion, shape retention, high absorbency, and breathability. The manufacturing of flax fibre consumes about 20% less energy than glass fibre and 10% less energy than carbon fibre. The EOL flax materials can be used for making composites for automotive, construction and thermal insulation applications.

2.2.6.7 Bamboo fibre

As an agricultural crop in many Asian countries, bamboo is considered to be the fastest growing plant in the grass family, can grow 1–4 inches per day and it is sustainable. The bamboo fibre is resilient and durable, with high breaking tenacity, better moisture-wicking properties and better moisture absorption properties, compared to cotton (Majumdar et al., 2010). The other advantages of bamboo fibre compared to traditional cotton are no use of pesticides or fertilizers and no use of fresh water during its growth. In addition, the EOL bamboo clothing can be disposed in an ecofriendly manner. Furthermore, bamboo fibre can be blended with other fibres such as hemp, cotton, polyester, and spandex to produce a range of blended fabrics. Bamboo fibre can be used in the design and development of general apparel clothing including polymer composites (Ghavami, 2005; Majumdar et al., 2010). There has been rapid growth in the application of bamboo in the polymer composites due to its biodegradability and mechanical properties (Ghavami, 2005; Rao et al., 2010).

Bamboo fibre is considered as regenerated fibre in the rayon group. The fibres can be produced by both the chemical and mechanical processes. The chemical process (viscose route) of bamboo fibre extraction is almost similar to the production of viscose fibre, which involves steps such as chip preparation, cellulose extraction, xanthation by carbon disulphide (CS_2), ageing, purification and fibre spinning. The xanthation process using CS_2 causes acute toxicity and can cause serious health effects such as heart attack, liver damage, blindness and psychosis (Vanhoorne et al., 1995). Hence, the chemical process of bamboo extraction involving CS_2 xanthation is not sustainable. The chemical process (lyocell route) can also be used to extract fibre from bamboo, which is considered to be sustainable. In addition, the mechanical process of bamboo fibre extraction, which involves enzyme retting followed by washing to extract the fibre (similar to flax fibre extraction), causes less environmental and health impact, hence considered being sustainable.

2.2.6.8 Polylactic acid fibre

Polylactic acid (PLA) fibre is a linear thermoplastic aliphatic polyester fibre prepared from renewable sources such as sugar beet, corn starch and sugarcane (Avinc and Khoddami, 2010). PLA fibre can be used as compost at the EOL as it degrades rapidly. From energy consumption, CO_2 emissions and end of life options, PLA is superior to many petroleum based polymers. PLA was initially promoted for single use packaging applications, given the key benefit of short life cycle due to its compostable nature. The production and use of PLA fibres has the following benefits (+signs) compared to the synthetic fibres such as polyester (Farrington et al., 2005).

+ The PLA fibres are produced from renewable natural resources; hence, there is no problem of depletion of the limited availability of synthetic raw materials.
+ The PLA fibre production process consumes less energy and there are less gaseous emissions compared to the synthetic fibres.

The production of PLA has some negative impacts, which has been marked as (−signs).

− The commercial production of raw material for PLA fibres needs a large amount of land, which are currently being used for cultivation of food for people.
− The foodstuffs (corn, sugarcane, or sugar beet) will be diverted from the mouths to the fibre manufacturing industries.
− Although the EOL garments of PLA are biodegradable, their rejection into landfill will result in generation of the GHG such as methane (CH_4), land and water toxication due to leaching of colour, and generation of human carcinogens.
− As there is limited availability of the PLA fibre, the cost is higher compared to other fibres such as polyester.
− Although PLA mimic polyester in many aspects, the melting point of PLA is lower than that of polyester, which restricts its applications at high temperature or even transfer printing of PLA is difficult.

- The properties of fabric made from PLA will change as the fibres get weakened in chemical processing with the combined use of water and temperature.

Due to many negative points than positives, the commercialization of PLA fibre is yet to be achieved.

2.2.6.9 Lyocell fibre

Lyocell is a rayon fibre prepared from eucalyptus trees, which is extremely fast-growing resource and requires low water and virtually no pesticide input to grow. It is considered to be an ecofriendly textile fibre (a rayon fibre) prepared using dry-jet wet spinning. Lyocell resembles viscose in the same group of rayon fibres in addition to resembling other cellulosic fibres such as cotton and flax. However, the environmental impact of manufacturing process of Lyocell is least among the rayon fibres. The solvent, amine oxide used to dissolve cellulose for fibre spinning can be recovered up to 98% and reused. In addition, the solvent is nontoxic and non-hazardous. The other benefits of Lyocell includes wrinkle resistance, softness, biodegradability, renewable source of raw material (i.e., Eucalyptus trees), bleach free process (as the fibre is white and clean), and less water and energy consumption during manufacturing. The only problem of Lyocell is the wetting of the fibre, which leads to slight loss of the mechanical strength and fibrillation (i.e., small fibrils separating from the main fibre or filament).

2.2.6.10 Soybean protein fibre

Soybean protein fibre is derived from the hulls of soybeans, which is a manufacturing by-product. It is a renewable plant source and can be produced in copious quantities at low cost. Although the fibre was developed in 1950s, there had not been much progress on the commercialization. The fibre is gaining importance due to the current awareness of sustainable material from renewable sources. The fibre is manufactured by a novel bioengineering technology from the soybean cake. The clothing prepared from soya fibres has softness similar to cashmere and elastic handle, but it is less durable than cotton and hemp (Vynias, 2011). Soya fibre has not only the lustre like silk but also has the excellent comfort properties similar to cotton. The fibre is antibacterial, UV resistant, similar to cotton in moisture absorbency and better than cotton in transporting moisture, which is the main contributing factor for comfort.

2.2.6.11 Milk protein or casein fibre

The milk protein fibre commonly known as "Casein" fibre is produced from milk protein (Rijavec and Zupin, 2011). The milk protein fibre possesses positive properties of both the natural and synthetic fibres. They have a glossy appearance like silk and possess natural long lasting antibacterial properties. The fibres have excellent air permeability, breathability, and water transport properties, hence, very comfortable. The fibre contains about 15 types of amino acids and pH is similar to skin, which helps to maintain healthier skin. Casein fibres have wide applications in intimate garments, children's garments, active wear, uniforms, apparel clothing and women's clothing. Casein fibre is biodegradable, ecofriendly and prepared from 100% renewable resource; hence, considered as one of the best sustainable fibres.

2.2.6.12 Lenpur fibre

Lenpur is a cellulosic fibre prepared from wood pulp of special trees such as white pine, clipping, which is a renewable source. The advantages of Lenpur fibre over other cellulosic fibres are its softness, higher moisture absorbency, higher thermal range, natural antiodour and antimicrobial properties (Dissanayake and Perera, 2016). Lenpur is compatible with many other textile fibres, hence can be easily blended. The yarns produced by ring spinning are smooth and can be directly used in knitting machines. Due to its comfort properties, Lenpur can be used as a replacement of cotton fibre. It can also be used to produce high end luxury fashion items.

2.2.6.13 SeaCell fibre

SeaCell fibre is produced from seaweed, which is always considered to have health and medical benefits. SeaCell fibre can nourish the skin by actively exchanging the material between fibre and the skin. The fibre contains iron and is an incredible source of iodine, which is important for thyroid function (Yimin, 2007). Seaweed is also a source of vitamin B12, a deficiency of which can lead to a type of anaemia. It's a rich source of antimicrobial nutrients and has anti-inflammatory properties. These properties make the SeaCell fibre an inherently healthy source for skin-friendly apparel clothing.

2.2.6.14 Chitin and chitosan fibre

The use of chitin and chitosan dates back to the 19th century by Prof. Henri Braconnot and Prof C. Rouget, respectively (Muzzarelli, 2013). Chitin is produced from animal sources such as moulds, yeasts, fungus, insects, shrimps and crabs, whereas Chitosan is only present in some types of fungi (Hirano et al., 2000). However, commercial chitosan is derived from chitin by alkali deacetylation process at high temperature. Both the chitin and chitosan fibres are prepared by wet spinning using a solvent, as their melting point exceeds the degradation temperature. Chitin fibre is highly crystalline and possesses oriented structure. The thermal properties of chitin and chitosan fibre are similar to other cellulosic fibres. They can be used for absorbent sutures, wound dressing materials, and mixed with other apparel fibres for providing antibacterial properties. Chitin and chitosan fibres possess antibacterial properties and biodegradable.

2.3 Consumer behaviour and sustainability

In 2013, there was an outcry when a textile factory building in Bangladesh collapsed and killed workers. It highlighted the dangers workers are facing in developing

Introduction to sustainable fibres for fashion and textiles 43

countries when producing fashion and textile products that are exported to the developed countries. There is an increasing awareness that the fashion and textile industry is not sustainable in its current form due to these incidents and high consumption of products. Consumer consumption has been increasing for years even though there are concerns about its sustainability. This can be attributed to the missing link between attitudes and behaviour. Having a sustainable attitude does not necessarily translate to a change in the purchasing behaviour even though attitudes can be used to predict the consumer behaviour.

One of the reasons that has been suggested to low purchasing behaviour of sustainable products is that sustainability is perceived to be driven by individuals who are perceived to be pushing their own selfish interests rather than that of the broader societal interests. Some consumers use sustainability as one of the criteria they consider when buying products. The quality of a sustainable product has an influence on the purchasing behaviour of the consumer. High quality promotes it whereas low quality does not. A high-quality product must have high durability and a longer life than a low quality product. Higher prices of sustainable products act as barriers when consumers compare them to conventional products and the purchasing behaviour of the consumer will be influenced by the price.

Lifestyles and appearance are important to some people and the quality is not on the top of the list of issues they consider when buying products. This can be seen in fast fashion which poses serious challenges in achieving sustainability due to the ever-changing fashion trends. As soon as the fashion trend disappears, products are disposed of. Education is important because consumers have enough information about the benefits of sustainable products which can persuade them to change their unsustainable established values. Changing established values requires the willingness on the part of the consumer to change. It is a long process which requires making the information available and listening to the concerns and needs of the consumers (Greco and De Cock, 2021; Jacob et al., 2018; Nayak et al., 2020).

Biocompatibility with the human skin is another important factor some consumers consider when buying products and this is where the organic fibres can play an important role. Clothing made of organic fibres is considered a niche market found mostly in small and medium shops and the lack of presence in big retailers means the majority of people might not know where to find them. Large retailers are at an advantage as they have resources at their disposal to market and sell their products unlike small and medium companies. Organic products with long life can reduce the consumption of new goods and waste (Jacob et al., 2018).

2.4 Sustainable designing

Designers are encouraged to take into consideration the environmental impact of their clothing and textiles products along the value chain. They are also encouraged to consider the energy required to produce a product and the waste that will be generated when a product reaches its end of life. The amount of fibre materials should be reduced where possible. Products must be made of fibre materials that are flexible, durable and long lasting. Flexible products that are elastic will accommodate the changes in the body of the wearer and avoid disposal. Designers must design products that are biodegradable by using fibres and chemicals that are non-toxic and biodegradable.

Natural fibres and dyes are biodegradable whereas the dyes and fibres from petrochemicals are not. The non-biodegradable synthetic polymer fibres should be replaced with biodegradable synthetic polymer fibres where possible. Fibre materials and the dyes must be durable to withstand repeated washing and ironing. Designing clothing that can be worn in all the seasons will prolong longevity. Customization of the product whereby a customer is also involved in the design of the product that will satisfy the customer's needs and the customer will be more attached to the product and keep it longer which will reduce overconsumption or disposal. Designing products that can be easily disassembled can encourage recycling and reuse. Conventional products are not designed to be easily disassembled and it is time-consuming and expensive which discourages recycling and reuse attributes (Claxton and Kent, 2020; Dissanayake and Weerasinghe, 2021).

2.5   Summary and future directions

Both natural and man-made fibres have an impact on the environment. Natural fibres impact is less than man-made fibres due to the fact that they are from renewable sources and are biodegradable and in addition they use less energy. The high environmental impact of man-made fibres is due to the fact that they require more energy to produce and are non-biodegradable. Recycling plays an important role in reducing waste that would end up landfills and incineration. Recycling blends like polycotton by using heat is destructive as polyester is a thermoplastic while cotton is damaged by heat. The heat melts polyester but destroys cotton. Using chemicals to dissolve the fibres affects the quality of fibres and when the chemical waste is discarded it damages the environment.

Consumer purchasing behaviour is complicated as indicated by the fact that a consumer who is conscious about sustainability and environmental issues will not necessarily buy sustainable products as there are other factors that have an influence such as fashion trends and willingness to change. Designers can help by designing environmentally friendly products. It is important that they select raw materials and processes that are sustainable and cause no or less harm to the environment.

Natural and man-made fibres and dyes are going to continue to be part of the fashion and textile industry. The natural fibres and dyes can be expected to be more dominant in the future as they cause less damage to the environment and are from renewable sources which make them biodegradable and biocompatible. The man-made fibres and dyes cause more damage to the environment and the pressure to improve their recyclability and reuse in order to minimize their negative impact is also going to continue. As recycling technology improves especially for the separation of blended fibres, the recycle, reuse and upscale is going to improve. Synthetic organic fibres made from non-conventional natural sources are also going to come to the market once the challenges of low yields are solved. This will present a new source of fibre materials that has the same characteristics as natural fibres.

Acknowledgements

The authors gratefully acknowledged various sources for granting permissions to reuse figures used in this chapter.

Funding
The author disclosed receipt of the following financial support for the research, authorship, and/ or publication of this article: This work is based on the research supported in part by the National Research Foundation of South Africa (grant-specific unique reference numbers (UID) 104840.

References

Avinc, O., Khoddami, A., 2010. Overview of poly (lactic acid)(PLA) fibre. Fibre Chemistry 42, 68–78.

Bwana, T.N., Amuri, N.A., Semu, E., Elsgaard, L., Butterbach-Bahl, K., Pelster, D.E., Olesen, J.E., 2021. Soil N_2O emission from organic and conventional cotton farming in Northern Tanzania. Science of the Total Environment 785 (147301).

Clancy, G., Froling, M., Peters, G., 2015. Ecolabels as drivers of clothing design. Journal of Cleaner Production 99, 345–353.

Claxton, S., Kent, A., 2020. The management of sustainable fashion design strategies: an analysis of the designer's role. Journal of Cleaner Production 268 (122112).

Dai, J., Dong, H., 2014. Intensive cotton farming technologies in China: achievements, challenges and countermeasures. Field Crops Research 115, 99–110.

Das, P.P., Chaudhary, V., 2021. Moving towards the era of bio fibre based polymer composites. Cleaner Engineering and Technology 4 (100182).

Dissanayake, G., Perera, S., 2016. New approaches to sustainable fibres. In: Sustainable Fibres for Fashion Industry. Springer.

Dissanayake, D.G., Weerasinghe, D., 2021. Towards circular economy in fashion: review of strategies, barriers and enablers. Circular Economy and Sustainability 1–21.

Farrington, D.W., Lunt, J., Davies, S., Blackburn, R.S., 2005. Poly (lactic acid) fibers. Biodegradable and Sustainable Fibres 6, 191–220.

Fletcher, K., 2013. Sustainable Fashion and Textiles: Design Journeys. Routledge.

Franck, R.R., 2001. Silk, Mohair, Cashmere and Other Luxury Fibres. Woodhead Publishing Limited, Kidlington, UK.

Ghavami, K., 2005. Bamboo as reinforcement in structural concrete elements. Cement and Concrete Composites 27, 637–649.

Gholamzad, E., Karimi, K., Masoomi, M., 2014. Effective conversion of waste polyester– cotton textile to ethanol and recovery of polyester by alkaline pretreatment. Chemical Engineering Journal 253, 40–45.

Greco, S., De Cock, B., 2021. Argumentative misalignments in the controversy surrounding fashion sustainability. Journal of Pragmatics 174, 55–67.

Guo, Z., Eriksson, M., de la Motte, H., Adolfsson, E., 2021. Circular recycling of polyester textile waste using a sustainable catalyst. Journal of Cleaner Production 283 (124579).

Hassan, M.M., Carr, C.M., 2019. A review of the sustainable methods in imparting shrink resistance to wool fabrics. Journal of Advanced Research 18, 39−60.

Hirano, S., Zhang, M., Chung, B.G., Kim, S.K., 2000. The N-acylation of chitosan fibre and the N-deacetylation of chitin fibre and chitin−cellulose blended fibre at a solid state. Carbohydrate Polymers 41, 175−179.

Hong, F., Guo, X., Zhang, S., Han, S., Yang, G., Jonsson, L.J., 2012. Bacterial cellulose production from cotton-based waste textiles: enzymatic saccharification enhanced by ionic liquid pretreatment. Bioresource Technology 104, 503−508.

Imran, M.A., Ali, A., Ashfaq, M., Hassan, S., Culas, R., Ma, C., 2019. Impact of climate smart agriculture (CSA) through sustainable irrigation management on Resource use efficiency: a sustainable production alternative for cotton. Land Use Policy 88 (104113).

Ion, S., Voicea, S., Sora, C., Gheorghita, G., Tudorache, M., 2021. Sequential biocatalytic decomposition of BHET as valuable intermediator of PET recycling strategy. Catalysis Today 366, 177−184.

Islam, M., 2020. Mapping environmentally sustainable practices in textiles, apparel and fashion industries: a systematic literature review. Journal of Fashion Marketing and Management 25 (2), 1361−2026.

Jacob, K., Petersen, L., Horisch, J., Battenfeld, D., 2018. Green thinking but thoughtless buying? An empirical extension of the value-attitude-behaviour hierarchy in sustainable clothing. Journal of Cleaner Production 203, 1155−1169.

Jiang, Z., Zhang, N., Wang, Q., Wang, P., Yu, Y., Yuan, J., 2021. A controlled, highly effective and sustainable approach to the surface performance improvement of wool fibres. Journal of Molecular Liquids 322 (114952).

Kebler, L., Matlin, S.A., Kummerer, K., 2021. The contribution of material circularity to sustainability—recycling and reuse of textiles. Current Opinion in Green and Sustainable Chemistry 32 (100535).

Luttikholt, L.W.M., 2007. Principles of organic agriculture as formulated by the international federation of organic agriculture Movements. NJAS—Wageningen Journal of Life Sciences 54, 347−360.

Majumdar, A., Mukhopadhyay, S., Yadav, R., 2010. Thermal properties of knitted fabrics made from cotton and regenerated bamboo cellulosic fibres. International Journal of Thermal Sciences 49, 2042−2048.

Mancini, F., Termorshuizen, A.J., Jiggins, J.L., van Bruggen, A.H., 2008. Increasing the environmental and social sustainability of cotton farming through farmer education in Andhra Pradesh, India. Agricultural Systems 96, 16−25.

Matlin, S.A., Mehta, G., Hopf, H., Krief, A., Kebler, L., Kummerer, K., 2020. Material circularity and the role of the chemical sciences as a key enabler of a sustainable post-trash age. Sustainable Chemistry and Pharmacy 17 (100312).

Morton, W.E., Hearle, J.W.S., 2008. Physical Properties of Textile Fibres. Elsevier and Woodhead Publishing, Kidlington, UK.

Muthu, S.S., 2014. Roadmap to Sustainable Textiles and Clothing: Eco-Friendly Raw Materials, Technologies, and Processing Methods. Springer.

Muthu, S.S., Li, Y., Hu, J.Y., Mok, P.Y., 2012. Quantification of environmental impact and ecological sustainability for textile fibres. Ecological Indicators 13, 66−74.

Muzzarelli, R.A.A., 2013. Chitin. Elsevier.

Myers, D., Stolton, S., 1999. Organic cotton: from field to final product.

Navone, L., Moffitt, K., Hansen, K.-A., Blinco, J., Payne, A., Speight, R., 2020. Closing the textile loop: enzymatic fibre separation and recycling of wool/polyester fabric blends. Waste Management 102, 149–160.

Nayak, R., Houshyar, S., Patnaik, A., Nguyen, L.T., Shanks, R.A., Padhye, R., Fegusson, M., 2020. Sustainable reuse of fashion waste as flame-retardant mattress filing. Journal of Cleaner Production 251 (119620).

Nibouche, S., Guerard, N., Martin, P., Vaissayre, M., 2007. Modelling the role of refuges for sustainable management of dual-gene BT Cotton in West African smallholder farming systems. Crop Protection 26, 828–836.

Park, S.H., Kim, S.H., 2014. Poly(ethylene terephthalate) recycling for high value added textiles value added textiles. Fashion and Textiles 1 (1).

Patil, S., Reidsma, P., Shah, P., Purushothaman, S., Wolf, J., 2014. Comparing conventional and organic agriculture in Karnataka, India: where and when can organic farming be sustainable? Land Use Policy 37, 40–51.

Patnaik, A., Mvubu, M., Muniyasamy, S., Botha, A., Anandjiwala, R.D., 2015. Thermal and sound insulation materials from waste wool and recycled polyester fibres and their biodegradation studies. Energy and Buildings 92, 161–169.

Provin, A.P., de Aguiar Dutra, A.R., Machado, M.M., Cubas, A.L., 2021. New materials for clothing: rethinking possibilities through a sustainability approach—a review. Journal of Cleaner Production 282 (124444).

Rao, K., Murali, M., Mohana Rao, K., Prasad, A.V.R., 2010. Fabrication and testing of natural fibre composites: vakka, sisal, bamboo and banana. Materials & Design 31, 508–513.

Rijavec, T., Zupin, Z., 2011. Soybean protein fibres (SPF). In: Recent Trends for Enhancing the Diversity and Quality of Soybean Products. InTech.

Shafie, S., Kamis, A., Ramli, M.F., Bedor, S.A., Puad, F.N., 2021. Fashion sustainability: benefits of using sustainable practices in producing sustainable fashion designs. International Business Education Journal 14 (1), 103–111.

Shen, M., Huang, W., Chen, M., Song, B., Zeng, G., Zhang, Y., 2020. (Micro) plastic crisis: unignorable contribution to global greenhouse gas emissions and climate change. Journal of Cleaner Production 254 (120138).

Stenton, M., Houghton, J.A., Kapsali, V., Blackburn, R.S., 2021. The potential for regenerated protein fibres within a circular economy: lessons from the past can inform sustainable innovation in the textiles industry. Sustainability 13 (2328).

Sumner, M.P., 2015. Sustainable apparel retail (chapter 8). In: Blackburn, R. (Ed.), Sustainable Apparel: Production, Processing and Recycling. Elsevier and Woodhead Publishing, Kidlington, UK, pp. 199–217.

Sun, X., Wang, X., Su, F., Tian, M., Qu, L., Perry, P., Ovens, H., Liu, X., 2021. Textile waste fibre regeneration via a green chemistry approach: a molecular strategy for sustainable fashion. Advanced Materials 33 (48), 2105174.

Textile Exchange, 2021. Fibre and Materials Market Report 2021. https://textileexchange.org/wp-content/uploads/2021/08/Textile-Exchange_Preferred-Fibre-and-Materials-Market-Report_2021.pdf. (Accessed 17 December 2021).

Tiza, T.M., Singh, S.K., Kumar, L., Shettar, M.P., Singh, S.P., 2021. Assessing the potentials of bamboo and sheep wool fibre as sustainable construction materials: a review. Materials Today Proceedings 47 (14), 4484–4489.

Tonk, R., 2021. Natural fibres for sustainable additive manufacturing: a state of the art review. Materials Today Proceedings 37, 3087–3090.

Tshifularo, C.A., Patnaik, A., 2020. Recycling of plastics into textile raw materials and products (Chapter 13). In: Nayak, R. (Ed.), Sustainable Technologies for Fashion and Textiles. Elsevier and Woodhead Publishing, Kidlington, UK, pp. 311−326.

Vanhoorne, M., De Rouck, A., De Bacquer, D., 1995. Epidemiological study of eye irritation by hydrogen sulphide and/or carbon disulphide exposure in viscose rayon workers. Annals of Occupational Hygiene 39, 307−315.

Vynias, D., 2011. 'Soybean fibre: a novel fibre in the textile industry. In: Soybean-Biochemistry, Chemistry and Physiology. InTech.

Wagner, M.M., Heinzel, T., 2020. Human perceptions of recycled textiles and circular fashion: a systematic literature review. Sustainability 12 (24), 10599.

Wu, X.J., Li, L., 2019. Sustainability initiatives in the fashion industry. In: Beltramo, R., Romani, A., Cantore, P. (Eds.), Fashion Industry—an Itinerary between Feelings and Technology. IntechOpen Limited, London, UK.

Wu, J.-M., Liu, R.-H., 2013. Cost-effective production of bacterial cellulose in static cultures using distillery wastewater. Journal of Bioscience and Bioengineering 115 (3), 284−290.

Yimin, Q., 2007. 'Structure and properties of Seacell fiber. Journal of Textile Research 11, 033.

Part Two

Sustainable natural fibres

Organic cotton and BCI-certified cotton fibres

Ashvani Goyal and Mayank Parashar
Department of Textile Technology, The Technological Institute of Textile & Sciences, Bhiwani, Haryana, India

3.1 Introduction

Cotton is the major fibre crop grown worldwide and India leads production of cotton among the different cotton growing countries. The demand for cotton in textile sector is increasing due to its inherent fibre properties. In recent years, the awareness among the world community about ill effects of toxic chemicals and pesticides used in cultivation and processing on health is increasing due to which the people are looking for organic products of different commodities. Cotton is one of the agricultural crops grown organically. The guidelines set for organic cotton cultivation have to be strictly followed for its certification. The practices such as selection of biotic and abiotic resistant varieties, use of compost, vermicompost and organic manures, intercropping with green manures, crop rotation, use of biofertilizers, bio-pesticides and botanicals, etc. form integral components in organic cotton farming. The use of locally available biological inputs significantly reduces the cost of cultivation. Thus, organic cotton cultivation is a systematic approach for maintaining biological diversity and bringing sustainable yield and income (Mageshwaran et al., 2019).

Organic cotton farming is done without use of synthetic pesticides and chemical fertilizers and the only additives come in the form of manures, while soil quality is controlled by crop rotation. The impact on the environment is therefore reduced drastically, producing clean and safe cotton while creating a sustainable cycle (http://www.ajsosteniblebcn.cat/the-life-cycle-assessment-of-organic-cotton-fiber_38172.pdf). However, readers need to keep in mind that cotton is only one crop grown on an organic farm in rotation (or intercropping) with a number of other crops. These other crops, like pulses, maize, sorghum, wheat, chillies, vegetables, and sugarcane, are also important for cash income, for home consumption or for fodder purposes (Eyhorn et al., 2005).

Organic cotton is attracting a lot of attention at present following some impressive growth in recent years. This situation raises questions and challenges, which the industry must attempt to answer. The Farm Development Program of Organic Exchange has been working for the past 3 years both to promote the production of organic cotton and to understand and shape how it is produced and what are the key components of an organic cotton system (Ferrigno and Lizarraga, 2008).

Extrapolating from experience in industrialized countries, some critics argue that organic farming systems cannot produce sufficient yields to ensure food security and to provide enough income for smallholder farmers in developing countries. While the economic and ecological impact of organic farming systems has been studied extensively in temperate zones, little research has been done on the performance of organic farming in tropical regions. Recent case studies highlight the potential of organic farming for poverty reduction and more sustainable livelihoods in developing countries but up to now these claims have lacked scientific evidence (Drinkwater et al., 1998; Lotter, 2003; Mader et al., 2002; Offermann and Nieberg, 2001; Reganold et al., 1993; Stolze et al., 2000; Tilman et al., 2002).

3.2 Cotton fibre

Cotton fibre has many desirable properties making it a major fibre for textile applications. It combines strength with good absorbency, making it a comfortable and durable apparel fabric. Mankind first learnt to utilize cotton more than 5000 years BC in India and the Middle East. Its use spread to Europe via the Greeks after the invasion of India by Alexander the Great. Modern cotton manufacture began in England in the eighteenth century and rapidly spread to the United States, resulting in a huge increase in production and international trade. It was not until the emergence of man-made fibres in the twentieth century that cotton was displaced as the most important textile fibre. It is still the most widely used natural textile fibre with over 25.2 million tons produced annually. Consumption of cotton is still growing at a rate of 2% per annum (Agrawal, 2005).

Cotton, the seed hair of plants of the genus Gossypium, is the purest form of cellulose available in nature. After flowering, an elongated capsule or boll is formed on the cotton plant in which the cotton fibres grow. Once the fibres have completed their growth cycle, the capsule bursts and fibres emerge. A cotton capsule contains about 30 seeds. Each seed contains around 2000–7000 fibres. Depending on the cotton type and growing conditions, the colour of the fibre is usually creamy white or yellowish. Cotton fibre is mostly composed of cellulose. Under 10% of the weight of the raw fibre consists of waxes, protein, pectate and minerals (Hsieh, 2007).

The length of different kinds of cotton fibre varies from 22 to 50 mm, and the diameter from 18 to 25 µm. The higher quality fibres are known as long staple fibres or extra-long staple. There are four main commercial species of cotton from the genus Gossypium (Agrawal, 2005):

- *G. arboretum* (Middle and Far East),
- *G. herbaceum* (Middle and Far East),
- *G. hirsutum* (America),
- *G. barbadense* (America and Egypt).

3.2.1 Cotton fiber structure

Cotton fibres have a multilayered structure that has been studied for nearly a century. The structure of the primary cell wall of the cotton fibre, and particularly the outer surface layer (the cuticle), has a major influence on fibre properties, processing and use (Degani et al., 2004). Cotton fibre has a fibrillar structure which consists of a primary wall, a secondary wall and a lumen (Losonczi, 2004). Under a microscope a cotton fibre looks like a twisted ribbon or a collapsed and twisted tube (Fig. 3.1). These twists are called convolutions: there are about 60 convolutions per centimetre. The convolutions give cotton an uneven fibre surface, which increases inter-fibre friction and enables fine cotton yarns of adequate strength to be spun.

The cross-section of a cotton fibre is often described as being kidney-bean shaped. The outermost layer, the cuticle, is a thin film of mostly fats and waxes. It has the waxy layer surface with some smooth grooves. The waxy layer forms a thin sheet over the primary wall that forms grooves on the cotton surface. The primary wall comprises non-cellulosic materials and amorphous cellulose in which the fibrils are arranged in a crisscross pattern. Owing to the non-structured orientation of cellulose and non-cellulosic materials, the primary wall surface is unorganized and open. This gives flexibility to the primary wall, which is required during cell growth. The basic ingredients responsible for the complicated interconnections in the primary wall are cellulose, hemicelluloses, pectin, proteins and ions. The secondary wall, in which only crystalline cellulose is present, is highly ordered and has a compact structure with the cellulose fibrils lying parallel to one another (Agrawal, 2005).

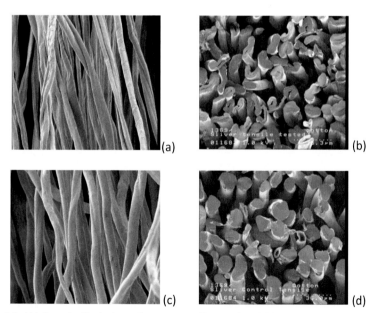

Figure 3.1 (A) Longitudinal view of raw cotton fibre (twisted ribbon structure); (B) cross-sectional view of raw cotton fibre (flat, elongated or kidney bean shaped); (C) longitudinal view of mercerized cotton fibre (rod like structure); and (D) cross-sectional view of mercerized cotton fibre (round or oval shape).

3.3 Organic cotton

Organic cotton by definition refers to any cotton that is cultivated and certified according to the organic agriculture standards. It includes cotton that is grown, harvested and processed without the use of chemical fertilizer, pesticide or herbicide, growth regulator or defoliant (Mageshwaran et al., 2019).

The crop is mostly rain-fed and does not pose any threat to human health, cattle, birds and insects or the environment as a whole. It represents an alternative method of farming that is self-sustainable. It uses many natural techniques of plant growth, pest and weed control, mulch and compost as fertilizer, intercropping for weed control and food crises management, and crop rotation for soil nutrition and soil moisture retention management (Textile Exchange, 2016a, b; Dorugade/Satyapriya, 2009).

The International Federation of Organic Agriculture Movements (IFOAM), which is the representative body for organic agriculture worldwide, defines 'organic agriculture' according to four principles (Ton, 2007):

> *The principle of health*: Organic agriculture should sustain and enhance the health of soil, plant, animal, human and planet as one and indivisible.
> *The principle of ecology*: Organic agriculture should be based on living ecological systems and cycles, work with them, emulate them and help sustain them.
> *The principle of fairness*: Organic agriculture should build on relationships that ensure fairness with regard to the common environment and life opportunities.
> *The principle of care*: Organic agriculture should be managed in a precautionary and responsible manner to protect the health and well-being of current and future generations and the environment.

3.3.1 Why organic cotton?

Organic cotton still only occupies a tiny niche of far less than 1% of global cotton production. However, the number of farms converting to organic cotton and the number of projects is constantly increasing. At present, organic cotton cultivation is reported in the following countries:

- Africa: Benin, Burkina Faso, Egypt, Mali, Mozambique, Senegal, Tanzania, Togo, Uganda, Zambia, Zimbabwe
- Asia: China, India, Kyrgyzstan, Pakistan
- South America: Argentina, Brazil, Nicaragua, Paraguay, Peru
- Middle East: Turkey, Israel
- Europe: Greece
- USA
- Australia

There are a number of reasons to grow cotton organically. The negative impacts of conventional cotton farming on the environment and health are obvious and well known. Some people may say: 'Why should I care about chemicals in cotton growing? We do not eat cotton'. But if you look at the fact that around 60% of the cotton weight harvest is cotton seed that is processed to edible oil and cattle feed, you realize that the bigger part of cotton production enters the human food chain. We also know that the pesticides sprayed on cotton do not only affect the target pest. Beneficial insects and other animals are killed, too, so that pests that formerly were of minor importance now have become a major problem (for example, whitefly and aphids). In some areas, the ground water has become so polluted with chemicals that people need to buy their drinking water from outside. In addition, many of the farmers and labourers spraying the pesticides face health problems that cause them to miss a lot of work and have additional costs for medical treatment. There are many cases in India where farmers have even died after applying chemical pesticides(Eyhorn et al., 2005).

Today, the emphasis is on environmentally friendly production practices and 'sustainable' production through the entire textile chain. Some of the reasons used to support their contentions are that conventional cotton production greatly overuses and misuses pesticides/crop protection products that have an adverse effect on the environment and agriculture workers and that conventionally grown cotton fibre/fabrics/ apparel has chemical residues on the cotton that can cause cancer, skin irritation and other health-related problems to consumers (Myers, 1999; Yafa, 2005; Organic Exchange, 2008; Patagonia, 2008; Hae Now, 2008).

3.3.2 Comparison of organic cotton with conventional cotton

Organic cotton production systems utilize biological substances rather than chemicals which are major focus of growing farming systems. Table 3.1 highlight the difference

Table 3.1 Comparison between conventional and organic cotton (Eyhorn et al., 2005).

	Conventional cotton	**Organic cotton**
Environment	• Pesticides kill beneficial insects • Pollution of soil and water • Resistance of pests	• Increases bio-diversity • Eco-balance b/w pests and beneficial insects • No pollution
Health	• Accidents with pesticides • Chronic diseases (cancer, infertility, weakness)	• No health risks from pesticides • Healthy organic food crops
Soil fertility	• Risk of declining soil fertility due to use of chemical fertilizers and poor crop rotation	• Soil fertility is maintained or improved by organic manures and crop rotation
Market	• Open market with no loyalty of the buyer to the farmer • Dependency on general market rates • Usually, individual farmers	• Closer relationship with the market partner • Option to sell products as organic at higher prices • Farmers usually organized in groups
Economy	• Higher production rates • High financial risk • High yields only in good years	• Lower costs for inputs • Lower financial risk • Satisfying yields once soil fertility has improved

between organic and conventional cotton fibre. The use of inorganic fertilizers is prohibited for cultivation of organic cotton and replaced by organic manure including farmyard and green manure, composite, cotton seed meal, fish meal, cake, leather meal or gypsum, etc. Also, the herbicides and pesticides of the botanical origin are usually used such as ipomeas, neem cake, etc. for organic cotton production while chemically synthesized pesticides, insecticides and herbicides are used for conventional cotton production. The use of chemical defoliants is also avoided in harvesting of organic cotton. Besides its significance in clothing chain, organic cotton is also a major component of food chain whereas the pesticide residues from seeds of conventionally grown cotton resides into the fatty tissues of animals and thus contaminate the meat and dairy products and pose a risk to consumers.

Organic cotton has reduced cost of production as it eliminates the cost for the use of agrochemicals for its cultivation (Fig. 3.2). However, the yield of organic cotton is less than that of conventional cotton and so organic cotton fabric costs much higher. Organic cotton production prohibits the use of toxic chemicals so not only farmer enjoys a healthy environment, but it also prevents environmental pollution. Soil also remains fertile as its composition is not destroyed by chemicals. Organic cotton production also eliminates the need for the treatment of water contaminated with chemicals as organic treatment process require just simple nontoxic dyes instead of hazardous chemicals, for example, chlorine, toxic finishes, or bleach, etc. (Patagonia, 2008; Hae Now, 2008).

3.3.3 History of organic cotton production

The certified production and consumption of organic cotton dates back to the early 1990s, when pioneers in the United States and Turkey started to create markets for

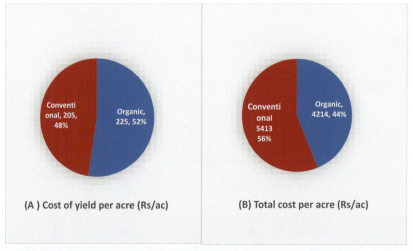

Figure 3.2 Cost of cultivation per acre (Rs/ac) for organic and conventional cotton cultivation (Raj et al., 2005).

cotton that was grown as a rotational crop on certified organic farms. The first organic cotton textiles brought to the market consisted of a limited range of 100% certified organic cotton products, which were sold in a small number of dedicated shops—usually natural and health food stores. They were primarily marketed for their ecological characteristics, rather than for their quality, design or fashionable appeal (Ton, 2007).

In late 1992, some environmentally motivated textile and clothing designers and companies launched the sale of more fashionable ecological textiles, later known as the 'eco-look' in fashion. Ecological textiles were now also for sale in fashionable shops such as Esprit and Hennes & Mauritz (H&M), in addition to the continuing sales in health food and natural textile shops. Products on offer tended to be either 'ecru' or dyed with soft 'natural' colours. Environmental awareness about fibre, textile and clothing production was created among consumers and in the industry. However, there was also confusion about the value of the different environmental claims found in the marketplace (e.g., 'natural' or '100% hand-picked cotton'). The design, quality and colour range of organic cotton items improved significantly in the second part of the 1990s. The range of yarns and fabrics available expanded, which widened the offer and the quality of the organic cotton textiles and clothing for sale. Supply of organic cotton fibre was in excess though, while overall demand stagnated. Several large United States-based companies involved in organic cotton usage at the time, such as Levi's and the Gap, withdrew from organic cotton use (Ton, 2007).

By 2000, in the United States, large companies operating internationally became increasingly concerned about the value of their brand and image, following public concern about social issues such as child labour and the working conditions in sweatshops. Some brands became aware that involvement in organics might help them to increase or restore their brand value and image. They were not keen though on paying significantly more to farmers for certified organic cotton than for conventional cotton. Supplying large brands would also require fibre volumes which were enormous for the newly emerging organic cotton market. Furthermore, organic cotton actors (in the United States in particular) were aware that the provision of large volumes of fibre to only a few large companies would generate a high level of dependency and risk. A solution was found in the development of so-called blending programs, in which brands engage in the use of a small percentage of organic cotton fibre (e.g. starting with 3%–5%) in their products. At the spinning mill, the organic cotton fibre is mixed into conventional cotton yarn or into other yarn types. By 2006, many large and medium-sized textile and clothing companies had followed the example of Nike and others and launched organic cotton conversion programs. The demand for organic cotton is increasing rapidly, with 100% organic cotton items now showing up in regular fashion fairs such as Magic (United States), Première Vision (France) and the London Fashion Week (United Kingdom) (Ton, 2007).

3.3.4 Current status of organic cotton

Production of organic cotton currently takes place in 22 countries across the world, covering the same regions as conventional cotton. However, the vast majority (87%)

of production takes place in just three countries: India, Turkey and Syria. The total land area globally confirmed by research is 161,000 ha. The extrapolated total area in organic cotton production is close to 200,000 ha. At least 177,678 confirmed farmers (with 70% of projects reporting data) are involved of whom nearly 18% are women. Total organic cotton farmer numbers are estimated at 217,000 (Organic Exchange, 2008). Growth in production has been until recently ad-hoc, meaning that while organic certification is generally to one of the EU (2092/91) or US (NOP) legal standards, the organization of production and farmers is not to any recommended or accepted set of practices (Ferrigno and Lizarraga, 2008).

India is currently the largest producer of organic cotton in the world, followed by China, Turkey, the USA and others (Textile Exchange, 2016a,b). Commercial organic cotton is now grown in more than 22 countries worldwide, including the sub-Saharan countries like Africa, Uganda, Tanzania and others (Afari, 2010). From a decent production of about 40,000 tons in 2005–06, organic cotton production in India went up to 241,697 tons, accounting for 81% of global organic cotton production in 2009–10. But due to the decrease in cotton fibre prices, production decreased by 3.8% in 2015–16. However, India is still the pioneer in organic cotton production, contributing around 74% of the world share in 2015. According to Statista and the US Department of Agriculture (2018), India still leads the world in organic cotton production at 5,748,000 metric tons, followed by China at 4,790,000 metric tons and USA at 2,680,000 metric tons. It shows a positive trend towards the adoption of organic cotton farming majorly by Indian farmers. It also indicates that India is facing a major challenge in the world market due to the rapid production growth of China, USA, Pakistan, Brazil and other countries, leading to reduced exports from India (Textile Exchange, 2017).

3.3.5 Organic cotton standard production practices

Organic cotton production system is a complex system as it requires stable environment and further farmers have required to manage the stable farm conditions to sustain and optimize the yield. This requires a heavy investment to maintain soil fertility. Organic cotton is majorly produced in the areas which grow cotton as the major cash crop or its cash crop (Memon, 2012). A systems approach to organic production involves the integration of many practices (cover crops, strip cropping, grazing, crop rotation, etc.) into a larger system. In order to market a crop as organic, a grower must be certified through a third party. This process involves several on-farm inspections and paying a certification fee. Organic production begins with organically grown seed. If certified organic seed cannot be located, untreated seed may be used as long as it is not derived from genetically modified (GM) plants (Mitchell, 1988).

Organic cotton production practices in India are an amalgamation of traditional wisdom and modern scientific knowledge. Management of ecosystem and use of local natural resources are inherent to all the organic cotton programs. Today, organic cotton encompasses a range of farming systems including those known as biological, low-input, organic, regenerative, green ecofriendly or sustainable. These alternate systems involve crop rotation, bio-control based Integrated Pest Management, conservation tillage, crop diversification including cover cropping and green-manuring, genetic improvement, disease/pest prevention and soil health amelioration. Further, the sound organic cotton production packages are developed based on the following principles (Mageshwaran et al., 2019):

- Promotion of natural processes nutrient cycling, N fixation, cotton plant-pest-natural energy integration of these complementary processes into the production system.
- Avoiding the use of those off-farm inputs that are a potential threat to the soil and aquatic environment, as well as health of farmer, farm animal and consumer.
- Allowing the biological/genetic potential of the variety/hybrid to be expressed fully with minimum interventions.
- Improved farm management to conserve soil, water, energy, and biological diversity and on-farm agro-wastes.
- A fertile soil is a pre-requisite for organic cotton production.
- The organic C content of soil should be improved and stabilized at such a level that the anticipated production levels do not decline it.
- Crop rotation with legume, cover cropping, green manuring, compost (vermicompost, Trichocompost, FYM), bio-mulch, bio-fertilizer are generally employed to improve fertility status.
- Soil amendments and naturally mined permitted (or regulated) chemicals can be employed to supplement native fertility
- Weed management is primarily achieved through preventive techniques (selection of perennial weed-free field, clean seeds, completely decomposed compost/FYM, crop rotation, cover cropping, mulching, smother crop, etc.) and soil solarization. Cultural, mechanical and manual method can be employed to supplement preventive measures.
- Pest management is achieved through the selection of pest tolerant variety, conservation of natural enemy and inundate release of predator/parasite pathogen.
- Since promotion on locally available on-farm resources is implied in all organic production system, no clear-cut package of practices can be prescribed. Farmers use their innovative power and in consultation with any certifying agency decide the inputs used.
- ICAR-CICR, Nagpur has developed a package of practices for organic cotton cultivation. The farmers can contact the Institute for more information. The biological inputs used in organic cultivation such as green manuring, organic liquid preparation, organic manure, biocontrol agent, and biofertilizer and biopesticide.

3.3.6 Growing organic cotton—a system approach

Converting a farm to organic production does not simply mean replacing chemical fertilizers and pesticides with organic ones. Organic cotton must be grown in a diverse and balanced farming system that also includes the other crops. Instead of troubleshooting, organic farmers should try to prevent problems and avoid substitutes to conventional inputs as far as possible. This requires a thorough understanding of nutrient and pest management and the ability to continuously observe and learn. To get satisfactory yields and income in organic cotton farming it is necessary to adopt a number of integrated measures in a system approach, ensuring that the interaction among soil, plants, environment and people is well balanced. The 'ingredients for success' all need to be applied together (Eyhorn et al., 2005):

- Suitable measures to improve and maintain soil fertility.
- Establishment of crop rotation and crop diversity; fostering natural balance
- Selection of varieties suitable to the conditions (soil, availability of irrigation, market requirements).
- Appropriate types and amounts of manures at the right time.
- Timely crop management such as intercultural operations, weeding and irrigation.
- Careful monitoring of the crop and sufficient protection against pests according to the concept of economic threshold level.
- Timely and proper picking of the cotton.
- Sufficient documentation for inspection and certification.
- Capacity building and experimenting for continuous improvement.

Following are the steps to grow cotton from seeds:

Step 1. Planning and growing

For growing organic cotton, people work with nature and to enhance the production, the organic farmers use the systems which are biologically based. Organic farmers are trying to manage the hazardous weeds and insects to keep the ecological balance and to secure the environment (Muthu, 2014).

The soil—Healthy soil is necessary for organic farming. The soil not only acts as a growing medium for plants, but also a living system for keeping the soil healthy and productive. Synthetic fertilizers can be replaced by compost, efficient nutrient recycling, cover crops and frequent crop rotation (Muthu, 2014). The organic matter in soil increases the biological activity of the soil by promoting activity of earthworms and microbes, which continuously work toward improving the soil fertility. The strategy such as crop residue management, intercropping application of compost, green manuring and crop rotation improve the organic matter content in the soil (Mageshwaran et al., 2019).

Weed control—Weeds spread can be controlled by hoes and other mechanical implement, rotation of crops, efficient use of irrigation water, intercropping, using mulches and also by adjusting the densities of plants and panting dates (Abrantes et al., 1978; Qamar et al., 2015; Shabbir et al., 2014; Elahi et al., 2011a,b; Aaliya et al., 2016). Careful observation of weed populations and the use of shallow soil cultivation (hoes, weeders), combined with selective hand weeding, usually allow the experienced organic cotton farmer to 'keep on good terms' with weeds (Eyhorn et al., 2005).

Pest control—Farmers can reduce the likelihood of insects, birds or mammas that cause damage to crop by biological diversity. The predator insects, crop rotation, biological pesticides like neem oil and intercropping can be used to control the pests by the farmer (Abrantes et al., 1978; Francis and Clegg, 2020; Puspito et al., 2015; Husnain, 2015). The best strategy in pest and disease management in organic cotton cultivation is to maintain diverse balance in farm ecosystem through increasing the population of natural enemies. The other strategies to support plant health are: diverse cropping system, intercropping, balanced nutrition, appropriate water management and employment of biocontrol system (Mageshwaran et al., 2019).

Step 2. Harvesting

Organic cotton is usually picked by hand without using machinery, defoliants or chemicals, especially in developing countries and it reduces the waste (Choudhury, 2015).

Step 3. *Cleaning and ginning* There are many processes involved in the manufacturing of cotton fibre into fabrics and garments, that is, cleaning, spinning, knitting, weaving, dyeing, cutting, assembly and finishing. Fibre producers receive processed plants and then remove plant material and waste by dividing and carding the lint. Waste is the mixture of soil, leaves, stems and lint. Cotton is an important source of food and it consists of 40% fiber and 60% seed by weight. After ginning, the fibre is processed in textile mills, while the by-products and seeds are used as food source (McGarry, 2009; Lewis et al., 1997). The cotton grown organically can be used for production of organic food and it is also valuable in clothing chain (Claudio, 2007).

3.3.7 Sustainable organic cotton

Organic cotton, to be a sustainable method of producing fibre, needs to be more than just 'environmentally friendly'. It needs to be productive, offer decent returns to farmers, and in a growing world with limited returns, needs to be efficient in terms of land use and offer opportunities to be more than just a fibre production system, that is, it needs to be an efficient system for producing other crops and offering other value to its farmers and to the wider world, such as opportunities to 'clean' air and water and reduce carbon footprints, for example. To be productive, organic cotton production needs to take place on fertile soils and in 'perfect' conditions where yields can easily match those from other production methods. In cases where organic cotton plays the role of a social safety net, 'capturing' farmers who can no longer afford to produce using other methods, this may be modified slightly (Ferrigno and Lizarraga, 2008).

Organic cotton has been criticized for offering lower yields and being water intensive, although both these points are far from borne out by the reality. Even where some evidence exists, the story is far from simple. For example, lower yields may be more related to the socio-economic status of the farmers or to areas where all yields are low because of water shortages, climatic conditions and change or general environmental degradation. With regards to water, most organic cotton is produced in rain-fed systems, while projects in areas with irrigation have often invested heavily in efficient technologies such as drip irrigation (Ferrigno and Lizarraga, 2008).

However, on a global level we can point towards some ingredients of a sustainable system, despite the diversity of production environments globally. These can be divided between the system within the farm (soil, land availability, inputs, water availability, labour, and so on) and the wider farm environment (market conditions, prices, regulatory issues, support services, organization and logistics and infrastructure, policy) (Ferrigno and Lizarraga, 2008).

Organic cotton production requires the involvement of many different sectors, and often, partnerships between public and private sectors, NGOs and business stakeholders as well as farmers. This makes it a very complex sector, more so because of the diversity of ecological, environmental, cultural, social, economic and political contexts in which farming takes place. Making sense of this complexity to understand what makes production sustainable requires some understanding of the production systems and farmers profiles (e.g., marginal systems, population pressure, fragmentation, smallholders, resource poor, etc.) and some analysis of successful business models and their objectives, reports, indicators, and factors influencing these from outside such as markets and policy (Ferrigno and Lizarraga, 2008).

3.3.8 Future prospects of the organic cotton market

In order to develop a better idea of the future prospects of the international market for cotton products, a holistic view of the entire fibre-to-clothing value chain is needed. Here we examine the arguments for and against expanding production of organic cotton, at each stage in the chain, as well as the opportunities for the industry (Ton, 2007).

3.3.8.1 Arguments in favour of expanding organic cotton (Ton, 2007)

Production

- Organic cotton is in demand. Production and trade increased 70% per year on average between 2001 and 2006 and have more than doubled annually since 2004.
- Cotton is an important rotational crop for many organic farmers in the world. Its cultivation will increase with growth of the market for organic produce.
- The additional production costs of organic cotton may be limited in systems with multiple organic crops, while overhead costs can be shared between various crops.
- Organic cotton production also has potential in production systems where cotton is the main cash crop or the sole cash crop. Here, market prices, access to information and production efficiency are important parameters in farmers' decision-making about whether to convert to organic production.

Processing

- Many spinning and textile mills around the world are involved in organic cotton processing. This favours economies of scale in processing.
- Blending organic cotton at some minor percentage at the level of spinning is an effective way to increase fibre demand at low cost.
- Infrastructure for the knitting of organic cotton is well established, more so than for weaving, because of the lower minimum quantities required per production run. Brands and retailers generally start with the production and sale of knitted items when first engaging in organic cotton usage.

Retail

- The concept of 'organic cotton' is successfully being marketed to brands and retailers in the fashion industry as being part of their policies for CSR.

- The new involvement of large brands and retailers increases the number of points of sale exponentially, making organic cotton items available to consumers in the usual points of purchase for textiles and clothing. Organic cotton items are increasingly found in regular sale channels such as high-street fashion shops, department stores, and supermarkets.

3.3.8.2 Arguments against expanding organic cotton (Ton, 2007)

Production
- The demand for organic cotton fibre currently outstrips supply. Supply growth is lower than demand growth. About half of global organic cotton is produced by two single projects—one in Turkey and one in India. This points to a fragile market.
- Easy options for production expansion were available a few years ago, when the market for organic cotton was much smaller than the market for other crops in the organic production system. Cotton could then relatively easily be added as a rotational crop. This is less the case today.
- Organic agriculture provides technical challenges to ensure appropriate yields and income. Conversion to organic takes time, knowledge and expertise.
- During the conversion to organic agriculture farmers are usually not rewarded with a price premium for their in-conversion produce. Organic farmers face significant financial risks in conversion.

Processing
- Organic cotton fibre, yarn, fabrics and garments cannot be distinguished from conventional ones, and generally not even from GM cotton, other than through documentation about production lots and volumes.
- The additional costs of blending organic cotton at low percentages have thus far not been rewarded by brands and retailers through price premiums. The involvement of spinning mills has been more or less imposed upon them by brands and retailers.

Retail
- Organic cotton demand currently outstrips supply.
- Many brands and retailers do not advertise to the general public their involvement in organic cotton, because the organic quality of the product is only an additional feature in consumers' purchasing decisions.

3.3.8.3 Opportunities for organic cotton

Production
- Organic cotton production has potential in areas where cotton is the main cash crop or the sole cash crop, as long as market prices, access to information and production efficiency are ensured.
- Organic cotton projects in countries in the South may aim for participation in fair trade to receive a higher price for most or part of their produce.
- Fair trade cotton growers have a higher probability of becoming organic fair trade producers than conventional producers, because of their more frequent linkage with consumer markets. Fair trade cotton eliminates the use of the most toxic and dangerous cotton chemicals in production (Ton, 2007).

Processing

- Organic cotton demand will continue to grow in the future, thus increasing the number of spinning and textile mills involved and enlarging the range of intermediate and end-products available to the industry and to consumers.
- With demand outstripping supply, organic cotton prices are likely to increase. This creates opportunities for processors to increase the price of organic cotton yarn, fabrics and garments beyond the additional production costs of organic cotton.
- Brands and retailers using organic cotton are likely to be among the first to welcome and implement globally agreed standards for organic textile processing. 26 Organic textile processing is a logical next step adding value to organic fibre use.

Retail

- The importance end-consumers attach to health and 'wellness' is likely to increase in time, to the benefit of organic agriculture and trade.
- Price differences between organic cotton items and conventional items may decrease, because of more efficiency in processing (higher-volume production runs, etc.), and following increased product availability. Organic cotton items may evolve in the marketplace from being a specialty item towards becoming commonly available goods.
- Consumer information about the organic nature of organic cotton items (for example through labels inside, hang-tags, consumer brochures or advertisements) is still in its infancy. New strategies and tools may be developed by brands and retailers to cash in on their involvement in organic cotton, improving their image and profile among consumers.

3.3.9 National obligatory standards for organic cotton and organic cotton certifiers

Certification is a prerequisite for a product that is to be sold as 'organic' cotton. Certification provides a guarantee that a specific set of standards has been followed in the production of the organic cotton. Cotton was first certified as organic in 1989/1990 in Turkey. Currently there are hundreds of private organic standards worldwide for all organic products. Organic standards have been codified in the technical regulations of more than 60 governments. There are at least 95 accredited certifying companies/organizations in the world dealing with cotton. EU regulations, International Federation of Organic Agriculture Movements (IFOAM) standards, and the US National Organic Standards (NOP) have helped to formulate organic farming legislation and standards in the world. Certifying companies develop their own standards but all are essentially comparable. Organic cotton producers have to commit to follow the standards set by the certifying organizations/companies, and this includes verification through visits by independent third parties. The certifying agency must be accredited, and recognized by buyers, and the system must be independent and transparent (Wakelyn and Chaudhry, 2009).

3.3.9.1 International Federation of Organic Agriculture Movements (IFOAM)

The IFOAM (2008) is an organic umbrella organization that was established in 1972 and unites member organizations in 108 countries. IFOAM implements specific projects to facilitate the adoption of organic agriculture, particularly in developing countries. The IFOAM adopted basic standards for organic farming and processing in 1998. IFOAM standards, revised over time, are not binding for any country/producer of organic agricultural products, but they do provide valuable guidelines for organic producers and processors. The IFOAM Organic Guarantee System enables organic certifiers to become 'IFOAM Accredited' and this allows their certified operators to label products with the IFOAM Seal, in conjunction with their own seal of certified production/processing.

3.3.9.2 European Union Regulation 2092/91, regulation of organic cotton (Europe)

The EU adopted Regulation 2092/91 in June 1991 for the organic production of agricultural products and labelling of organic plant products. It came into effect in 1993.

The logo assures that the product complies with the EU rules on organic production. EU regulation 2092/91 permits labelling a product with 'Organic Farming' if:

- at least 95% of the product's ingredients have been organically produced.
- the product complies with the rules of the official inspection scheme.
- the product has come directly from the producer or preparer in a sealed package.
- the product bears the name of the producer, the preparer or vendor and the name or code of the inspection body.

For labeling, the USDA NOP standards allow

- a product to be certified as '100% organic', if it is all organic or contains 95% organic ingredients.
- a product to be certified as 'made with organic ingredients', if it has at least 70% organic ingredients.

3.3.9.3 Japanese Agricultural Standard (JAS)

The production and processing of organic textiles sold in the Japanese market are regulated/certified by the Japanese Agricultural Standard (2001). These certified organic textiles can be identified with the JAS organic seal of the Japanese government.

3.3.9.4 Australian Organic Standard (AOS)

The Australian Organic Standard (AOS, 2006) outlines requirements for operators wishing to attain certification. This includes records required, inputs allowed, and minimum practices required. Ultimately, certification to the AOS allows certified operators to use the Australian Certified Organic bud logo which is the customers' guarantee of organic integrity. It is the only organic standard that gives accountability of organic status for the organic industry in Australia. In addition to organic criteria, the standard includes certification for biodynamic production systems, as well as criteria for the registration of agricultural input products.

3.3.9.5 Organic Exchange

The Organic Exchange (2008) is a non-profit organization in the USA committed to expanding organic agriculture, with a specific focus on increasing the production and use of organically grown fibres, such as cotton. It has sponsored work at the farm level in India and many countries in Africa and Latin America. The Organic Exchange is accredited by USDA NOP to certify organic cotton production. It also tracks world production of organic cotton. The Organic Exchange is funded through company sponsorships and revenues from its activities. It has an 'Organic Cotton' logo for use by its member companies to identify their products or family of products, from yarn to finished goods that contain organic cotton.

3.4 BCI (Better Cotton Initiative)

In view of these socio economic and environmental problems arising from conventional agriculture, environmentalists and ecologists have been emphasizing the promotion of an innovative cleaner alternative. The sustainability challenges facing agriculture can be addressed through innovation and, therefore, have sparked interest in sustainable innovation (Boons et al., 2013).

Alternative to chemical-intensive conventional cotton, Better Cotton Initiative (BCI) provides an opportunity to cotton farmers around the world to grow cotton in a way that is more environmental friendly. It improves the social and economic status of cotton growers by reducing the use of fertilizers, pesticides, and irrigation. BCI also involves farm practices promoting decent work standards and ensuring fibre quality. Thus, the objectives of BCI are to improve the environment and increase the income of cotton growers on sustainable bases. Better Cotton Standard System (BCSS) involves the adoption of better management practices in cotton cultivation, encouraging farming community to increase the quantity and quality of cotton yield, thereby leading to an increase in profit (BCI, 2017).

BCI is a multi-stakeholder initiative launched in 2005 with the objective of globally raising the sustainability standard of conventional cotton and making it available in large volumes to mainstream retailers at the same price as conventional cotton. The BCI exists to make global cotton production better for the people who produce it, better for the environment it grows in and better for the sector's future. The principles, covering a range of environmental and social aspects of cotton production, describe the broad areas that need to be addressed by the farmer for the production of Better Cotton. The Minimum Production Criteria represent the initial core requirements for farmers to grow Better Cotton. It has also been agreed to help farmers adopt practices consistent with the BCI Production Principles and BCI will coordinate a program of farmer support activities, delivered through experienced implementing partners. The program will enable knowledge sharing and skills development on Better Cotton, help smallholder farmers to organize and advocate more effectively, and facilitate equitable access to responsible financial services and to market as well. Under supply chain support, BCI will create a 100% Better Cotton bale, and connect the supply of

Better Cotton to demand, putting in place a supply chain system that facilitates the procurement of Better Cotton (Makhdum et al., 2011).

For achieving the objectives, BCI has launched capacity building programs through implementing partners for training farmers to grow cotton according to BCSS in cotton-growing regions of the world. In BCSS, farmers are registered and grouped into LGs (learning group) comprising of 30–40 farmers. LG meetings are organized on regular basis, and all farmers are trained on different aspects of BC principles by implementing partner. Each farmer is required to fill farmer field card containing information about all farm-related operations and complete information about input and output uses. To minimize pesticide exposure, promote water stewardship, care of soil health, promotion of decent work practices, care for natural habitats, and ensure fiber quality are six principles of better cotton. Under these principles, there are 24 criteria for which compliance of small farmers is checked. Farmers are assessed through Better Cotton Assurance Program by conducting second-party and third-party credibility checks. On the basis of the outcome of the second- and third-party credibility checks, license of producing better cotton is awarded to those farmers who follow the criteria of BCSS (BCI, 2014). So, a farmer who minimizes the pesticide exposure, ensures water use efficiency, takes care of soil health, promotes decent work practices, cares about natural habitats, and ensures quality of fiber is the better cotton farmer. BCSS has already been introduced in 21 cotton growing countries across the world, and approximately 14% of the global cotton is produced by 1.3 million licensed better cotton farmers (BCI, 2017).

3.4.1 Why BCI?

In view of social, environmental and economic problems arising from conventional agriculture, environmentalists and ecologists have been emphasizing the promotion of organic agriculture to make agriculture sustainable. Organic agriculture uses only organic fertilizers and pesticides to ensure the sustainability of agro-ecological systems (Samie et al., 2010). Organic agriculture, on the other hand, cannot meet the demands of an ever-increasing population for food and fibre because its output is often significantly lower than that of conventional agriculture. Organic agriculture is financially unfavourable for farmers that make labour and capital investments in the expectation of a positive financial return due to its poor yield and high labour costs (de Ponti et al., 2012; Rattanasuteerakul and Thapa, 2012). Moreover, a genuine organic agriculture has strict certification requirements entailing farmers to maintain all the record of inputs used and management practices for pest and disease control, which are difficult and costly tasks for them (Chongtham et al., 2010). Therefore, it is almost impossible to make organic agriculture financially better than conventional agriculture unless appropriate policy interventions enabling organic products to fetch premium price are made (Rattanasuteerakul and Thapa, 2012).

Conventional agriculture depends highly on external inputs, including seeds, pesticides, fertilizers and irrigation water (Rasul and Thapa, 2003). Cotton produced using high amounts of external inputs, including fertilizers and pesticides, and irrigation water has been referred to as 'conventional cotton'. Although cotton is making a significant contribution to national and rural economies, the intensive use of irrigation water, inorganic fertilizer and pesticides has impinged severely on environment, public health and financial return. Conventionally, cotton farmers are using high amounts of pesticide to protect crops from pests, insects and diseases (BCI, 2013).

The findings of scientific studies have revealed that crop yields and profits can be maintained by even reducing the amounts of inputs used (Abraham et al., 2014; Coulter et al., 2011). Despite reduced application of inorganic inputs and irrigation water, the yield of 'better cotton' was found to be 11%, 18% and 15% higher than the yield of conventional cotton in China, India and Pakistan, respectively. This was attributed to reduction of the use of pesticide, fertilizer and irrigation water. However, crop variety significantly influences the yield and financial return. The basic aim of promoting 'better cotton' by BCI was to reduce the use of pesticides, fertilizer and irrigation water without compromising on the yield, thereby increasing farmers' net income. At the same time, BCI could increase financial benefit by capitalizing on the ever increasing national and international demand for cleaner agricultural products (BCI, 2013).

3.5   Bt cotton

Bt cotton is a GM pest resistant plant cotton variety, which produces an insecticide to combat bollworm. Strains of the bacterium Bacillus thuringiensis produce over 200 different Bt toxins, each harmful to different insects. Most notably, Bt toxins are insecticidal to the larvae of moths and butterflies, beetles, cotton bollworms and flies but are harmless to other forms of life. The gene coding for Bt toxin has been inserted into cotton as a transgene, causing it to produce this natural insecticide in its tissues. In many regions, the main pests in commercial cotton are lepidopteran larvae, which are killed by the Bt protein in the GM cotton they eat. This eliminates the need to use large amounts of broad-spectrum insecticides to kill lepidopteran pests (some of which have developed pyrethroid resistance). This spares natural insect predators in the farm ecology and further contributes to noninsecticide pest management.

Bt cotton is ineffective against many cotton pests such as plant bugs, stink bugs, and aphids; depending on circumstances it may be desirable to use insecticides in prevention.

3.5.1   Bt cotton production versus organic cotton production

Bt seeds are GM seeds that are high-yielding and pest-resistant in nature. The GM seeds are made by the terminator technology that strips the seeds of further reproduction capabilities. Thus, once means that the farmer has to buy new seeds every year. These seeds will result in high yields only when they are well irrigated. The farming also requires a higher input of chemical fertilizers and pesticides (Yusuf, 2010). Also, Bt cotton is a cash crop and is mostly grown on a plantation scale. It is produced as a mono crop i.e., single crop plantation. This leads to excessive soil degradation, loss of surface water, ground water pollution and proliferation of insects, pests and disease-causing pathogens, like bollworms, aphids, jassids, thrips and whitefly (Gopalakrishnan et al., 2007).

Initially, Bt cotton cultivation led to higher yields, but gradually the cost of cultivation was impacted by the high cost of seeds, pesticides and fertilizers. Monsanto completely controlled the seed market in India since its entry in 2002 and hence cheap indigenous varieties of seeds were unavailable on the market (Orphal, 2005). The farmers were uneducated about the usage of pesticides for Bt cotton and indiscriminately used pesticides 20–30 times as against the requirement of 15 times. Thus, the pests developed resistance and reinfested the Bt cotton, thereby lowering production. New pests started attacking the fibre, necessitating an ever great number of pesticide applications (Shetty, 2004).

3.6 Application of organic cotton

(i) Organic Cotton is unquestionably the most skin-friendly, relaxing, and safe natural fibre. While conventional cotton can be unpleasant to baby skin at times, Organic Cotton never is. It's the perfect material for protecting and cleaning newborns, especially for clothing, bandages, covering and cleaning wounds, baby crib beddings, baby outfits, towels, and dozens of other items. It can also be used safely in surgical procedures where contamination from any source could be lethal. Organic Cotton Seed Oil, a by-product of organic cotton production, offers a wide range of applications in snacks and livestock feed. (https://www.organicfacts.net/organic-cotton.html)

(ii) Organic cotton has been introduced as an alternative to traditional cotton industries (due to environmental constraints and worldwide eco-friendly market dynamics). Many brands are now incorporating organic cotton lines in their clothing lines. Nike, Coop Switzerland, Otto, Patagonia, and others are examples of good enterprises. Now a day, organic cotton is a reliable selection. Recently it is reported in U.S, since few years, there is two times more increase in growth due to the organic and eco-friendly textile sales. Many companies are making plans to use either pure organically produces cotton or mixed it with small percentage of conventional cotton. The more use of domestic and international organic cotton is doing by many companies (Rana et al., 2014).

People are becoming more concerned about organic cotton cultivation as they become aware of the detrimental impact of conventional cotton on the environment and persons. Organic cotton which is also known as 'green cotton, environmentally friendly cotton and bio-logical cotton' has seen a drastic rise in its demand in the past few years. In the last few years, there have been many industries that are making products from organic cotton (https://www.yarnsandfibers.com/textile-resources/other-sustainable-fibers/natural-fibers-other-sustainable-fibers/organic-cotton/organic-cotton-fiber-consumption/what-are-the-end-uses-of-organic-cotton/).

- *Personal care*: Sanitary products like napkins, make-up removal pads, cotton puffs, ear swabs are being made from organic cotton because it is much safer for skin than conventional cotton.
- *Home furnishing*: products like towels, bathrobes, sheets, blankets, bedding, beds are also being manufactured using organic cotton.

- *Children's product*: The diapers and many other stuffed toys are being made out of organic cotton because organically grown cotton is safer for the skin as they avoid the use of synthetic fertilizers, pesticides, heavy metals, chlorine and chemicals dyes for its manufacturing.
- *Apparel*: Apparel made from organic cotton is gaining a lot of attention from people who are concerned about nature and the environment. Many international apparel brands like Nike, Patagonia, Levi's, The Gap, C&A, Pact, Inditex, Boll & Branch and designers like Mara Hoffman, Eileen Fisher, Stella McCartney, etc. are using certified organic cotton in their collections.

(iii) *Organic fabrics*—Organic cotton fabric is made from cotton fibres that have been grown on the surface of the Earth for at least 3 years and have not been treated with chemical pesticides. In comparison to other materials, organic cotton cloth smells fresh and has a smooth hand feel. Furthermore, allergies are not triggered by this eco-friendly fabric. Natural cotton seeds, rather than processed cotton seeds, are used to make organic cotton fibres. Organic cotton fibres minimize soil and water pollution through a natural process. Fabrics made of organic cotton are extremely absorbent. They are cool in the summer and warm in the winter. This makes the cloth appealing all year long. This eco-friendly clothing made up of organic cotton fabric does not offer a huge colour palette, as they avoid using chemical dyes (https://thedesigncart.com/blogs/news/organic-fabrics).

3.7 Way ahead

We should provide a viewpoint that the rest of the world recognises the importance of developing a green and sustainable corporate system. India isn't an exception in this regard. Cotton is the most widely used natural fibre, and it supports billions of people around the world. In many types of textile material, it affects the lives of billions of people. As a result, it might be positioned as a symbol of sustainability and risk-free living in order to build a greener, healthier, and more equitable society for all. We've also sought to highlight the growing demand for organic cotton among a variety of brands that want to make a difference in our lives. As a result, small farmer cooperatives, non-governmental organisations, government agencies, and other private fashion merchants may band together to develop sustainable systems founded on deep ecological principles (Özalp and Öran, 2014).

3.8 Conclusions

Organic cotton fibre production, which avoids the use of toxic chemicals, is an alternative to conventional cotton fibre production that benefits farmers, employees, domesticated animals and the environment. Organic cotton production must increase to meet the demands of brands and merchants in the textile and apparel industries. Cotton may soon become a driving force behind agricultural reform, with a viable market for organic cotton encouraging farmers to convert their land to organic cultivation. Demand growth must be sustained in order for this to happen. Cotton has a number of pests that must be treated organically and without the use of chemicals. Non-synthetic pesticide approaches are the only options for weed control. Defoliation is a challenging task with limited options for completion. Farmers also need help with organic cotton production; otherwise, their investment could be jeopardized. The government and NGOs should explain the facts and realities of chemical fertiliser and pesticide use through organized extension teaching and trainings. On the other hand, rather than taking a holistic approach to natural resource management, our agricultural progress has been focused on increasing production. Cotton that is of higher grade adds to soil health, environmental preservation, and fibre quality.

'BCI Cotton' is more input efficient in terms of both volumes of mostly inorganic inputs and irrigation water consumption per unit of land, as well as financial return per unit cost of input, resulting in a significant reduction in its production cost. This, together with the higher yield (thanks to the use of organic fertilisers) and lower price, made 'better cotton' a more profitable crop than traditional cotton. As a result, the use of 'better cotton' benefited both BCI partners and farmers financially. Furthermore, 'better cotton' helped to save resources such as land and water by using fewer chemical inputs and boosting the use of organic fertilisers, bio-pesticides, and hand weeding. It also reduced the hazards to farmers' health by reducing their exposure to inorganic pesticides. It is necessary to transition from the traditional cotton system to a more profitable and sustainable alternative cotton system, such as organic cotton and BCI cotton, in order to achieve better sustainability in cotton production.

References

Aaliya, K., Qamar, Z., Ahmad, N.I., Ali, Q., Munim, F.A., Husnain, T., 2016. Transformation, evaluation of gtgene and multivariate genetic analysis for morpho-physiological and yield attributing traits in *Zea mays*. Genetika 48 (1), 423−433.

Abraham, B., Araya, H., Berhe, T., Edwards, S., Gujja, B., Khadka, R.B., Verma, A., 2014. The system of crop intensification: reports from the field on improving agricultural production, food security, and resilience to climate change for multiple crops. Agriculture & Food Security 3 (1), 1−12.

Abrantes, I.M.O., IMO, A., DE Morais, M.M.N., 1978. Nemátodos e plantas hospedeiras identificados em Coimbra, Portugal, durante 1972-1977.

Afari-Sefa, V., 2010. Horticultural exports and livelihood linkages of rural dwellers in southern Ghana: an agricultural household modeling application. The Journal of Developing Areas 1−23.

Agrawal, P.B., 2005. The Performance of Cutinase and Pectinase in Cotton Scouring. thesis. University of Twente, the Netherlands.

AOS (AUSTRALIAN ORGANIC STANDARD) (2006), http://www.bfa.com.au/_files/AOS%202006%2001.03.2006%20w%20cover.pdf.

BCI, 2013. Harvest Report 2013. Better Cotton Initiative, Switzerland.

BCI, 2014. The Better Cotton Assurance Program; Applicable from 2014 Harvest Season.

BCI, 2017. Better Cotton Production Principles and Criteria.

Boons, F., Montalvo, C., Quist, J., Wagner, M., 2013. Sustainable innovation, business models and economic performance: an overview. Journal of Cleaner Production 45, 1−8.

Chongtham, I.R., de Neergaard, A., Pillot, D., 2010. Assessment of the strategies of organic fruit production and fruit drying in Uganda. Journal of Agriculture and Rural Development in the Tropics and Subtropics 111 (1), 23−34.

Choudhury, A.R., 2015. Development of eco-labels for sustainable textiles. In: Roadmap to Sustainable Textiles and Clothing. Springer, Singapore, pp. 137−173.

Claudio, L., 2007. Waste Couture: Environmental Impact of the Clothing Industry.

Coulter, J.A., Sheaffer, C.C., Wyse, D.L., Haar, M.J., Porter, P.M., Quiring, S.R., Klossner, L.D., 2011. Agronomic performance of cropping systems with contrasting crop rotations and external inputs. Agronomy Journal 103 (1), 182−192.

Degani, O., Gepstein, S., Dosoretz, C.G., 2004. A new method for measuring scouring efficiency of natural fibers based on the cellulose-binding domain-β-glucuronidase fused protein. Journal of Biotechnology 107 (3), 265−273.

De Ponti, T., Rijk, B., Van Ittersum, M.K., 2012. The crop yield gap between organic and conventional agriculture. Agricultural Systems 108, 1−9.

Dorugade, V.A., Satyapriya, D., 2009. Organic cotton. In: Man-Made Textiles in India, June 2009.

Drinkwater, L.E., Wagoner, P., Sarrantonio, M., 1998. Legume-based cropping systems have reduced carbon and nitrogen losses. Nature 396 (6708), 262−265.

Elahi, M., Cheema, Z.A., Basra, S.M.A., Ali, Q., 2011a. Use of allelopathic extracts of sorghum, sunflower, rice and brassica herbage for weed control in wheat (*Triticum aestivum* L.). International Journal for Agro Veterinary and Medical Sciences 5 (5), 488−496.

Elahi, M., Cheema, Z.A., Basra, S.M.A., Akram, M., Ali, Q., 2011b. Use of allelopathic water extract of field crops for weed control in wheat. International Research Journal of Pharmaceutical Sciences 2 (9), 262−270.

Eyhorn, F., Ratter, S.G., Ramakrishnan, M., 2005. Organic Cotton Crop Guide-A Manual for Practitioners in the Tropics. Research Institute of Organic Agriculture (FibL), Frick, Switzerland.

Ferrigno, S., Lizarraga, A., November 2008. Components of a sustainable cotton production system: perspectives from the organic cotton experience. In: Proceedings of the International *Cotton Advisory Committee, 67th Plenary Meeting*, Ouagadougou, pp. 17−21.

Francis, C.A., Clegg, M.D., 2020. Crop rotations in sustainable production systems. In: Sustainable Agricultural Systems. CRC Press, pp. 107−122.

Gopalakrishnan, N., Manickam, S., Prakash, A.H., 2007. Problems and Prospects of Cotton in Different Zones of India. Project Coordinator & Head, AICRP on Cotton, Coimbatore.

Hae Now, 2008. 'Why Choose Organic', Organic Cotton Clothing Hae Now Fair Trade. http://www.ajsosteniblebcn.cat/the-life-cycle-assessment-of-organic-cotton-fiber_38172.pdf. https://www.organicfacts.net/organic-cotton.html. https://thedesigncart.com/blogs/news/organic-fabrics. https://www.yarnsandfibers.com/textile-resources/other-sustainable-fibers/natural-fibers-other-sustainable-fibers/organic-cotton/organic-cotton-fiber-consumption/what-are-the-end-uses-of-organic-cotton/.

Hsieh, Y.L., 2007. Chemical Structure and Properties of Cotton. Science and Technology, Cotton, pp. 3−34.

Husnain, T., 2015. Transformation and transgenic expression studies of glyphosate tolerant and cane borer resistance genes in sugarcane (*Saccharum officinarum* L.). Molecular Plant Breeding 6 (12).

JAPANESE AGRICULTURAL STANDARD (2001) (http://www.maff.go.jp/soshiki/syokuhin/ hinshitu/organic/eng_yuki_59.pdf)

2008. IFOAM (INTERNATIONAL FEDERATION OF ORGANIC AGRICULTURE MOVEMENTS) (2008), http://www.ifoam.org/ and http://www.ifoam.org/about_ifoam/standards/norms/ibsrevision/ibsrevision.html.

Lewis, W.J., Haney, P.B., Reed, R., Walker, A., 1997. A total systems approach for sustainable cotton production in Georgia and the southeast: first year results. In: Beltwide Cotton Conferences (USA).

Losonczi, A.K., 2004. Bioscouring of Cotton Fabrics.

Lotter, D.W., 2003. Organic agriculture: a review. Journal of Sustainable Agriculture 21 (4), 59—128.

Mäder, P., Fliessbach, A., Dubois, D., Gunst, L., Fried, P., Niggli, U., 2002. Soil fertility and biodiversity in organic farming. Science 296 (5573), 1694—1697.

Mageshwaran, V., Satankar, V., Shukla, S.K., Kairon, M.S., 2019. Current status of organic cotton production. Indian Farming 69 (02), 09—14.

Makhdum, A.H., Khan, H.N., Ahmad, S., February 2011. Reducing cotton footprints through implementation of better management practices in cotton production; a step towards Better Cotton Initiative. In: Proceedings of the Fifth Meeting of the Asian. Cotton Research and Development Network, Lahore, Pakistan, pp. 23—26.

McGarry, D., 2009. Conservation agriculture as a sustainable option for the central Mexican highlands. Soil and Tillage Research 14—18.

Memon, N.A., 2012. Organic cotton: biggest markets are Europe and the United States. Pakistan Textile Journal 61 (3).

Mitchell Jr., C.C., 1988. New Information from Old Rotation. Highlights of agricultural research-Alabama Agricultural Experiment Station, USA.

Muthu, S.S. (Ed.), 2014. Roadmap to Sustainable Textiles and Clothing: Eco-Friendly Raw Materials, Technologies, and Processing Methods. Springer.

Myers, d, 1999. 'The problems with conventional cotton', Chapter 2. In: Myers, D., Stolton, S. (Eds.), Organic Cotton—from Field to Final Product, Intermediate Technology Publications. Intermediate Technology Development Group, London, pp. 8—20.

Offermann, F., Nieberg, H., 2001. Wirtschaftliche Situation ökologischer Betriebe in ausgewählten Ländern Europas: Stand, Entwicklung und wichtige Einflussfaktoren. Agrarwirtschaft; Zeitschrift für Betriebswirtschaft, Marktforschung und Agrarpolitik 50 (7), 421—427.

Organic exchange, 2008. Organic Cotton: Your Healthier Choice. http://www.organicexchange.org/health/intro.php.

Orphal, J., 2005. Comparative Analysis of the Economics of Bt and Non-Bt Cotton Production. Inst. für Gartenbauökonomie.

Özalp, B., Ören, M.N., 2014. Recent developments in WTO negotiations on agriculture and position of Turkey. In: Presidency of the Congress, p. 81.

Patagonia, 2008. Fabric: Organic Cotton.

Puspito, A.N., Rao, A.Q., Hafeez, M.N., Iqbal, M.S., Bajwa, K.S., Ali, Q., Husnain, T., 2015. Transformation and evaluation of Cry1Ac+ Cry2A and GTGene in Gossypium hirsutum L. Frontiers of Plant Science 6, 943.

Qamar, Z., Aaliya, K., Nasir, I.A., Farooq, A.M., Tabassum, B., Qurban, A., Husnain, T., 2015. An overview of genetic transformation of glyphosate resistant gene in Zea mays. Natural Science 13 (3), 80—90.

Raj, D.A., Sridhar, K., Arun, A., Lanting, H., Brenchandran, S., 2005. Case study on organic versus conventional cotton in Karimnagar, Andhra Pradesh, India. In: Second International Symposium on Biological Control of Arthropods, Davos, Switzerland, 12—16 September, 2005. United States Department of Agriculture, Forest Service, pp. 302—317.

Rana, S., Pichandi, S., Parveen, S., Fangueiro, R., 2014. Biodegradation studies of textiles and clothing products. In: Roadmap to Sustainable Textiles and Clothing. Springer, Singapore, pp. 83−123.

Rasul, G., Thapa, G.B., 2003. Sustainability analysis of ecological and conventional agricultural systems in Bangladesh. World Development 31 (10), 1721−1741.

Rattanasuteerakul, K., Thapa, G.B., 2012. Status and financial performance of organic vegetable farming in northeast Thailand. Land Use Policy 29 (2), 456−463.

Reganold, J.P., Palmer, A.S., Lockhart, J.C., Macgregor, A.N., 1993. Soil quality and financial performance of biodynamic and conventional farms in New Zealand. Science 260, 344−349.

Samie, A., Abedullah, Ahmed, M., Kouser, S., 2010. Economics of conventional and partial organic farming systems and implications for resource utilization in Punjab (Pakistan). Pakistan Economic and Social Review 245−260.

Shabbir, M.Z., Arshad, M., Hussain, B., Nadeem, I., Ali, S., Abbasi, A., Ali, Q., 2014. Genotypic response of chickpea (*Cicer arietinum* L.) for resistance against gram pod borer (*Helicoverpa armigera*). Advancements in Life Sciences 2 (1), 23−30.

Shetty, P.K., 2004. Socio-ecological implications of pesticide use in India. Economic and Political Weekly 5261−5267.

Stolze, M., Piorr, A., Häring, A.M., Dabbert, S., 2000. Environmental Impacts of Organic Farming in Europe. Universität Hohenheim, Stuttgart-Hohenheim.

Textile Exchange, 2016a. Cotton—India, Material Snapshot. Texas.

Textile Exchange, 2016b. Organic Cotton Market Report 2016. Texas.

Textile Exchange, 2017. Organic Cotton Market Report 2017 (Texas).

Tilman, D., Cassmann, K.G., Matson, P.A., Naylor, R., Polasky, S., 2002. Agricultural sustainability and intensive production practices. Nature 418, 671−677.

Ton, P., 2007. Organic Cotton: An Opportunity for Trade.

US Department of Agriculture, 2018: "Leading Cotton Producing Countries Worldwide in 2016/2017." Statista; at: https://www.statista.com/statistics/263055/cotton-production-worldwideby-top-countries/ (16 June 2018).

Wakelyn, P.J., Chaudhry, M.R., 2009. Organic cotton: production practices and post-harvest considerations. In: Sustainable Textiles. Woodhead Publishing, pp. 231−301.

Yafa, S., 2005. Big Cotton: How a Humble Fiber Created Fortunes, Wrecked Civilizations and Put America on the Map, Viking Penguin. a member of Penguin Group (USA) Inc., London, pp. 290−301.

Yusuf, M., 2010. Ethical issues in the use of the terminator seed technology. African Journal of Biotechnology 9 (52), 8901−8904.

Hemp, flax and other plant fibres

4

Ryszard Kozlowski[1] and Malgorzata Muzyczek[2]
[1]FAO/ESCORENA European Cooperative Research Network on Flax and other Bast Plants, Institute of Natural Fibres, Poznan, Poland; [2]Institute of Natural Fibres and Medicinal Plants, FAO/ESCORENA European Cooperative Research Network on Flax and other Bast Plants, Poznan, Poland

4.1 Introduction—natural fibres, yarns, fabrics and knitting for fashion

According to the European Confederation of Flax and Hemp (CELC) data, the most important natural fibres for textile manufacturing and fashion are among others cotton, flax, viscose, hemp, ramie and bamboo. There are estimates that on the textile market new products will account for more than 30% of profits in the nearest future and the sales from new products will grow by 37%, while only 54% of sales will result from the products offered up to now.

World production of all apparel and textile fibres (Fig. 4.1) reached 110 million tonnes in 2018, and natural fibres accounted for 29% of the total (down from 41% in 2008).

The decline in the share of natural fibres in the total fibre production in the last decade is the result of the exponential growth in polyester production.

Unfortunately, according to the last announcement, global warming connected with polyester microfibers coming from washing machine with effluent are the biggest microplastic pollution issue on the earth (air and oceans (Bomgardner, 2017, 16—17).

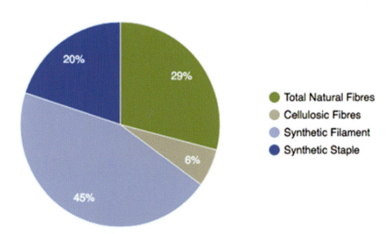

Figure 4.1 World fibre production (2018).

Sustainable Fibres for Fashion and Textile Manufacturing. https://doi.org/10.1016/B978-0-12-824052-6.00017-2
Copyright © 2023 Elsevier Ltd. All rights reserved.

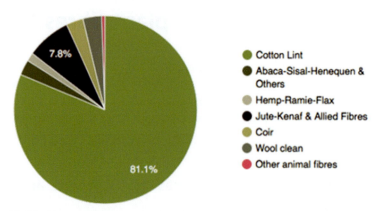

Figure 4.2 World natural fibres production (2018).

Of the world total production, natural fibres accounted for 32 million tonnes of production during 2018 (Fig. 4.2), an increase of less than two million tonnes in 10 years. The share of natural fibres in the world fibre production fell from 41% in 2008 to less than 30% in 2018. Cotton accounted for 81% of the natural fibre production by weight in 2018, jute accounted for 7%, while coir and wool accounted for 3% each (Townsend, 2019).

Flax is a rare product, which represents less than 1% of all textile fibres consumed worldwide. 80% of the world production of scutched flax fibres are originated from Europe and France is the world leader.

In Fig. 4.3, world fibre production in 1980–2025 is presented (Yang, 2014). Total fibre production is expected to grow 3.7% per annum to 2025. The demand for fibres grows steadily by about 3% annually, which is correlated with population growth and increasing prosperity.

World fibre production has excellent correlation with GDP (Growth Domestic Product), which we use as a basis for projections to 2025, as presented in Fig. 4.4.

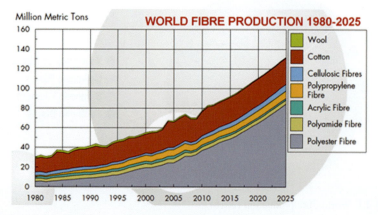

Figure 4.3 Total world fibres production in 1980–2025.

Hemp, flax and other plant fibres

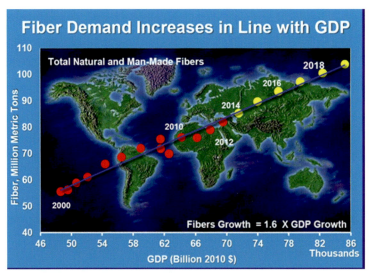

Figure 4.4 World fibre production with GDP (Growth Domestic Product).

The production of natural fibres is expected at the level of about 38 million tons/year in the middle of the 21st century. It is estimated to grow 3% annually to reach 122 million tonnes in 2025 of total fibre production (Kozlowski and Muzyczek, 2017, 215–236). Various goods are produced from natural fibrous plants: woven, knitted, technical and nonwoven textiles, eco-friendly composites.

All market segments within textile industry are expected to grow, with the highest growth rate (absolute value and percent value) for technical textiles. Following a period of dynamic development for man-made fibres, a trend of a slow increase of natural fibres production is observed (Fig. 4.5).

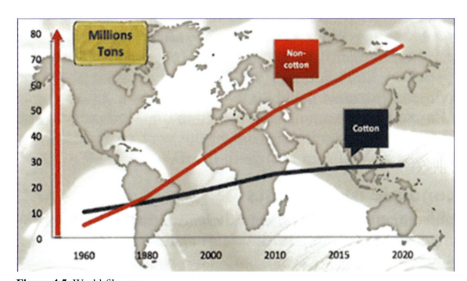

Figure 4.5 World fibre use.

The global fibre market driving factors responsible for the growth of fibre market are (Oerlikon, 2013, 37)

- ➤ abundant availability,
- ➤ cost-effectiveness,
- ➤ a wider scope of application,
- ➤ population increase,
- ➤ wealth
- ➤ eco response.

The production of natural fibres depends on the availability of the cultivable land and favourable water conditions, which is changing faster than ever in the 21st century because of global warming and its effects. While the man-made fibres are derived from organic raw materials, which are expected to be available virtually on the unlimited basis in the coming decades and which will cost significantly cheaper than natural fibre.

Apart from this, the other factor that ensures the unlimited availability of fibres is the recycling of man-made fibres as against natural fibres and the freedom of designing these fibres in research and production to alter properties to make fibres suitable for other applications (Fibre2Fashion, 2020).

Increasing environmental concerns, coupled with volatile prices associated with conventional fabrics, have spurred the adoption of eco fibres in the global market. Favourable regulations on manufacturing bio-based products, coupled with stringent environmental sanctions on synthetic polymers, are also expected to have a positive impact on the growth of the market.

The global eco fibre market size is expected to reach 58.29 billion USD by 2027, expanding at the CAGR(Compound Annual Growth Rate), of 4.6% from 2020 to 2027, according to a new report by Grand View Research, Inc. Rising concerns regarding the harmful environmental impacts of using synthetic fibres have been contributing to the increased demand for organic eco fibre.

The use of chemicals and polymers in the manufacturing of synthetic fabrics is projected to limit its demand in textiles. Rising awareness regarding sustainable textile production to meet the environmental and social aspects, such as lowering pollution, is projected to fuel the demand for eco fibre across the globe over the forecast period (Guest Contributor Grand View Research, 2020).

New technologies with fewer manufacturing steps will improve the cost efficiency of non-wovens compared to woven textiles. Technical textiles are suitable for other industries and will substitute traditional materials such as steel, cement and wood as they are light, flexible, durable, cost-effective, and multi-functional materials (Maity et al., 2014, 365—390).

4.2 The sustainability aspects of natural fabrics and knitting from flax, hemp, ramie, curaua, bamboo, pineapple fibres. Example of apparels

Natural fibres are a renewable resource *par excellence*. Natural fibres are carbon neutral, as they absorb the same amount of carbon dioxide they produce. They are easily biodegradable, and they do not left in the environment catastrophic microfibres coming from polyester. Unlike their synthetic counter parts, natural fibres are produced from renewable resources. Natural fibres can be extracted both from the plants and the animals. The plant-based fibres also known as vegetable fibres can be extracted from various parts of the plant such as seed (e.g., cotton, akund, kapok, coir) bast (e.g., flax, jute, hemp, ramie, kenaf) and leaf (e.g., sisal, abaca). The vegetable fibres are also called as cellulosic fibre as cellulose is the main component of the fibres. Hence, they are named according to the part they are extracted from. Among all the natural fibres, plant-based fibres extracted from the bast of the plant are of special interest due to their sustainability approach while harvesting.

BAST FIBRES are termed as 'soft fibers' with the following plants within the group: flax, hemp, jute, ramie, kenaf, urena and nettle. Bast fibres are extracted from the stalks of dicotyledonous plants. The bast bundles of fibres are fixed to the central woody part by gum (mainly pectins) and the bundles determine the strength to the stalk of the fibrous plants (Wasif and Singh Vivek, 2005; Blackburn, 2005; Kozlowski et al., 2012a, 200−222).

4.2.1 Flax

Flax (*Linum usitatissimum* L.) or linen is the oldest textile fibre, the earliest trace of its use dates from 8000 BCE. One hundred per cent linen knitted, transparent Pharaoh wife apparel can be found in the National Museum in Cairo, Egypt. In 2016, the Oxford University got from Egypt a shirt older than 2000 years; the age of this shirt was confirmed by the method of radiocarbon by measuring the carbon isotope C-14. This shirt was made from famous Egyptian linen fibres.

Next, the Phoenician merchants spread the flax across Europe.

The modern phase in the processing and utilization of flax in Europe began in the 17th century, especially in Ireland, England, Holland, Belgium, France, the territory of Poland, Germany and Russia. Later, in the 19th and 20th centuries, the production expanded to South America, Argentina, Brazil, Chile, Peru, even to Japan. However, at the end of the 18th century, lower costs of the processing of cotton contributed to the fall of the flax industry and the rapid expansion of the cotton and man-made fibres.

The colour of the bast fibres such as flax depends upon the type of retting and conditions in retting. For example, the water retted fibre can range from white to yellow and the dew retted fibre can be grey to dark silver colour. The features of flax include clockwise twist similar to cotton, polygonal cross-section of ultimates with thick walls and small lumina. The feature of flax fibre which is common with other bast fibres is the presence of dark dislocations, often in the form of an 'X' under microscope that are roughly perpendicular to the longitudinal axis of the fibre. The diameter of elementary flax fibres is within the range of 15−22 μm. It has polygonal cross section of the ultimates which are angular with medium sized lumina. The cross and longitudinal section of flax fibre by scanning microscope are presented in Fig. 4.6.

For the good quality of flax, pulling has to be done at the stage of light green maturity. Retting (degumming of flax) is the process of loosening the bonds between fiber bundles and surrounding them pectins. This process can be done in water tanks or

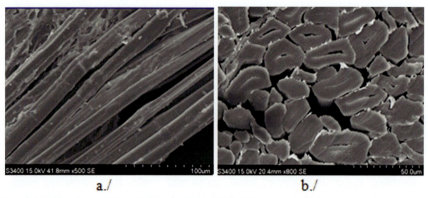

Figure 4.6 The cross and longitudinal section of flax fibre.

rivers (now it is not used in Europe) and on the fields (dew retting). Practically, the two methods are used: water retting and dew retting. Water retting consumes 20–30 m^3 warm water (temp. 30–33°C) per 1 tonne of straw water retted straw in an anaerobic process is characterized with fibres that have the odour of an unsaturated fatty acid, mainly n-butyric acid. In the case of dew retting on the fields, there is a risk connected with agroclimatic conditions (not sufficient humidity or too much rain and problems with drying and frequently over-retting of fibres) (Kozlowski, 1990, 251–261; Kozlowski et al., 2020).

The new emerging potential of genetically modified (GM) natural fibres provides them with better performance, including the yield of major products such as cellulosic fibres (in the case of lignocellulosic plants) and carbohydrates to obtain, in statu nascendi, in plants, polyhydroxyalkanoates (PHA) such as polyhydroxybutyrate (PBA) (natural polyester) (Sabbir and Tasneem, 2014, 9494–9499). GM fibrous plant can be resistant to herbicides and also to environmental stress such as salinity and drought. There are attempts to apply GM techniques to increase the biomass of fibrous plants and control the lignin and pectin levels.

The degumming method of bast fibrous plants is presented in Fig. 4.7 (Kozlowski and Rozanska, 2020). The osmotic degumming method is very promising. The fibres obtained as the result of osmotic degumming are more delicate, whiter and finer, and they can be used for producing better textile products (yarn, fabric, knitting or nonwoven).

Bast fibres have to be harvested and processed in a special way to make them finer, more homogeneous and with additional functionality (Kozlowski et al., 2012a, 200–222). There is the long-expected opportunity for flax and hemp to diminish the still existing gap between flax versus other fibres. This opportunity is fast coming.

4.2.2 Hemp

Hemp (*Cannabis Sativa*), an annual plant, grows in temperate climatic conditions in Europe, Asia, North and South America. Hemp is one of the bast fibres known to

Hemp, flax and other plant fibres 81

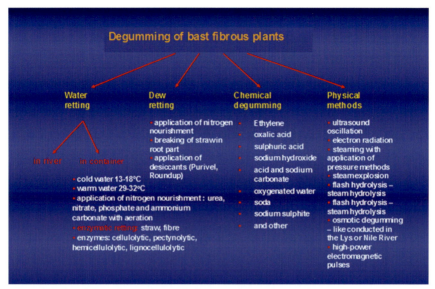

Figure 4.7 Directions in fibres degumming and treatment.

ancient Asians long before Christ (Kozlowski et al., 2005a, 207−227; Kozłowski et al., 2005b, 36−87). There are the two types of hemp: industrial hemp with the low level of gamma tetra hydrocanabinol (THC)—below 0.2/0.3% and marihuana hemp planted mainly for drugs—marihuana. It can be distinguished differences between monoecious and dioecious cultivars as well.

Hemp is characterized by allelophatic phenomena which is very important in cleaning the cultivated soil and eliminating the necessary use of plant protection chemicals. Results of research conducted at the Institute of Natural Fibres in Poland indicate that flax and hemp plants can also be used for extracting heavy metals: lead (Pb), copper (Cu) and zinc (Zn), from the polluted soil after cultivation on the degraded land.

The technology of hemp harvesting is a very specialized area, which, to be successful, requires the extensive knowledge of the final product. Tebeco company in cooperation with the biggest processor of industrial hemp in the Czech Republic, 'Canabia a.s.' developed the unique technology to harvest industrial hemp called 'Alpha' (Bednar, 2007, 127−129).

Hemp fibres have high tenacity, hygroscopicity and quick moisture transport, high strength and low homogeneity, depending on agrotechnological conditions, for example sowing density (for textiles about 60 kg/ha), time of harvesting and the cultivar type.

Hemp fibres, for the purpose of textile production, are nowadays mainly imported, from China and are mostly produced by chemical treatment in NaOH. In view of sustainable production, this process is not feasible due to its high consumption of water, chemicals and energy. Alternatives are, for example, the steam explosion process or enzymatic separation. The main task of enzymes in the separation process of bast fibres is to catalyse the dissolving of pectin species connecting the single fibre in order to reduce the size of the fibre bundles (Fischer et al., 2006, 39–53).

Hemp fibres are used for clothing, technical products, such as ropes, composites, and as an upholstery material, carpet underlay, nonwovens, household textiles, such as towels, bed linen and tablecloths. Because yarn in hemp fabrics swells and enlarges its diameter in wet conditions, hemp is used for making canvas (Horn, 2020).

4.2.3 Clothing from linen and hemp

Linen clothing is characterized by hydrophilic properties does not cause the increase of reactive oxygen species and the oxidative stress of human organism in opposite to the garments made of synthetic fibres (Zimniewska et al., 2006, 17–21).

Linen garment does not cause the increase of reactive oxygen species and oxidative stress. Sleeping in linen bedding ensures deeper sleep, quicker human body regeneration and better rest—the human immune system grows stronger (Zimniewska and Kozlowski, 2004, 851–858).

The lowest body temperature and an increase of immunoglobulin A during the sleep-in linen bedding proved that such a raw material is the most positively influencing human rest and sleep quality (Fig. 4.8) (Kozlowski et al., 2016).

The current trend in textiles is to move away from traditional uses and look for increased complexity and 'intelligence' in terms of applications. Intensive research is in progress, following a multi-directed scheme where key sectors can be distinguished: textiles structures, functionality and composites.

In the 21st and 22nd century, flax and hemp future fibres are very thin, clean and strong cellulosic fibres, including bast nanocellulosic fibres, environmentally friendly, sustainable and fully biodegradable.

The progress in the area of breeding, agricultural technology and processing of fibrous plants opens also future prospects for the cost-effectiveness of their production and the application of lignocellulosic fibrous plants in the apparel sector industry without the risk of global pollution (Muzyczek, 2020).

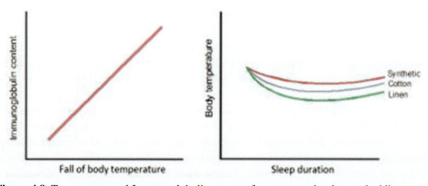

Figure 4.8 Temperature and Immunoglobulin content for cotton and polyester bedding.

Semi-products such as roving, yarns, knitting and fabrics after special treatment are widely used in different branches of economy. New techniques of processing and functionalization become open to a new area in modern transport and building industry as well. Various goods can be produced from natural fibrous plants: woven, knitted, technical and non-woven textiles, eco-friendly composites.

Great attention will be paid to environmental pressures on existing chemical, enzymatic processes and the development of new functional finishes to further enhance performance, especially in the apparel sector. Nowadays, sustainability is a very hot topic in the fashion world (Kozlowski et al., 2020). Fashion is expected to remain one of the most lucrative application segments (Figs 4.9 and 4.10).

Figure 4.9 Linen apparel designed by Jola Zalecka for INFs (Institute of Natural Fibres).

Figure 4.10 Hemp apparel designed by Rafal Berezowski for INFs (Institute of Natural Fibres).

Special treatment and functionalization methods such as degumming, enzymatic, ultrasound, plasma, corona, ultrasound, liquid ammonia, and flame retardant treatments, and protection against biodeterioration provide new promising features and properties of all natural fibres.

Nowadays we observe the development of many technical textiles like nonwovens, geo-mats, fibrous materials including 3D and 3D warp-knitted reinforcing materials. By adding a 3D to textile structure, we open doors for the development of a wide range of new products: geo-synthetics, insulation felts, composites. The 3D textiles currently are made by weaving and knitting, for example Laroche Napco, 3D Web Linker technology allows the manufacturing of three-dimensional (3D) linked fabrics at very high speeds and with a possibility of the simultaneous introduction of additional components like powder, fluids, profiles and foams, for example aerogel into the structures (Fischer, 2016, 8).

Some of bast fibres, for example flax and hemp, are susceptible to dyeing by natural dyes and are characterized by natural antibacterial properties and UV blocking thanks to nanolignin and its structure. They can be easily protected against flammability. They are excellent for 'healthy appearance' by decreasing the oxidative stress of the users.

Natural fibres and fabrics after special functionalization by MOF (*Metal-organic Frameworks*), POM (polyoxymetalates) and dendritic polymers will play an important role in the near future not only in defence and military apparels. This mentioned functionalization can protect the users, including soldiers, against biological, chemical, radiation weapons, also against cold and fire (Kozlowski and Muzyczek, 2017, 215−236).

Other flat flexible textile composites made of the blends of cellulosic fibres are dedicated to the production of packaging and decorative materials, preferably bags, roller blinds, lamp shades and packaging bags.

All market segments within the textile industry are expected to grow, with the highest growth rate (absolute value and per cent value) for technical textiles. Technical textiles are suitable for other industries and will substitute traditional materials such as steel, cement and wood, as they are light, flexible, durable, cost-effective and multifunctional materials. New technologies with fewer manufacturing steps will improve the cost efficiency of non-wovens compared to woven textiles (Maity et al., 2014).

Today's natural textiles not only keep the wearers 'cool' in a fashion sense but also protect them from 'the extremes' of modern life. However, anyone who pays attention to the fashion industry knows that we have an urgent sustainability problem on our hands. According to the Ellen MacArthur Foundation, if we stay on our current path, the fashion industry alone will capture a quarter of the entire world's carbon budget by 2050. To avert the climate catastrophe, we need a number of different creative approaches and interventions at multiple stages in the lifecycle of a garment and their recycling. But to change the fashion industry's trajectory on the deepest level, there is a strong case for going back to the fundamentals: the raw materials themselves. A brand's choice of raw materials has a greater impact on its environmental footprint than any other single element (Guest Contributor, 2021).

The natural fibres, including flax and hemp, are characterized by unique and important properties like air permeability, hygroscopicity, the capacity of increasing the moisture level, UV blocking properties (thanks to nanolignin), not causing an allergic effect (represented by a higher level of histamine in the human blood and other properties), biodegradability and safer behaviour in the flame and fire conditions in comparison with man-made fibres.

4.2.4 Properties of bast fibres

Chemical composition of flax and hemp in comparison to some others lignocellulosic fibres are presented in Table 4.1 (Kozlowski et al., 2012b). Mechanical properties of flax and hemp in comparison with some others lignocellulosic fibres are presented in Table 4.2 (Banerjee, 2012, 401−427).

4.2.5 Application of flax and hemp

A range of products can be manufactured from the bast fibres, flax and hemp. The main area application of flax in 2009 is presented on Fig. 4.11 (Muzyczek, 2009, 312−328).

In contrast, now 90% of European linen is destined for the textile market (60% for clothing, 15% for household linens, 15% for the furniture and lifestyle) and just 10% is dedicated to technical opportunities: eco-construction, insulation, automobile parts, sports, equipment, boating, stationery, surgery and health items (Nelen and Delbeke, n.d).

Table 4.1 Chemical composition of plant fibres in %.

Fibre	Cellulose	Lignin	Pectin and hemicelluloses
Abaca	60−80	6−14	13
Bamboo	26−43	21−31	−
Cabyua	80	17	−
Cotton	83−99	6	5
Curaua	70−80	13	−
Flax	64−84	0.6−5	19
Hemp	67−78	3.5−5.5	17
Henequen	60−78	8−13	4−28
Isora	75	23	−
Jute	51−78	10−15	37
Kenaf	44−57	15−19	−
Nettle	53−82	0.5	0.9−4.8
Pineapple	80	13	−
Pita	80	17	−
Ramie	67−99	0.5−1	22
Sisal	60−80	6−14	13
Coir	44	46	3.25

Table 4.2 Mechanical and physical properties of natural fibres.

Fibre	Length average [mm]	Diameter average [mm]	Tensile strength [N/tex]	Breaking extension [%]	Density [g/m³]	Initial modulus [N/tex]	Moisture absorption [%]
Flax	32	0.019	0.54	3.0	1.54	18.0	7
Jute	2.5	0.018	0.31	1.8	1.44	17.2	12
Ramie	120	0.040	0.59	3.7	1.56	14.6	6
Hemp	25	0.024	0.47	2.2	1.48	21.7	8
Sisal	3	0.021	0.30	3.0	1.45	15.0	11
Coir	0.7	0.020	0.31	1.8	1.24	17.2	10
Cotton	25	0.019	0.19–0.45	5.6–7.1	1.52	5.9	7–8

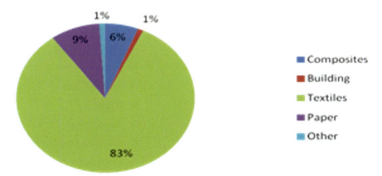

Figure 4.11 Area of flax application in 2009.

Application area of hemp in Europe (2013) is presented in Fig. 4.12 (Carus, 2016) and global market share for hemp in 2019 is presented in Fig. 4.13 (Guest Contributor Grand View Research, 2020).

The global industrial hemp market is expected to grow at the compound annual growth rate (CAGR) of 15.8% from 2020 to 2027 to reach USD 15.26 billion in 2027. The Asia Pacific dominated the industrial hemp market with a share of 32.6% in 2019. Asian countries, such as China, have a significant tradition of hemp production and exports, which has allowed these countries to dominate the global market in the recent past (Guest Contributor Grand View Research, 2020).

Recently, there has been an increasing trend in the functionalization of the bast fibres. The functionalization can help to achieve improved properties for various

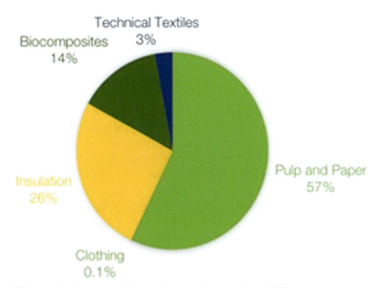

Figure 4.12 Application areas of hemp fibres in Europe from 2013.

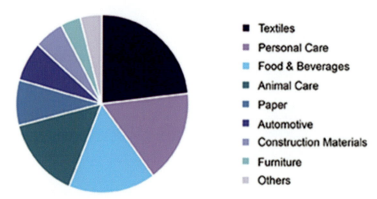

Figure 4.13 Global market share for hemp in 2019.

technical applications. The trends in flax fibre and hemp functionalization and finishing are presented below:

- improvement of hydrophylicity
- self-cleaning, bacteriostatic properties
- using natural dyestuffs
- ultrasound and osmotic degumming
- liquid ammonia treatment, like delicate mercerization
- fire retardancy
- bio-deterioration protection
- functionalization by dendrimers improvement of dyeability, antibacterial effect, special finishing fibres and fabrics in biomedical application
- functionalization by metal-organic frameworks (MOFs) application in military fabrics and other
- plasma and corona treatment
- nano Ag, nano–TiO_2, stain removal, UV barrier (especially by nanolignin activity),
- functionalization by reaction with polyoxymetalate (e.g., molybdenium polyoxy metalate).

Bast fibres are well known for their sustainability aspects. Growing and harvesting the bast fibres is more ecofriendly than the most common cellulosic fibre, cotton. Unlike cotton fibre, flax and hemp can be grown with very little irrigation or even with the rain water. It is worth to specify that the cotton fibres need about 20,000 L (approx.) of freshwater per kg of the fibres. Further, flax and hemp are grown with far less pesticides than cotton as they are resistant to many insects and pests. The yield of flax and hemp per acre of land is much higher than the cotton. Flax plants readily absorb the toxic greenhouse gas such as carbon dioxide (CO_2). For example, 1 ha of flax cultivation can absorb about 3.7 metric tonnes of CO_2 and release oxygen (O_2) to the air after photosynthesis. Hence, they can be considered as carbon neutral. Further, in flax or hemp cultivation, there is no waste generated as all the parts of the plant can be converted to useful products.

Similar to cotton clothing, flax clothing is comfortable due to high moisture absorbency and breathability of the fibres. The clothing made from flax is easier to take care and more durable than cotton clothing. Flax is stiffer, UV-resistant and dries faster that the cotton. Flax textiles and clothing are well-known for their good antibacterial and thermoregulation properties. The flax fibres can be as long as 1 m, whereas as the hemp can grow up to 5 m, which is much longer than cotton fibres (1.5−5 cm).

Flax can be used to produce a range of linen clothing for tropical climates such as trousers, shirts, skirts, bath towels, beach towels, bed linens, tablecloths and home furnishings. Further, flax and hemp fibres can also be used for twines, ropes and other high-performance textiles. Flax and hemp, if untreated with dye, are fully biodegradable. Summing up, the essential features of the bast fibres are

➤ comfort/healthful aspects of clothing,
➤ biodegradability and eco-response,
➤ carbon neutral,
➤ good antibacterial and thermoregulating properties,
➤ renewability,
➤ durability of clothing,
➤ social impact especially in rural areas of globe.

In spite of several advantages, the usage of flax is about 1% of all textile fibres consumed worldwide. This can be attributed some properties of the flax fibre such as stiffness, which needs to be fixed by various processing to make it softer. However, the demand for the flax textiles is increasing due to the increased awareness of consumers and fashion brands. For the better promotion of healthy natural fibres used in the apparel sector, there would be need of getting involved high talented fashion designers to work with natural fibres.

Research on clothing comfort has a fundamental meaning for the survival of human being and the improvement of the quality of life. Clothing comfort has substantial financial implications for the efforts to satisfy the needs and wants of consumers in order to obtain competitive advantages in the modern consumer market (Li, 2001, 1−138). Hence, research in the bast fibres has been significantly increased to make the fibres the future sustainable fibres. Various proposals to emphasize the comfort, health, UV protection, provided for the human body by natural fibres (see results of the relevant physiological research of INF) are as follows:

➤ oxidative stress,
➤ muscle tension,
➤ level of Alpha-globulin, and
➤ other very important physiological factors influencing the well feeling of uses.

4.3 Recycling of natural textiles as a sustainable solution

Recycling is the problem, the solving of which is to release the mankind of Earth catastrophes and the destruction of the climate. As mentioned before, even oceans of Earth are polluted by microparticles of plastic and fibres coming from pollution from the washing effluent of textiles. Nylon, polyester and acrylic are the synthetic fibres, which release microplastics to the ocean causing risk to the aquatic animals. The microplastic then enters to the human food chain, through aquatic food and create risks for human life.

Natural fibres in comparison with polymers and man-made fibres are sustainable and biodegradable. They are free from the problems of generating microplastics. Furthermore, they are renewable and biodegradable. Since the years 1990−2000, many announcements have put attention to the problem of recycling fabric waste (mainly waste of fibres polyester and polyamide, polyacrylonitrile, polyurethane) (Grosso, 1996, 21−30). There is a range of opportunities for recycling of products made from natural fibres such as flax and hemp. Various consumer goods such as men's and women's clothing, wipes, socks, nonwoven and knitwear product can be manufactured from recycles materials. However, the recycling market is not so wide now and it is gaining popularity in a sustainability driven fashion and textile market.

Now in 2021, the problem is growing. We know that recycling must be an integrated effort and the partnership with manufactures, consumers, retailers, recyclers and obviously governments. Study mentioned above has shown that if the recycled textiles materials were available, companies would be interested in their reproduction. Some limits to do this are the lack or weak: market, technology and proper equipment. Summing up, greatest barriers to entering the wide recycling of textiles are

➤ the lack of market,
➤ the lack of technology,
➤ the lack of equipment, and
➤ the lack of consumer awareness.

The lack of market and the lack of technology were identified as the most significant barriers to recycled textiles, sewn products and successful entering the marketplace. In the 20th century, many technologies connected with dominant fibres—polyester recycling—were developed (Hell, 2016, 26−29). There is a growing interest for thee recycling of natural fibre-based clothing at the end of life including linen and hemp. Many fashion brands are developing new products from the recycled natural fibres or blends with synthetic fibres.

In the last years, an urgent problem with medical waste connected with COVID 19 appeared.

A joint UK/India team led by Svensea University is developing a novel process called photoreforming by using the sunlight to simultaneously kill viruses and convert non-recycled waste into clean hydrogen fuel. This process works by using nanostructured semiconductors to drive the degradation of waste and pathogen with the sunlight and by generating hydrogen from plastic waste (Kuehnel, 2020; Starlinger, 2011, 8−10).

4.4 Future trends and further information and advice

Most of lignocellulosic fibrous plant raw materials, such as cotton, flax, hemp, jute, kenaf, sisal, ramie, curaua, pineapple, bamboo, coir, can be extracted, processed, modified (functionalized) and used in production of not only textiles (woven, knitted, nonwoven, technical and 3D textiles) but also as a reinforcement of more friendly composites.

Functionalization and special treatment such as enzymatic, ultrasound, plasma, corona, liquid ammonia, flame retardant treatments, as well as protection against biodeterioration provide new promising features and properties of all natural fibres, such as short and fine fibres including nano-cellulosic fibres and semi-products such as roving, yarns, and special knitting, also 3D structures and fabrics. The surface of natural fibres can be modified by many methods, for example graft copolymerization (Kozlowski and Wladyka-Przybylak, 2004).

Linen and hemp are a valuable raw material for apparel. According to data from the INF, linen and hemp have the required clothing physiology and comfortability to be integrated in the fashion industry and even have the chance to substitute cotton in the future (Zimniewska et al., 2004, 69–81). Furthermore, it is indicated that people need to be convinced about using more linen and hemp in apparel, by capturing the sustainability factor of the crop and communicating it publicly.

4.5 Conclusions

This chapter has discussed two of the most significant fibres, flax and hemp which are receiving increased attention in a sustainable global fashion market. Some of the physical properties, chemical properties and scope for sustainable applications are discussed. Various aspects of these bast fibres, which make them more sustainable, are also covered in this chapter. The latest news about the catastrophic role that the microfibres of man-made fibres coming from washing machines have in our globe (oceans, air, soil) bring a better chance for future natural fibres. Natural fibres can be a good raw material for the future fashion and textiles to avoid the problems of microplastics.

Summing up, the 21st century is the century of the coexistence and competition of natural fibres with man-made fibres, especially in the area of quality, sustainability, applications and the economy of their production. Researchers and industrials look for the best way for the coexistence and competition between man-made and natural fibres. Natural fibres are considered to be ecofriendly, biodegradable and safer compared with the synthetic fibres. Natural fibres such cotton, flax, hemp, jute, kenaf, ramie and bamboo can be extracted and made into a range of products (woven, knitted, nonwoven and 3D textiles) in safer and cleaner way. They can also be used in reinforcing materials and composites for technical applications. Finally, at their EOL, they can be recycled back to new products by recycling, which can eliminate the problems of waste ending in the landfill.

References

Banerjee, P.K., 2012. Environmental textiles from jute and coir, 12. In: Kozlowski, R. (Ed.), Handbook of Natural Fibres, vol. 2. Woodhead Publishing, pp. 401–427.

Bednar, P., 7-10 October 2007. Harvesting technologies for industrial hemp, tebeco—oral presentation. In: Proceedings of the 4th Global Workshop of the FAO/ESCORENA Innovative Technologies for Comfort, pp. 127–129. Arad (Romania).

Blackburn, R.S., 2005. Biodegradable and Sustainable Fibres. Woodhead Publishing in Textiles, Cambridge.

Bomgardner, M.M., January 9, 2017. The Great Lint Migration. Chemical&Engeeniering News, pp. 16–17.

Carus, M., Sarmento, L., April 2016. The European Hemp Industry: Cultivation, Processing and Applications for Fibres, Shivs, Seeds and Flowers. EIHA.

Fibre2Fashion, June 2020. Global Fibre Market Outlook—an Overview. Published. https://www.fibre2fashion.com/industry-article/8700/global-fibre-market-outlook-an-overview.

Fischer, H., Mussig, J., Bluhm, C., 2006. Enzymatic modification of hemp fibres for sustainable production of high quality materials: influence of processing parameters. Journal of Natural Fibers 3 (2/3), 39–53.

Fisher, G., 2016. Meeting the sustainability challenge with sustainable innovation. Fiber Journal 30 (6), 8.

Guest Contributor, 2021. Starting at the Start for Sustainable Fashion: The Argument for Focusing on Raw Materials. (Accessed 9 March 2021).

Guest Contributor Grand View Research, Published February 2020. Industrial Hemp Market Size, Share & Trends Analysis Report by Product (Seeds, Fiber, Shives), by Application (Animal Care, Textiles, Food & Beverages, Personal Care), and Segment Forecasts, 2020–2027. https://www.grandviewresearch.com/industry-analysis/industrial-hemp-market.

Grosso, M.M., 1996. Recycling fabric waste. The challenge industry. Journal of the Textile Institute 87, 21–30. https://fashionunited.uk/news/business/starting-at-the-start-for-sustainable-fashion-the-argument-for-focusing-on-raw-materials/2021030954240.

Hell, E., August 2016. Tenacity polyester fiber recycling: a challenge tackled with innovative technology. Fibreculture Journal 30 (4), 26–29.

Horne, M., 2020. Bast fibres:Hemp cultivation and production. In: Kozlowski, R., Mackiewicz-Talarczyk, M. (Eds.), Handbook of Natural Fibers. Processing and Applications, vol. 1. Elsevier Ltd, UK, pp. 163–196.

Kozlowski, R., 1990. Retting of flax in Poland. In: Sharma, H.S.S., Van Sumere, C.E. (Eds.), The Biology and Processing of Flax. M.Publications, Belfast, pp. 251–261.

Kozlowski, R., Wladyka-Przybylak, M., 2004. Chapter 14: uses of natural fibres reinforced plastics. In: Wallenberger, F.T., Weston, N.E. (Eds.), Natural Fibres, Plastics and Composites. Kluwer Academic Publishers, pp. 249–274.

Kozlowski, R., Rawluk, M., Barriga-Bedoya, J., 2005a. Ramie. In: Frank, R.R. (Ed.), Bast and Other Plant Fibres. Woodhead Publishing, Cambridge, UK, pp. 207–227.

Kozłowski, R., Baraniecki, P., Barriga-Bedoya, J., 2005b. Bast fibres (flax, hemp, jute, ramie, abaca). In: Frank, R.R. (Ed.), Bast and Other Plant Fibres. Woodhead Publishing, Cambridge, UK, pp. 36–87.

Kozłowski, R., Mackiewicz-Talarczyk, M., Muzyczek, M., Barriga-Bedoya, J., 2012a. Future of natural fibers, their coexistence and competition with man-made fibers in 21st century. Molecular Crystals and Liquid Crystals 556, 200–222.

Kozlowski, R., Mackiewicz-Talarczyk, M., Allam, A.M., 2012b. Bast fibres: flax. In: *Handbook Of Natural Fibers, Vol. 1 Types, Properties And Factors Affecting Breeding And Cultivation*. Woodhead Publishing Ltd, Cambridge, UK, pp. 56–113.

Kozlowski, R., Muzyczek, M., Zimniewska, M., Mackiewicz-Talarczyk, M., Barriga-Bedoya, J., April 25th–28th, 2016. Opportunities and challenges of natural fibres, production and application. Poznan, Poland. In: The 90th Textile Institute World Conference: Inseparable from the Human Environment. Hosted by the Institute of Natural Fibres and Medicinal Plants (INF&MP), pp. 21–29. Proceedings.

Kozlowski, R., Muzyczek, M., 2017. Introduction. In: Kozlowski, R., Muzyczek, M. (Eds.), Natural Fibers: Properties, Mechanical Behavior, Functionalization and Applications. Nova Science Publishers, pp. 215−236.

Kozlowski, R., Mackieicz-Talarczyk, M., Wielgus, K., Praczyk, M., Allan, M., 2020. Bast fibres: flax. In: Kozlowski, R., Mackiewicz-Talarczyk, M. (Eds.), Handbook of Natural Fibers. Processing and Applications, vol. 1. Elsevier Ltd, UK, pp. 93−156.

Kozlowski, R., Rozanska, W., 2020. Enzymatic treatment of natural fibres. In: Kozlowski, R., Mackiewicz-Talarczyk, M. (Eds.), Handbook of Natural Fibers. Processing and Applications, vol. 2. Elsevier Ltd, UK, pp. 227−244.

Kuehnel, M., 2020. Recycling medical waste to produce clean fuel. winter Advances Wales 95, 14.

Li, Y., 2001. The science of clothing comfort. Textile Progress 31 (1/2), 1−138 (The Textile Institute).

Maity, S., Prasad, G.D., Palash, P., 2014. A review of flax nonwovens: manufacturing, properties, and applications. Journal of Natural Fibers 11 (4), 365−390.

Muzyczek, M., 2009. The use of flax and hemp for textile end uses. In: Kozlowski, R. (Ed.), Handbook of Natural Fibers. Processing and Applications, vol. 2. Woodhead Publishing Ltd, Cambridge, UK, pp. 312−328.

Muzyczek, M., 2020. The use of flax and hemp in textile application. In: Kozlowski, R., Mackiewicz-Talarczyk, M. (Eds.), Handbook of Natural Fibers. Processing and Applications, vol. 2. Elsevier Ltd, UK, pp. 147−166.

Oerlikon, September 5, 2013. Manmade fibers segment analyst and media briefing. Remscheid 37.

Sabbir, A., Tasneem, F., February 2014. Polyhydroxybutyrate—a biodegradable plastic and its various formulations. International Journal of Innovative Research in Science, Engineering and Technology 3 (2), 9494−9499.

Starlinger, April 2011. Recycling technology offers innovative machinery solutions. Fiber Journal 25 (2), 8−10.

Townsend, T., August 22, 2019. Natural Fibres and the World Economy—July 2019. Published. https://news.bio-based.eu/natural-fibres-and-the-world-economy-july-2019/.

Wasif, A.J., Singh, V.L., 2005. Naturally coloured cotton growing awareness. Colourage Annual 89.

Yang, Q.M., 2014. Global Fibres Overview. Tecnon OrbiChem, Synthetic Fibres.

Zimniewska, M., Kozlowski, R., 2004. Natural and man—made fibers and their role in creation of physiological state of human body. Molecular Crystals and Liquid Crystals 418, 841−858.

Zimniewska, M., Witmanowski, H., Kozlowski, R., 2006. Clothing effect on selected parameters of oxidative stress. LENZINGER BERICHTE 85, 17−21.

Zimniewska, M., Kozlowski, R., Rawluk M, M., 2004. Natural vs. man-made fibres—physiological viewpoint. Journal of Natural Fibers 1 (2), 69−81. Materials Committee Meeting at APIC 2014 Pattaya.

Lotus fibre drawing and characterization

Ritu Pandey[1], Amarish Dubey[2] and Mukesh Kumar Sinha[3]
[1]Department of Textiles & Clothing, Chandra Shekhar Azad University of Agriculture & Technology, Kanpur, Uttar Pradesh, India; [2]Rajiv Gandhi Institute of Petroleum Technology, Jais, Uttar Pradesh, India; [3]Technical Textile Division, Ministry of Textiles, New Delhi, New Delhi, India

5.1 Introduction

Aquatic lotus (*Nelumbo nucifera* Gaertn.) fibre is one of the most luxurious sustainable fibres for the fashion industry (Gardetti and Mutthu, 2015). It is green and eliminates negative environmental impact because of no chemical intervention from fibre production to yarn and fabric manufacturing stage. Fabric made of lotus fibre has a natural fragrance, resiliency, hydrophilicity, and antibacterial properties. Unique characteristics of lotus fibre also include coolness in summer and warmness in winter that makes it one of the most significant textile fibres (Cheng et al., 2017). Lotus plant is an important cash crop for lotus farmers (Lin et al., 2019). Lotus farmers are benefited from the marketing of lotus flowers, seeds, and rhizomes. Plant leaves, stamens, and peduncles have uses in herbal tea, food, fish feed, and cosmetic products besides having cultural significance (Paudal and Panth, 2015; Tungmunnithum et al., 2018). A study in India reveals that farmers can earn USD 1070–1200/ha in a year by selling different plant parts (except fibre) of lotus (Sahu and Chandravanshi, 2018). Awareness about lotus fibre drawing and marketing may bring even more dividends to their profit. Although, the lotus plant is having several applications, yet some parts of lotus plant which may turn lotus as a bigger asset for its growers, is just waste; this is happening because of slackness in design innovation by its stakeholders. Lotus peduncles, after plucking flowers, are generally left as waste to dry alongside ponds and lakes throughout the year, whereas, they could be utilized for fibre production. Fibres from the lotus could be drawn from its peduncles as well as rhizomes. Practice of leaving peduncles as waste is common in larger parts of the world except in parts of south-east Asia where some people extract fibres manually for making fabric for donning lord Buddha in pagodas and to honour the elites (Zhao et al., 2018).

The fibre has found applications in fashion apparel. In one of the previous works, cellulosic microfibre from lotus is reported as one of the finest quality fibres suitable for current trendy garments. The fibre is comparable with the best quality cotton, silk, and pashmina without any chemical treatments (Pandey et al., 2020a). Lotus robes are preferred by the environmentally conscious consumers of Europe (Gardetti and Mutthu, 2015; Nayak et al., 2020; Pandey et al., 2020b).

5.2 Lotus cultivation

Lotus cultivation is traditionally done in China and India for over 7000 years (Zhu, 2017). Its cultivation is also widely observed in parts of Asia, Australia, Russia, north, southern, and Eastern America. National Garden of Aquatic Vegetable at Wuhan has conserved around 572 lotus accessions from different parts of the world. Lotus plants are classified into three types; rhizome, seed, and flower types (Table 5.1). Area under lotus cultivation in China is estimated to be more than 0.2 million ha, producing about three million tons of edible rhizomes and 15,000 tons of dry seeds annually. Lotus plants coexist with rice plantation, fish, crab, and shrimp (Guo, 2009). Lotus propagation is through seeds, rhizomes, and root slips. Seeds are first soaked in water for four to five days. Sprouted lotus seeds are later transferred to ponds 5 cm deep in the soil, or soil containers submerged in water. Rhizomes planted in soil, underwater give rise to several lotus peduncles that grow out tall passing the water surface. Sandy loam, clay, and animal manure are recommended medium for lotus culture. Plants flourish well in acidic soil with 4.6 pH. Nutrient requirements in the soil are nitrogen, phosphorus, and potassium in 18:18:21 ratio. The process from lotus cultivation to product development is depicted in Fig. 5.1.

5.3 Lotus fibre drawing

Lotus fibre production process is depicted in Table 5.2. Fibres are drawn out from the peduncles as well as rhizomes of the lotus plant (Fig. 5.2);
(a) Fibre from lotus peduncle

The lotus peduncle consists of a single flower which blooms in temperate region in different parts of the world. Plucking of flowers for cultural uses renders the plant peduncle useless. Plant peduncles bereft of flowers are rooted out later as a drive to clean water bodies. Tender green peduncles contain innumerable circular rings of fine and clean filaments drawn out with physical force using different methods (Fig. 5.2A).

Table 5.1 Classification of lotus plant based on DNA markers (Guo, 2009).

Types of lotus plants	Characteristics/ number of accessions'
Rhizome lotus	Higher plant height and yield, few or negligible flowers, edible rhizomes/ 310
Seed lotus	Higher number of carpels and seeds with better seed quality compared to rhizome lotus' seeds/ 33
Flower lotus	Higher flower petals count, ornamental lotus flower with more than 1000 petals as the stamen and pistils transform into flower petals, low plant height, thin rhizome/ 229

Lotus fibre drawing and characterization 97

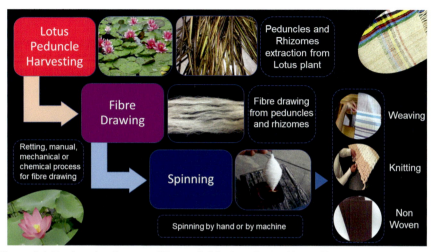

Figure 5.1 Product development (fibre to finished product) of lotus fibre.

Table 5.2 Lotus fibre production process.

Lotus propagation	Peduncle harvest	Fibre drawing			
		Peduncle			Rhizome
		Manual	Mechanical	Retting	Manual
Seed	Manual pulling	Hand drawing	Machine drawing	Water	Folding/breaking
Rhizomes	Mechanized uprooting	Biological, chemical, biochemical		Steam explosion	
Root slip		Irradiation		Chemical	

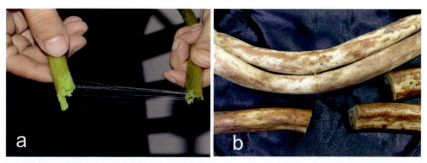

Figure 5.2 Fibre drawing from lotus. (A) Fibre drawing from lotus peduncle (B) Fibre drawing from lotus rhizome.

(b) Fibre from Lotus rhizome

Thick snappy rhizomes are edible and widely used to prepare snacks and preservative food items. Folding of rhizomes expose valuable lotus fibres lying along its length. The quantity of fibres obtained from single lotus rhizomes is lesser than fibres drawn from a single peduncle (Fig. 5.2B).

5.3.1 Manual drawing of lotus fibres

Manual drawing of lotus fibres from peduncles is carried out in two ways (Method I):

Method I involves twisting and breaking the peduncle's end in 50 mm sections one by one and manual twisting. Breaking and pulling apart from peduncle sections let the fibres drawn out from the peduncles. Peduncles containing natural spinneret in sclerenchyma and vascular tissues lead the way for the extraction of lissome ells of lotus fibres arranged in circular ring form that stretches up to 1400 mm lengths. The lotus fibres drawn out from the sets are helical and do not stretch further.

Method II of manual fibre drawing involves cutting off peduncles into 70–90 mm lengths. Peduncle's ends are twisted gently by hand to break open from the mid-point of the cut lengths, and slowly pulling both ends in opposite directions. The process draws out the fibres arranged lengthwise along the full length of the peduncle.

5.3.2 Mechanical extraction

The mechanized system to draw out lotus fibres from peduncles consists of a pair of jaws in-line but facing each other mounted on a shaft and designed to rotate slowly at about 10 s cm^{-1} in opposite direction simulating the hand twisting. About 10–12 pieces of 70–90 mm sections of the peduncles are gripped in the jaws of the device in one operation (Fig. 5.5).

In this art of fibre extraction, lotus peduncles are cut into the sections of 70–90 mm using blades of the machine and then 10 such cut sections are fixed on 10 separate pairs of moving jaws of the machine. Jaws containing peduncle sections are first slightly twisted in opposite directions by holding jaws and then pressed to move apart at a speed of 10 s cm^{-1}. The process breaks the thorny epidermis along with soft inner tissues and stretches causing drawing off fine long fibres. The fibres drawn out are collected on a velvet board kept underneath (Pandey et al., 2020a). The process obviates stress involved in manual drawing and reduces the duration of fibre drawing when compared to the manual method of fibre drawing. Extracted fibres are clean and free from any entangled vegetative materials (Method II).

There are several other methods of fibre drawing from lotus peduncles that usually involve the use of either Method I or II (Chen and Gan, 2008).

5.3.3 Biological extraction

The biological extraction method employs dipping of the peduncles in water. With the aid of bacteria in water, peduncles get soften. Fibres are drawn out from soft peduncles employing either Method I or II.

5.3.4 Biochemical extraction

Semi-cellulosic chemicals and different enzymes are used to dip the peduncles followed by Method I or II to draw out the fibres.

5.3.5 Chemical extraction

Lotus peduncles are submerged in dilute sodium hydroxide solution for several hours. Subsequently, peduncles are washed and vascular tissues of the softened epidermis are separated manually to draw out fibres using Method I. Drawn fibres are dried at 50 °C (Cheng et al., 2017).

5.3.6 Steam explosion

Lotus peduncles are bio-fermented to extract the crude fibres prepreged with surrounding vascular tissues. Extracted fibres are heated to put them into steam explosion under pressure in a closed vessel to separate fibres from vegetative matter. The resultant fibre is dried before use. Alkali treatment prior to steam explosion yields better separation of fibres from the plant vegetative matter (Yuan et al., 2013).

5.3.7 Retting

Lotus peduncles are submerged in water for water retting (Pandey, 2016) followed by fibre extraction (Zhang and Li, 2011). Fibre extraction through retting yields coarse fibres as compared to hand-drawn fibres (Table 5.3).

5.3.8 Irradiation

Plant peduncles are cut into sections and treated with NaOH followed by microwave irradiation for 20 min. Fibres are squeezed from peduncles and rinsed to remove vascular impurities (Cheng et al., 2017).

5.4 Fibre physical properties

5.4.1 Fibre moisture

The higher amount of moisture in fibre corresponds to increasing softness in fabric (Fuzek, 1985). Lotus fibre moisture is about 11.3%−12.32% which is more than in cotton, pina (Pandit et al., 2020), and most of the other natural fibres. Fine microstructure, low crystallinity, and the presence of high number of hemicelluloses contribute to high moisture regain of lotus fibre (Pan et al., 2011). Moisture absorption and release rate of lotus fibre is higher than flax and cotton fibre (Liu et al., 2020; Mishra et al., 2016; Pandey et al., 2020a). Higher moisture absorption and desorption ensures the escape of body perspiration into the atmosphere and recommended for providing a high level of comfort during wear.

Table 5.3 Physical characteristics of lotus fibre reported in various researches.

Fineness (tex)	Length (mm)	Tenacity (g/d)	Elongation (%)	Moisture content (%)	References
1.8−2.2	40−150	1.7−1.8	5.72−5.80	12.35	Chen and Gan (2008)
0.091	100−200	2.5	2.6	12.3	Pan et al. (2011)
2.59	−	1.9	6.68	−	Zhang and Li (2011)
0.16	−	4.36	2.1	−	Yuan et al. (2012)
0.16	−	3.9	2.75	−	Yuan et al. (2013), Chen et al. (2012)
−	−	−	2.6	12.3	Gardetti and Mutthu (2015)
0.09−0.18	35.2−58.0	0.8−2.4	4.46	−	Cheng et al. (2017)
0.22	120−1400	0.57−2.8	1.95	11.3	Pandey et al. (2020a)
0.13	−	−	−	−	Liu et al. (2020)

5.4.2 Physico-mechanical properties

Lotus fibre is fine, lustrous, clean, and off-white in colour (CIELAB L value = 79.46). The fibre is having a flat strip but helical structure (Fig. 5.3A) with round cross section (Chen et al., 2012; Pandey et al., 2020a). Fig. 5.3A shows an image of scanning electron microscope (SEM) of lotus fibre which was drawn out manually. The image depicts that the fibre is having flat strip structure, which is showing the soft nature of fibre. Lotus fibres are long with an average length of 467 mm. Creep length of lotus fiber is better in wet state (0.10 mm) than at ambient temperature (0.13 mm) and dry state (0.21). Fibre strength is higher at dry state, whereas, elongation improved at wet state. Fibre strength of lotus fibre is reported to be greater than cotton at ambient temperature and in the state of wet (Yuan et al., 2012). Stretching and fibre drawing process, storage duration of harvested peduncles, and the particular portion of peduncles determine lotus fibre length to some extent besides plant nutrition and growth factors. Fibres from decayed peduncles, over-stretching, and fibres from lower or submerged parts of the plant, break the fibres causing a shorter length of fibres as compared to fresh and green peduncles. Fibre colour of the putrescent peduncle turns darker in hue. Table 5.3 presents the lotus fibre characteristics determined by several researchers. The Highest tenacity and elongation reported for lotus fiber is 4.36 g/d (Yuan et al., 2012) and 6.68% (Zhang and Li, 2011) respectively. Variation in physical properties of the fibre, as evident in Table 5.3, may be attributed to variation in method of fibre extraction (Chen and Gan, 2008) and fibre stretching to draw out fibres (Yuan et al., 2012).

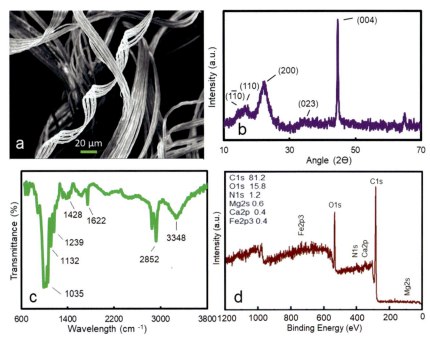

Figure 5.3 Characterization of lotus fibre. (A) SEM image of extracted lotus fibre (B) XRD plot of lotus fibre (C) FTIR spectrum of lotus Fibre (D) XPS Broad spectrum of lotus fibre.

5.5 Chemical analysis of lotus fibre

5.5.1 Chemical composition of lotus fibre

Lotus fibre is mainly composed of cellulose (70.18%). The X-ray photoelectron spectroscopy (XPS) elemental analysis also confirms the presence of cellulose as a major constituent in lotus fibres (Pandey et al., 2020a). Other components present in the lotus fibre are hemicellulose, lignin, pectin, and lipids (Cheng et al., 2017). Lotus fibre consists of 17 different amino acids (1.94%) (Yuan et al., 2012). Weak peaks of nitrogen, magnesium, calcium, and iron present in the fibre might have been obtained from soil (Fig. 5.3D). The XPS broad spectrum shows the relative presence of carbon (81.2%), oxygen (15.8%), nitrogen (1.2%), and other elements with different stages and amount.

5.5.2 Crystal structure and crystallinity

Fibre has cellulose I structure, with 48 % and 60 % crystallinity and orientation respectively (Chen et al., 2012; Yuan et al., 2012). The X-ray diffraction (XRD) picture of lotus fibre shows four crystalline peaks at 15.9 degrees, 22.6 degrees, 34.7 degrees, and 45.2 degrees respectively (Fig. 5.3B). The peaks pattern shows the characteristic of the crystal polymorph I of cellulose (as there is no double peak at 22.6 degrees). The two peaks at 15.9 degrees suggest that the lotus fibre contains higher cellulose content, like cotton or flax. It can also be noticed here that hemicellulose and lignin (non-cellulosic substances) might be the merging of these two distinct peaks into a single one. The peak 15.9 degrees corresponds to (101) crystallographic plane, peak 22.6 degrees corresponds to (200) plane, and the peak at 34.7 degrees and 45.2 degrees corresponds to (023) and (004) planes.

5.5.3 Chemical structure

Fourier transform infrared spectroscopy analysis (FITR) spectrum of lotus fibres show four broader peaks visible in Fig. 5.3C for lotus fibre. The recorded FTIR spectrum of the lotus fibre contains the typical bands of cellulose, lignin, and hemicellulose (Fig. 5.3C). The strong peak present at 3348 cm^{-1} shows the hydroxyl (OH) groups of cellulose, lignin, and water (Pandey et al., 2020c). Distinguish peak present at 2852 cm^{-1} representing the stretching vibration of the C−H bond, which could be present due to cellulose and hemicellulose content in lotus fibre. The peak at 1622 cm^{-1} might be related to the presence of water in the fibres. The 1428 cm^{-1} peak is associated as an absorption band for CH_2 symmetric bending of the cellulose. The peak at 1239 cm^{-1} and 1132 cm^{-1} are related to bending vibrations of the C−H and C−O groups, respectively (Pandey et al., 2021). These peaks might be present because of aromatic rings of cellulose polysaccharides. A high intense peak present at 1035 cm^{-1} is connected with the CO stretching vibrations of the polysaccharide in cellulose.

5.6 Comparison of lotus fibre with cotton fibre

The texture and appearance of the lotus fibre is very much similar to cotton fibre. As shown in Fig. 5.4A, the cotton fibre spectrum is almost similar to the lotus fibre spectrum. The C−O (carbonyl) stretching peak in lotus fibre is present at 1035 cm^{-1}, which is similar to peaks observed in cotton FTIR with enlarged stretching. The diffused peak of C−O in lotus suggests lower molecular weight of lotus fibre than cotton. A similar trend exists for CH (1241 cm^{-1} is) and OH (3348 cm^{-1}) stretching of lotus and cotton fibres (Fig. 5.4A). The peak at 2258 cm^{-1} for cotton fibre, which represents stretching vibration of C−H bond, is almost similar to lotus fibre at the same wavelength. It signifies that lotus fibre and cotton fibre are chemically similar, yet constituents may vary (Chung et al., 2004). Lower intensity peaks of lotus fibre indicate lotus and cotton are chemically similar but chemical constituents and composition are less as compared to cotton fibre. Similar to cotton, lotus fibre is resistant to alkali and disintegrates in concentrated acid (Pandey et al., 2020a).

Similar to FTIR, the XRD spectrum is also overlapping with each other for lotus fibre and cotton fibre (Fig. 5.4B). Yet the cotton fibre XRD peaks are more prominent as compared to lotus fibre, which signifies the higher crystallinity nature as compared

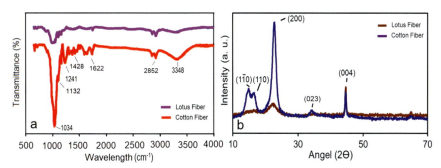

Figure 5.4 A material characteristic comparison between manually drawn out lotus fibre and cotton fibre. (A) FTIR patterns of lotus fibre in comparison with cotton fibre (B) XRD patterns of lotus fibre in comparison with cotton fibre.
No Permission Required.

to lotus fibre. The crystalline peak at the plain (200) is more prominent in cotton than in the lotus (Fig. 5.4B). Diffused peaks, as observed in most of the crystal planes in lotus fibre, also support the statement that the lotus fibre is having less crystallinity than cotton fibre. However, crystalline peak (004) at 45.2 degrees is more or less similar in both the fibres. In general, cotton fibre is having higher degree of crystallinity than lotus fibre.

5.7 Application of lotus fibre for commercial product

Application of lotus fibres for commercial products is listed as:

Chemical free food packaging material
Comfort moisture management
Apparel with wrinkle recovery property
Biomedical due to high absorption, healing in wounds
Substitute of cotton

Fibre drawing from lotus in Asia is an ancient practice prevalent in 5000 BCE (Pandey et al., 2020a). Lotus robe is one of the most traditional and significant fabric with healing properties, that has found references in Buddhist literature Jinatthapakasani and Thai Ramayan (Fraser-Lu and Thanegi, 2016). The earliest references of lotus fibre drawing and robe production for elites in Myanmar were reported by Balfour (1885). Use of lotus fibre started in Cambodia in 1910 when a team of weavers prepared lotus garments for abbots. The appreciation received from the meritorious monks encouraged the weavers to draw, and prepare lotus robes for highly revered monks. The weavers' group was later renamed Daw Kya Oo (Madame Lotus Egg) to honour their achievement. Two sets of lotus robes prepared in a year were reserved only for part of abbots' offerings and not a commercial activity (Aishwariya and Thammima, 2019). The ancient tradition of drawing of lotus fibre and preparation of lotus robe continued due to awareness about the hazardous outcome of fast fashion to the mankind and universe. 0.3 t of lotus peduncles are required to produce 1000 g of lotus fibres (Pandey et al., 2020a). One lotus fibre scarf is representative of 1000 lotus flowers, since 1000 lotus peduncles are required to produce a single scarf (Myint et al., 2018). Lotus fibre drawing and product formation are depicted in Fig. 5.5. Various lotus fibre artists working to promote sustainable and vegan culture in the fashion world are listed in Table 5.4. Lotus fibre fabric can be a toxic-free substitute in food sector for making tea bags and food packaging materials. The tea bag material currently in use is chemically synthesized nylon and polyethylene terephthalate, which releases billions of micro- and nano-plastics in steeping process affecting human behaviour and developmental process (Hernandez et al., 2019).

Figure 5.5 Lotus fibre drawing: From lotus plucking to mechanical fibre drawing.

Table 5.4 Lotus textile merchandizing around the world.

Artist/brand/tradename	Products	Country
Samatoa	Fibre, accessory, vegan leather, garments	Cambodia
Loro Piana	Apparel	Italy
NoMark Lotus	Men's shirt	India
Kyar Chi	Scarf, shawl, accessory	Myanmar
Phan Thi Thuan	Fabric weaving	Vietnam
Bijoy Shanti	Yarn, bags, masks, accessories	India
YGN collective	Lotus textiles	Myanmar

5.8 Lotus inspired design culture

Lotus has the most inspirational survival instinct; a 1300 years old lotus rhizome was able to germinate in Liaoning, China (Shen-Miller et al., 2002). Lotus flower used for cultural uses is truly a symbol of courage; from besmirched bottom to bright beams by overcoming hurdles of the water currents. Divine flower buds symbolizing perfect wisdom, and serenity comes out of the muddy surface and opens only after embracing the bright sun. Not only it emerges faultless but also repels grime from its surface. This property of repelling dirt has inspired many researchers to mimic the same on interface surface material (Gao and McCarthy, 2006; Shmueli and Leikovich, 2015). Intelligent textiles with self-cleaning features are possible due to inspiration drawn from the lotus flower and leaves (Chinta et al., 2013; Chen et al., 2019). Apart from stain guard fabric surface, lotus bioinspired composite material (Wu et al., 2014) and wearable graphene e-textile for human health environmental hazards sensor (Cheong et al., 2021) has also been achieved. Fabrication of hydrophobic multi-bore hollow fibre (MBF) membrane imitating lotus roots were found to have improved mechanical properties than single-bore and rectangular MBF for membrane distillation applications (Wang and Chung, 2012). Lotus shaped core design is used to develop negative curvature anti-resonant fibre for data transmission applications (Nawazuddin et al., 2017). Lotus is one of the oldest and most beautiful motifs in Persian carpets and embroidery (Siddiqui, 2012). Many old building architecture is inspired by the eight-petaled inner calyx of lotus flower petals. A few such examples are the great Egyptian pyramid (Seyfzadeh, 2018), and Ranakpur Jain temple (India) ceilings (Fig. 5.6A−C). The inverted conical lotus-shaped structure of the ceilings is said to have maintained the airflow of the building and keep the premises cool. Lotus is also prevalent in Gandhara sculpture and Zoroastrian architecture. A 15th CE Lord Vishnu installation at Sotheby's is

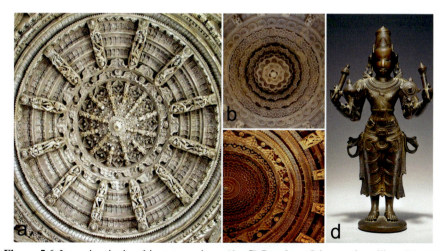

Figure 5.6 Lotus inspired architecture and art. (A−C) Ranakpur Jain temple ceiling (D) 15th CE Lord Vishnu installation at Sotheby's.

depicted holding lotus buds' peduncles in his both hands (Fig. 5.6D). Hindu Deities and lord Buddha are often projected on a lotus seat. Pushkalavati (The Lotus city) now called Prang, settled between the river Swat and Kabul founds mention by Hiuen Tsang in seventh CE (Siddiqui, 2012).

5.9 Conclusion

Lotus fibre is a potential substitute for cotton. It is recommended that fibre can be used in producing blended yarn with sustainable fibres. Apart from this, lotus fibre is biodegradable, cost-effective, and sustainable to the environment. Fibre is highly lustrous and thus can be used as a substitute for viscose rayon in the fashion segment. The mechanical extraction process is easy and suitable for the large-scale production of lotus fibres. Lotus peduncles are just waste for lotus growing farmers and this application of fibre production can improve their monetary benefits in multiple folds. Properties of the lotus fibre are comparable to normally established fibres (silk, wool, cotton, pashmina, viscose) and suitable for numerous applications in the field of fashion to technical textiles. Most importantly being eco-friendly, hand-drawn, and hand-spun yarn without chemical intervention, it is the most suitable and hygienic fibre for medical textiles and food grain packaging.

References

Aishwariya, S., Thammima, S., 2019. Sustainable textiles from lotus. Asian Textile Journal 28 (10), 56–59.

Balfour, E., 1885. Cyclopedia of India and of Southern and Eastern Asia. White Pebble, Inc. GluedIdeas.

Chen, D.S., Gan, Y.J., 2008. Lotus Fibers and Preparation. Journal of Minjiang University, p. 5.

Chen, D.S., Gan, Y.J., Yuan, X.H., 2012. Research on structure and properties of lotus fibers. Advanced Materials Research 476-478, 1948–1954. https://doi.org/10.4028/www.scientific.net/AMR.476-478.1948.

Chen, R., Wan, Y., Wu, W., Yang, C., He, J.H., Cheng, J., Jetter, R., Ko, F.K., Chen, Y., 2019. A lotus effect-inspired flexible and breathable membrane with hierarchical electrospinning micro/nanofibers and ZnO nanowires. Materials & Design 162, 246–248. https://doi.org/10.1016/j.matdes.2018.11.041.

Cheng, C., Guo, R., Lan, J., Jiang, S., 2017. Extraction of lotus fibres from lotus stems under microwave irradiation. Royal Society Open Science 4, 170747. https://doi.org/10.1098/rsos.170747.

Cheong, D.Y., Lee, S.W., Park, I., Jung, H.G., Roh, S., Lee, D., Lee, T., Lee, S., Lee, W., Yoon, D.S., Lee, G, 2021. Bioinspired lotus fiber-based graphene electronic textile for gas sensing. Cellulose 29, 4071–4082. https://doi.org/10.21203/rs.3.rs-1001747/v1.

Chinta, S.K., Landage, S.M., Swapnal, J., 2013. Water repellency of textiles through nanotechnology. International Journal of Advanced Research IT Engineering 2 (1), 36–57.

Chung, C., Lee, M., Choe, E.K., 2004. Characterization of cotton fabric scouring by Ft-Ir ATR Spectroscopy. Carbohydrate Polymers 58, 417–420.

Fraser-Lu, S., Thanegi, M., 2016. History of lotus. https://samatoa.lotus-flower-fabric.com/eco-textile-mill/lotus-fabric. (Accessed 27 October 2016).

Fuzek, J.F., 1985. Absorption and desorption of water by some common fibers. Industrial and Engineering Chemistry Product Research and Development 24 (1), 140−144.

Gao, L., McCarthy, T.J., 2006. "Artificial lotus leaf" prepared using a 1945 patent and a commercial textile. Langmuir 22 (14), 5998−6000.

Gardetti, M.A., Mutthu, S.S., 2015. The lotus flower fiber and sustainable luxury. In: Gardetti, M., Mutthu, S. (Eds.), Handbook of Sustainable Luxury Textiles and Fashion. Environmental Footprints and Eco-Design of Products and Processes. Springer. https://doi.org/10.1007/978-981-287-633-1_1.

Guo, H.B., 2009. Cultivation of lotus (*Nelumbo nucifera* Gaertn. ssp. nucifera) and its utilization in China. Genetic Resources and Crop Evolution 56 (3), 323−330.

Hernandez, L.M., Xu, E.G., Larsson, H.C., Tahara, R., Maisuria, V.B., Tufenkji, N., 2019. Plastic teabags release billions of microparticles and nanoparticles into tea. Environmental Science and Technology 53 (21), 12300−12310. http://doi/10.1021/acs.est.9b02540.

Lee, S.W., Jung, H.G., Roh, S., Lee, D., Lee, T., Lee, S., Lee, W., Yoon, D.S., Lee, G., 2021. Bioinspired Lotus fiber-based graphene electronic textile for gas sensing. https://doi.org/10.21203/rs.3.rs-1001747/v1.

Lin, Z., Zhang, C., Cao, D., Damaris, R.N., Yang, P., 2019. The latest studies on Lotus (*Nelumbo nucifera*)-an emerging horticultural model plant. International Journal of Molecular Sciences 20 (15), 3680. https://doi.org/10.3390/ijms20153680.

Liu, Y., Wang, Y., Yuan, X.H., Prasad, G., 2020. The function of water absorption and purification of lotus fiber. Materials Science Forum 980, 162−167.

Mishra, S., Pandey, R., Singh, M.K., 2016. Development of sanitary napkin by flax carding waste as absorbent core with herbal and antimicrobial efficiency. International Journal of Science, Environment and Technology 5, 404−411.

Myint, T., San, D.K.N., Phyo, U.A., 2018. Lotus fiber value chain in Myanmar. In: Project Report, Helvetas Myanmar. A study conducted on behalf of the Regional Biotrade Project, Myanmar.

Nawazuddin, M.B.S., Wheeler, N.V., Hayes, J.R., Sandoghchi, S.R., Bradley, T.D., Jasion, G.T., Slavík, R., Richardson, D.J., Poletti, F., 2017. Lotus-shaped negative curvature hollow core fiber with 10.5 dB/km at 1550 nm wavelength. Journal of Lightwave Technology 36 (5), 1213−1219.

Nayak, R., Panwar, T., Nguyen, L.V.T., 2020. Sustainability in Fashion and Textiles: A Survey from Developing Country. Sustainable technologies for fashion and textiles, pp. 3−30. https://doi.org/10.1016/B978-0-08-102867-4.00001-3.

Pan, Y., Han, G., Mao, Z., Zhang, Y., Duan, H., Huang, J., Qu, L., 2011. Structural characteristics and physical properties of lotus fibers obtained from *Nelumbo nucifera* petioles. Carbohydrate Polymers 85 (1), 188−195.

Pandey, R., 2016. Fiber extraction from dual-purpose flax. Journal of Natural Fibers 13 (5), 565−577. https://doi.org/10.1080/15440478.2015.1083926.

Pandey, R., Sinha, M.K., Dubey, A., 2020a. Cellulosic fibers from Lotus (*Nelumbo nucifera*) peduncle. Journal of Natural Fibers 17 (2), 298−309.

Pandey, R., Pandit, P., Pandey, S., Mishra, S., 2020b. Solutions for sustainable fashion and textile industry. In: Pandit, P., Singha, K., Srivastava, S., Ahmad, S. (Eds.), Recycling from Waste in Fashion and Textiles: A Sustainable and Circular Economic Approach. Scrivener Publishing LLC, pp. 33−72.

Pandey, R., Jose, S., Sinha, M.K., 2020c. Fiber extraction and characterization from Typha domingensis. Journal of Natural Fibers 1–12. https://doi.org/10.1080/15440478. 2020.1821285.

Pandey, R., Jose, S., Basu, G., Sinha, M.K., 2021. Novel methods of degumming and bleaching of Indian flax variety Tiara. Journal of Natural Fibers 18 (8), 1140–1150. https://doi.org/ 10.1080/15440478.2019.1687067.

Pandit, P., Pandey, R., Singha, K., Shrivastava, S., Gupta, V., Jose, S., 2020. Pineapple leaf fibre: cultivation and production. In: Jawaid, M., Asim, M., Tahir, P., Nasir, M. (Eds.), Pineapple Leaf Fibers. Green Energy and Technology. Springer, Singapore. https://doi.org/ 10.1007/978-981-15-1416-6_1.

Paudal, K.R., Panth, N., 2015. Phytochemical profile and biological activity of *Nelumbo nucifera*. Evidence-Based Complementary and Alternative Medicine 789124, 1–16. https://doi.org/10.1155/2015/789124.

Sahu, R., Chandravanshi, S.S., 2018. Lotus cultivation under wetland: a case study of farmers innovation in Chhattisgarh, India. International Journal of Current Microbiology and Applied Science 7, 4635–4640.

Seyfzadeh, M., 2018. Hemiunu used numerically tagged surface ratios to mark ceilings inside the great Pyramid hinting at designed spaces still hidden within. Archaeological Discovery 6, 319–337. https://doi.org/10.4236/ad.2018.64016.

Shen-Miller, J., Schopf, J.W., Harbottle, G., Cao, R.J., Ouyang, S., Zhou, K.S., Southon, J.R., Liu, G.H., 2002. Long-living lotus: germination and soil γ-irradiation of centuries-old fruits, and cultivation, growth, and phenotypic abnormalities of offspring. American Journal of Botany 89 (2), 236–247.

Shmueli, E., Leikovich, A., 2015. Substrate Having a Self-Cleaning Anti-reflecting Coating and Method for its Preparation. U.S. Patent 9,073,782.

Siddiqui, K.S., 2012. Significance of Lotus depiction in Gandhara art. Cell 321, 2870169.

Tungmunnithum, D., Pinthong, D., Hano, C., 2018. Flavonoids from *Nelumbo nucifera* Gaertn., a medicinal plant: uses in traditional medicine. Phytochemistry and pharmacological activities. Medicines 5 (4), 127. https://doi.org/10.3390/medicines5040127.

Wang, P., Chung, T.S., 2012. Design and fabrication of lotus-root-like multi-bore hollow fiber membrane for direct contact membrane distillation. Journal of Membrane Science 421, 361–375.

Wu, M., Shuai, H., Cheng, Q., Jiang, L., 2014. Bioinspired green composite lotus fibers. Angewandte Chemie International Edition 53 (13), 3358–3361.

Yuan, X.H., Gan, Y.J., Chen, D.S., Ye, Y.J., 2012. Analysis on mechanical properties of lotus fibers. Advanced Materials Research 476-478, 1905–1909. https://doi.org/10.4028/ www.scientific.net/amr.476-478.1905.

Yuan, B.Z., Han, G.T., Pan, Y., Zhang, Y.M., 2013. The effect of steam explosion treatment on the separation of lotus fiber. Advanced Materials Research 750, 2307–2312.

Zhang, H.W., Li, Y.L., 2011. Lotus fiber production and its property. Advanced Materials Research 146, 93–96.

Zhao, M., Yang, J.X., Mao, T.Y., Zhu, H.H., Xiang, L., Zhang, J., Chen, L.Q., 2018. Detection of highly differentiated genomic regions between Lotus (*Nelumbo nucifera* Gaertn.) with contrasting plant architecture and their functional relevance to plant architecture. Frontiers of Plant Science 9 (1–14), 1219. https://doi.org/10.3389/fpls.2018.01219.

Zhu, F., 2017. Structures, properties, and applications of lotus starches. Food Hydrocolloids 63, 332–348.

Macrophyte and wetland plant fibres

Ritu Pandey[1], Mukesh Kumar Sinha[2] and Amarish Dubey[3]
[1]Department of Textiles & Clothing, Chandra Shekhar Azad University of Agriculture & Technology, Kanpur, Uttar Pradesh, India; [2]Technical Textile Division, Ministry of Textiles, New Delhi, New Delhi, India; [3]Rajiv Gandhi Institute of Petroleum Technology, Jais, Uttar Pradesh, India

6.1 Introduction

Importance of fibre yielding plants is next to cereals. Natural cellulosic fibres are non-hazardous, non-polluting and sometimes they act as health cures (Kozlowski, 2009). Since flora and fauna resources are depleting there is a constant need to explore new textile fibre materials with immense potential to reduce the burden on existing textile fibres that are in short supply, expensive, and becoming rare. Advancements in the field of synthetic fibres caused an abrupt end to local sustainable practices of natural fibres' extraction and yarn formation for making fabric and carpets.

Problems arising in safe disposal of synthetic fibres is prompting present world to move towards eco-friendly processes involving natural fibres and slow fashion because of its biodegradable and sustainable product development support (Pandey et al., 2020a). The designers are also breaking old trends of using natural fibres traditionally and creating a new trend to utilize the natural fibres in a very innovative way such as jute for generating high-performance carbon tube (Dubey et al., 2017), silk cocoon for generating nitrogen-doped graphene (Jangir et al., 2018), corn hair as an energy generation tool (Kumar et al., 2018), cocoon as a soft magnetic memory (Roy et al., 2016), sanitary napkins from flax spinning waste (Mishra et al., 2016) and many more. It is expected that aquatic macrophytes will also find their use in such new innovative ways in the future. Therefore, under-utilized aquatic peduncles generated fibre is very important for producing the next generation of smart textiles. However, wetland macrophytes are being used continuously in different parts of the world due to their traditional significance. Macrophytes are of economic importance in fibre production, weaving, basketry, and making different handicraft utility articles. A review of their identification, classification, extraction methods, physicochemical properties, commercial, and environmental aspects has been presented in this chapter.

6.2 Classification of macrophyte and wetland plants

Classification of macrophytes of fibre importance including definition and fibre size, as suggested by United States' National Technical Committee for Wetland Vegetation (NTCWV), is presented in Table 6.1 (Lichvar et al., 2012, 2016) and Fig. 6.1. Macrophytes such as lotus (*Nelumbo nucifera*), water lily (Nymphaeaceae), water hyacinths (*Eichhornia crassipes*) grow above underwater soil and nourish from water, soil, and air. Upon drying of water beds, macrophytes' are unable to survive on soil and dries up. Facultative wetland (FACW) or riparian plants are generally hydrophytes, which grow on wetlands, but are capable of taking nourishment required for growing on upland or dried-up fields (Lichvar et al., 2016). Wetland plants of fibre importance, presented in the chapter are typha species, milkweed (*Calotropis*), munja (*Saccharum munja*), rice (*Oryza sativa*), wild sugarcane (*Saccharum spontaneum*), reed plant (*Phragmites australis*), coconut (*Cocos nucifera*), money plant (*Epipremnum aureum*), and Poplar (*Populus*). The potential and general characteristics of macrophytes and wetland plants of fibre importance are presented in Table 6.2.

Table 6.1 Classification of aquatic/wetland plants according to US NTCWV definition.

Indicator status of wetland/ upland plants (abbreviation)	Designated as/ qualitative description	Fibre yielding wetland/ macrophytes	
Obligate (OBL)	Hydrophyte/always an aquatic	Submerged	Pondweeds
		Floating	Water hyacinths
		Floating-leaved	Water lily
		Emergent	Lotus, typha
Facultative wetland (FACW)	Hydrophyte/aquatic but may grow in upland	Reed plant, typha	
Facultative wetland (FAC)	Hydrophyte/aquatic and upland both	Wild cane, money plant, poplar (populus species)	
Facultative upland (FACU)	Nonhydrophyte/may grow in aquatic	Munja, rice	
Upland	Nonhydrophyte/ wetland in some regions	Milkweed, miscanthus, coconut	

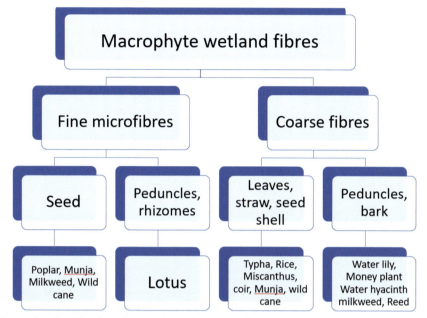

Figure 6.1 Classification of macrophyte fibres according to fineness.

6.2.1 Lotus fibre

Lotus (*Nelumbo nucifera* Gaertn.) is an emergent macrophyte and details for the fibre are explained in Chapter 5. Plant peduncles and rhizomes contain long strands of fine microfibres that are drawn manually and mechanically. The fibre can be obtained throughout the year since the plant flourishes the year around, and only blooming of the flowers is restricted for few months. Lotus fibres are fine, lustrous, wavy, and off white to white in colour (Pandey et al., 2020b). Lotus fibre is characterized by a large specific surface area, high surface energy and strong moisture absorption property. The fibre surpasses cotton in comfort properties as it cools and dries quickly (Liu et al., 2020).

6.2.2 Water hyacinth

Water hyacinth (*Eichhornia crassipes*), an offset floating on water bodies belong to the family of Pontederiaceae. It is a perennial aquatic macrophyte that flourishes in tropical and subtropical regions. The plant is considered as a ferocious weed that has spread out to every part of the world. The plant is highly invasive, making it difficult for other useful macrophytes to survive and also causes clogging up of waterways throughout canals and river basins. The menace has been so alarming that sometimes machines are installed to get rid of water hyacinth plants to maintain water quality. Efforts to turn waste weeds to worth have resulted in greater attention than ever before about water hyacinth for utility articles. Water hyacinth peduncles measuring 0.1–1 m in

Table 6.2 Macrophytes potential and general characteristics.

Fibre	Obtained from	Harvesting season	Habitat	Distribution	Considered as	Characteristics/possible uses
Lotus	Peduncle	January–December	Obligate	India, South-East Asia	Ornamental	Long microfibre/food packaging, apparel
Water hyacinth	Peduncle	January–December	Obligate	India, South-East Asia, Sri Lanka	Invasive weed	Course wavy fibre/bags, furniture
Water lily	Peduncle	January–December	Obligate/Rainforest	Americas, Asia, Europe, Africa	Ornamental	Course fibrils
Typha	Leaf and seed	January–December April–October	Obligate/Riparian, sandy loam	India, Asia, Australia, Europe	Invasive weed	Strong fibre/cord, composite, paper
Milkweed	Stem and seed	January–December May–June	Wetland/Upland	India	Weed	Strong course fibre fine white soft fibre
Munja and wild cane	Stem and seed	January–December July–October	Grassland river bank/skeletal soil rocky soil	India	Noxious weed	High tenacity/cord, automobile
Reed plant	Peduncle and seed	January–December	Halophyte/alkaline soil	Americas, Africa, Asia, Australia	Invasive weed	Strong fibre rope/Roof thatching, flute, aroma disperser
Rice	Straw	October–November	Wetland/upland	Asia	Food grain	Strong fibre/mats
Coir	Seed shell	January–December	Upland	Asia, Middle East, USA, Australia	Fruit	Course fibre/Mats, mulching
Money plant	Peduncle	January–December	Obligate/upland	Australia, Asia, Pacific Islands	Ornamental vine	Strong fibre/Carpet, composite
Poplar	Seed	May	Riparian/forest crop	North and South America, Europe, Asia, New Zealand	Industrial timber	Fine fibre, fibre fill, apparels

length, is water retted, boiled and chemically treated to obtain useful fibre (Chonsakorn et al., 2019). Fibres extracted from water hyacinth and polyester blends are reported to be spun in 10—12 Ne yarn count, which is suitable in apparel and home furnishings. Non-woven sheets of pure WH sanitary napkins are a cost-effective and useful product. As a cottage industry paper making, rope, basket, bags, fibreboard, mats, briquettes are the successful applications of water hyacinth fibres besides biogas production and sewage water treatment (Haider, 1989).

6.2.3 Water lily

Water lily (Nymphaeaceae), an ornamental aquatic hydrophyte, is native of Southeast Asia. Water lily and water hyacinth are among few known aquatic fibres that are being used for useful articles. Lily and hyacinth fibres' extraction is by mechanical as well as retting method using water, chemicals, and enzymes. Presence of fine filaments as well as dense coarse fibrils in plant metaxylem has been reported (Schneider et al., 2009). Beautiful water lily flowers are also used for cultural practices and décor similar to lotuses. Water lily with bright red, pink, yellow, violet, and blue colours surpass the lotus in beauty and often mistaken for lotus. However, the distinguishing features of both the flowers are differences in their flower stamen and peduncle appearances. Water lily stamen are laminar shaped, whereas, lotus stamen is filamentous and surrounds the large receptacle (Fig. 6.2). Water lily peduncles are reddish, thicker and softer than lotus.

Figure 6.2 Divine flower decor of water gardens; distinguishing features of the three aquatic macrophytes are their flower stamen, peduncle cross-sectional features, thickness, and colour (A) Lotus (B) Water lily (C) Water hyacinth.

6.2.4 Typha

Typha (*Typha domingensis Pers.*) is a wetland perennial grass belonging to the Typhaceae family. It is an aquatic weed that invades waterlogged sites for thriving and is widespread universally (El-Aimer, 2013). Typha plant length range between 1.5 and 4.8 m, containing 4—10 leaves (Elhaak et al., 2015). The plant yields two different fibres obtained from its leaves and seedpod. Typha leaf fibres (TLF) are coarse and stiff as compared to typha seed fibres (TSF) obtained from typha inflorescence. Plant leaves have a composite structure to handle harsh environmental conditions (Rawlatt and Morshead, 1992). Typha species are rightly termed ecosystem engineers due to their survival potential on water and upland equally. Not only it survives on dried-up fields, but also contributes to soil nutrient enrichment (Alvarez and Becares, 2006). Long typha leaves are green, convex, pliable, and rich in aerenchyma tissues that are water repellent like lotus leaves. Plant inflorescence is used for human consumption as well as for healing wounds. Plant leaves are water retted for 15 days to obtain long fibres. Seed fibres are detached from the plant stem-head by hand. Fine and soft TSF (Fig. 6.3C and F) are used for stuffing in cushions, toys, and paper pulp (Pandey et al., 2020c), whereas leaf fibres (Fig. 6.4F) are used to prepare composite materials for building, automobile panels, insulation boards, biofuels, and diversified products. Traditional uses of typha leaves are elephant feed, roof thatching, and stuffing in saddle pads for elephants, fruit baskets, fibre, paper, and ropes (Uddin et al., 2006). The Typha plants are self-propagating; therefore, it is a low cost, abundantly available source of valuable fibre.

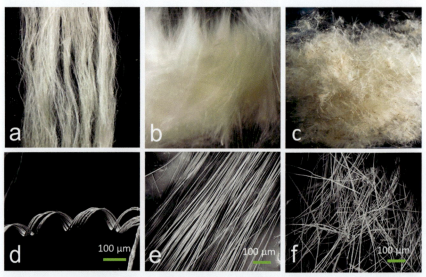

Figure 6.3 (A—C) Fine macrophyte fibres (D—F) Scanning electron microscope (SEM) images at 100 magnitude and 100 μm range; (A and D) Lotus fibre (B and E) Milkweed fibre (C and F) TSF.

Macrophyte and wetland plant fibres

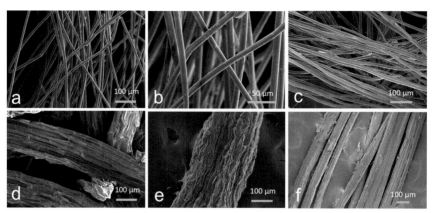

Figure 6.4 (A—C) Scanning electron microscope (SEM) images of fine wetland seed microfibres (200 X) (D—F) SEM images of course macrophyte fibres (200 X); (A—B) Wild sugarcane (C) Munja fibre (D) Water hyacinth fibre (E) Water lily fibre (F) TLF.

6.2.5 Milkweed

The advent of ASEAN vision 2025, the UN 2030, and ambitious sustainable resolution (Su.Re) projects has prompted the researchers, industrialists, and all stakeholders to search for alternative sources for obtaining natural and sustainable fibres (Pandey et al., 2020a). Milkweed (*Calotropis gigantea*) and (*Calotropis procera*), a perennial, is one such plant that has caught the attention of a few researchers (Karthik and Murugan, 2013) and a certain entrepreneur (Indian Creates Vegan Wool, 2020). Milkweed is a perennial that flourish near water drains and on sandy loam soil as a waste weed plant. Plant prospers all over India even without proper care provided to its nutritional needs. Plant leaves and flowers have medicinal as well as cultural value. Plant seedpod contains silky white and very fine floss (Fig. 6.3B and E) similar to cotton and silk, whereas stem yields coarse fibres on retting (Pandey, 2016) for textiles and biocomposites. Fibre has fineness, good tenacity (Table 6.3), high lignin content and low density (Table 6.4). Seed fibre is highly lustrous like synthetic viscose fibre and it is due to high wax content. The hollow structure of the fibre is useful for cold weather clothing by trapping air. Moisture absorption comfort, and whiteness value of milkweed fibre is higher than cotton. Scanning electron microscope (SEM) image in Fig. 6.3E shows the morphological structure as smooth and uniform fibre. This fine micro-fibre with renewability, availability, and low cost certainly has the potential to be the substitute for synthetic fibres.

6.2.6 Munja

Saccharum munja Roxb. grass (SMG) of the Gramineae family is a native of India and distributed in entire north India and parts of Asia. Tall perennial and wild plant grow on wasteland on their own year after year. The plant is self-propagating, and adapt to harsh environmental conditions such as high temperature and drought (Pandey et al.,

Table 6.3 Physico-mechanical properties of macrophyte and wetland plant fibres.

Fibre	Fibre length (mm)	Fineness (denier)	Elongation (%)	Tenacity (g/d)	Moisture (%)	References
Lotus	476	1.98	1.95	13.41	11.3—12.32	Pandey et al. (2020b)
Water hyacinth	300—500	64.48—95.0	3.53—4.8	0.90—0.33	12.1—16.1	Present study; Chonsakorn et al. (2019)
TLF	65—500	49.5	3.3	4.25	7.8	Pandey et al. (2020c)
Milkweed	4—25	1.19	2.56	4.12	11.3	Present study
Munja	3—7	65.93	1.3	9.79	—	Present study
Reed	—	—	0.8—1.4	—	—	Pandiarajan and Kathiresan (2018)
Rice	25—80	27	1.67—2.19	3.45	—	Reddy and Yang (2006); Kim and Lim (2009)
Money plant	—	—	1.38—4.24	—	—	Maheshwaran et al. (2018)

Table 6.4 Chemical constituents of macrophyte and various wetland plant fibres.

Fibre	Cellulose (%)	Lignin (%)	Wax (%)	Ash (%)	Moisture (%)	Density (g/cc)	References
Lotus	77.42	10.73	–	–	9.3	1.18	Zhang and Li (2011)
TLF	68.35	17.6	0.37	5.71	7.8	3.13	Pandey et al. (2020c)
TSF	73.46	9.88	0.93	5.9	9.6	1.29	Pandey et al. (2020c)
WH	50.38	2.25	–	–	12.1	–	Chonsakorn et al. (2019)
Munja	70–72	18–20	–	–	0.9	–	Singh et al. (2017)
Milkweed	53.0	15.0	4.0	1.2	10.0	0.8	Karthik and Murugan (2013)
Rice straw	37–64	8–15.8	8–38	5.0	9.8	–	Ready and Yang (2006); Kim and Lim (2009); Abas and Abd-Rased (2016)
Wild cane	53.45	11.7	1.3	5.6	2.1	–	Vijay et al. (2019)
Reed plant	33–64.56	10.84–25	–	–	–	–	Kobbing et al. (2013); Pandiarajan and Kathiresan (2018)
Coir	43–48.3	30.4–45	–	–	0.8	1.2	Biagiotti et al. (2004); Mathura and Cree (2016)
Money plant	66.34	14.01	0.37	4.61	7.44	0.65	Maheshwaran et al. (2018)

2015). SMG fibre has traditional uses in rural areas due to its durability and capacity to withstand rain. Plant stem and leaves are used to weave basket mats, hand fan (Fig. 6.5B–E), furniture, roof thatching, and such utility articles. Recently SMG handicrafts have been introduced as an indigenous industry in Prayagraj, India as part of 'One District One Product (ODOP)' scheme to support the munja-specific industrial hub and its marketing. SMG fibre (Fig. 6.5C) is used to make strong rope for weaving furniture, and pulp for the paper and textile industry. Seed-bursts from feathery SMG flowers yield microfibres (Fig. 6.4C) that is traditionally used as fibre fill in saddle pads. Search for non-polluting sources has compelled the researchers to explore under-utilized strong natural fibres that can be substituted for synthetic fibres. It is one of the most under-utilized fibre that offers great avenues for industrial applications and commercial proposition. Alkali treated composite fabrication from SMG cane has a good impact factor comparable to composite from synthetics (Singh et al., 2017). Low density and high strength make the munja fibres suitable for use in biocomposites and panel board fabrication. SMG Fibre is also a good source of furfural and pyrolytic carbon for cooking fuel use (Vasudevan et al., 1984).

6.2.7 Wild sugarcane

Wild sugarcane (*Saccharum spontaneum* L.) is a perennial tall grass in the family of the Poaceae and it is one of the most under-utilized fibres. A waste weed, the plant grows near waste and drainage dump in tropical countries. The plant, considered

Figure 6.5 Munja and wild cane fibre/straw and its utility articles (A) Wild sugarcane grass in the field (B) Munja straw (C) Munja fibre (D) Hand air blowing fan (generally used in remote villages) (E) Munja bowl for traditional uses (F) Herbal tea made of wild sugarcane grass stem (G) Shoes made of wild sugarcane grass.

highly invasive, is propagated by seed, rhizomes, and stem cutting. Similar to munja, the wild cane has religious significances and is used for making traditional crafts for household use (Fig. 6.5D–G). Plant stems and clams are used for making roof thatching and greaseproof currency respectively. Plant Seed bears gleaming white fibres (Fig. 6.5A) that have fineness (Fig. 6.4A, B) and good strength (Pandey et al., 2015). Moisture absorbance of soft lightweight fibres is 13.2%. Another species *Saccharum bengalense* fibres obtained from plant stems are a source of strong fibre (Vijay et al., 2019) and sustainable energy (Rawal et al., 2018).

6.2.8 Reed

Reed plants (*Phragmites australis*) naturally growing to the height of 2–6 m height is one of the widely distributed halophyte of temperate and tropical latitudes. The plant form reed beds across freshwater lakes, rivers and ponds edge. The long sticks of green plantlets with dense inflorescence or bristles on top is commonly used as household broom. Mature hollow plant stems are the source of strong fibre, musical instruments (flute), panel board, insulation material, paper pulp, ropes, mat, renewable energy, and biofuel (Szijártó et al., 2009; Köbbing et al., 2013; Pandiarajan and Kathiresan, 2018).

6.2.9 Rice

Rice (*Oryza sativa*) is an important food grain crop. The crop ranks third in worldwide production next to sugarcane and maize. Plant by-product straw is used for cattle feed, and manure. Ash derived from combusted rice straw is naturally alkaline in nature and traditionally used in cleaning utensils. However, a larger chunk of straw is burnt in the fields itself causing pollution and smog during winters. The solution to this long-overdue problem due to straw burning pollution is to utilize the plant straw for extracting cellulosic fibres. Rice straws are one of the annually renewable and low-cost sources of cellulosic fibres that are used for reinforced plastics, composites, paper pulp, chemicals, enzymes, and bioenergy (Kim and Lim, 2009; Abas and Abd-Rased, 2016). Fibres derived from rice straw are long with high strength and elongation suitable for textiles and high-value applications (Reddy and Yang, 2006). Rice husk is utilized as filling material in composite, nano-cellulose, and fertilisers (Ramamoorthy et al., 2015).

6.2.10 Coconut

Lignocellulosic coconut fibre is obtained from the seed shell of the coconut (*Cocos nucifera* L.) palm. Coconut, a vascular seed palm has upland status in North America and FACU in the Caribbean and Hawaii. Fibre is resistant to saline water and used in a multitude of ways like handicraft articles, rope, mats, and composites. Green and brown ripe coconut yield fibres with different physical performance. Fibres from brown mature coconut are stiff and high in tenacity as compared to green coconut (Mathura and Cree, 2016). The massive availability and renewability of coconut are making it possible to explore diversified uses of the fibre. Coir peat, a lightweight coconut fibre by-product is used as a substitute for soil in horticultural gardens (Biagiotti et al., 2004).

6.2.11 Money plant

Money plant, golden pothos, and silver vines are the common names of *Epipremnum aureum* of the monocotyledonous Araceae family (Fig. 6.6A). It is one of the most common house plants that grows indoor as well as outdoor, both as hydrophyte and nonhydrophyte with dynamic vigour and vitality. Studies indicate that mere watching of money plant acts as stress-buster in old age (Hassan et al., 2019). The stem of the plant is aerenchymatic that consists of long fibres lengthwise (Fig. 6.6B–C) and is suitable for composite manufacturing. Rich in cellulosic content, the fibre has a crystallinity index of 49.33% and 1.38%–4.24% strain rate (Maheshwaran et al., 2018). Money plant is an air purifier, electricity generator (Mathuriya et al., 2015), and heavy metal absorber (Chehrenegar et al., 2016; Baron et al., 2018).

6.2.12 Poplar

Poplar tree species are widely spread across the globe and popularly known as a forest and timber crop. In order to utilize fast growing population of Populus, its wood was fabricated in composite manufacturing for building construction (He et al., 2016). Seed capsules of the tree contains cottony balls. Fig. 6.7 shows white cottony Himalayan poplar (*Populus ciliata*) fibres that surpass cotton fibres in whiteness (Table 6.5), and softness. Perhaps this could be the reason behind its common name as cottonwood. A single populus tree produces 40 million seeds that are attached to cottony strands with an average length of 5.42 mm. Poplar fibre fineness is comparable to the finest cotton and kapok fibre. Vast availability, whiteness ($L = 78.28$), and fineness (10.1 μ) of poplar fibre may be useful for filling in cushions and in blends with cotton in apparel and hosiery sector.

Figure 6.6 Money plant and fibre (A) Money plant vine (B) Water retted money plant peduncle containing fibres (C) Manually extracted long length of money plant fibres.

Macrophyte and wetland plant fibres

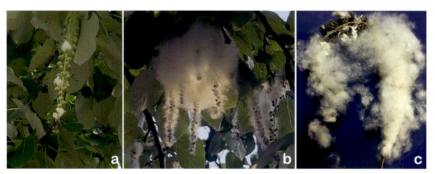

Figure 6.7 Himalayan poplar (*Populus ciliata*) (A) Poplar tree with seed capsule (B) Bursting of seed capsule into fibrous danglers (C) Dried dehiscent seed fibres.

6.3 Fibre morphology

Micrograph images of the macrophyte and wetland fibres are presented in Figs 6.3 and 6.4. Macrophyte fibres such as lotus are comprised of fine helical (Fig. 6.3D) microfibres, whereas, water hyacinths, lily, coconut, reed, and rice provide course fibres (Fig. 6.4). Milkweed, wild sugarcane, munja, and typha yield fine seed as well as course bast fibres.

6.4 Physicomechanical properties

Important physical parameters such as fibre length, tenacity, elongation and fineness values of lotus, WH, SMG, rice, reed, coconut, and wild sugarcane macrophyte fibres are presented in Table 6.3. High tenacity of course macrophyte fibres make them suitable for canvas fabric, mattresses, carpets, durable composites, and reinforced panels. In contrast, macrophyte seed fibres are shorter and finer. Lotus and milkweed seed fibres' colour values exhibited higher whiteness and brightness as compared to course macrophyte bast fibres (Table 6.5). WH Fibre has wool-like waviness, good elongation, and reasonable strength. The stress—strain curve of WH in Fig. 6.8 shows fibre elongation between 2.5% and 8%.

6.5 Chemical composition

Macrophyte fibres mainly consist of celluloses, lignin, pectin, and lipids. Cellulose, lignin, and moisture content of each fibre differs from each other due to differences in plant maturity and cell wall growth characteristics of the macrophyte fibres. Silky, shiny appearance, and wax content make the fine milkweed and TSF hydrophobic in comparison to course cellulosic fibres. Lower lignin content of water hyacinth fibre is useful in paper production (Table 6.4). The moisture content of macrophyte fibres is

Table 6.5 Colour values of aquatic macrophyte and wetland plant fibres.

Fibre	Whitness index (CIE-76/10DEG)	Brightness index (Tappi 525)	Yellowness index (YI_D1925)	Fluorescence	L	a	b
Lotus	−60.02	36.35	45.8	36.96	79.47	2.3	22.02
WH	−76.92	22.42	47.1	22.5	64.93	3.94	18.25
TLF	−53	31.0	30.0	–	–	–	–
Milkweed	0.25	56.52	28.58	57.3	88.32	0.19	14.78
Munja	−62.98	26.97	43.44	27.34	68.83	3.8	17.4
Himalayan poplar	17.83	47.49	14.46	47.91	78.28	0.33	6.6

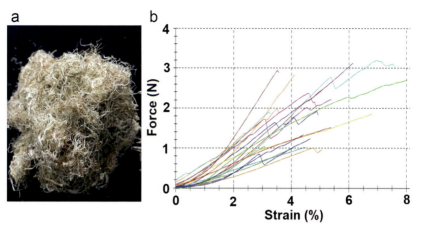

Figure 6.8 Water hyacinth fibre with its stress vs strain curve. (A) Raw natural water retted fibre (B) Stress-strain curve of water retted fibre.

comparable to common cellulosic fibres (Ramamoorthy et al., 2015). Alkali treatment improves cellulose and reduces the lignin content of cellulosic fibres (Kim and Lim, 2009; Zhang and Li, 2011; Li and Fu, 2015; Pandey et al., 2021).

6.6 Application of macrophytes in effluent treatment

Water purity directly correlates to the vital health of humans and all living beings (Khetan and Collins, 2007). Macrophytes and riparian flora are important for protecting water quality, erosion prevention and temperature control in aquatic as well as terrestrial ecosystems. Aquatic macrophytes enhance water quality due to their ability to absorb excessive loads of nutrients and effluents from an aquatic ecosystem. Freshwater fishes get oxygenation, food, and refuge from aquatic macrophytes (Sahu and Chandravanshi, 2018). Macrophytes manage the water as well as nearby land temperature, soil, and air quality. The Macrophytes with bryophytes are essential components for watercourse ecosystems and physical substrate for fish, habitat, and periphyton. One can evaluate the waterbody's health just by observing depth, density, diversity, and types of macrophytes present in the waterbody. Their presence reduces the flowing water velocity, increases sedimentation, and reduces turbidity. Due to this property, worldwide the macrophytes are considered as an indicator of water and sediment quality. Money plant has proven its potential for phytoremediation of ibuprofen and lead absorption from lead-based paints (Chehrenegar et al., 2016; Baron et al., 2018). Wetland plants such as SMG prevents soil erosion and fertiliser runoffs (Singh et al., 2017). SMG binds soil due to its self-propagation and extensive root system (Pandey et al., 2015). Typha absorbs salts in wastewater and chemical effluents contributing to toxic free soil and water. Typha, also a biofuel generating plant (Satya and Maiti, 2013), constitute of lignoproteins and some amount of nitrogen, which fertilize the soil-enriching the surrounding ecosystem (Brinkmann et al., 2002). The absence of macrophytes in the waterbody is an indication of qualitatively compromised water ecosystem and environmental hazards. In absence of macrophytes, the water body has excessive turbidity, herbicides, and salinization which is detrimental to fish and aquaculture. The abundance of macrophytes in the waterbody shows good habitat structure, fishability, recreational use and nutrient dynamics. Yet, the overabundance of macrophytes might harm ecosystem health. The excess nutrients in the water body called eutrophication may initiate the vanishing of submerged macrophytes as the light could not reach properly because of the overgrowth of phytoplankton.

6.7 Conclusion

Aquatic and wetland plants are precious resources of the planet. The study provides basis for exploiting new aquatic macrophyte fibres for technical textile applications. Fibres derived from plants surviving in water bodies are numerous and abundantly available. Fibres are fine as well as course and thus can be useful for different purposes ranging from basic garments to carpet, bags, biocomposites and paper production. All the wetland fibres presented in this chapter are discarded waste weed (except rice and coconut) that grows in marshy areas where land is not suitable for high-value crops. The utilization of environmentally sustainable natural wetland fibres for industrial applications thus opens the path for farmers to increase their earnings. Aquatic macrophytes and wetland fibres have the advantage of lower density, low cost, and easy biodegradability. Awareness and utilization of the same will minimize raw material costs and manage waste-generated pollution and also help in increasing the income of the resource-starved and marginal farmers. In summary, the macrophytes and wetland fibres have enormous potential for industrial and technical textile applications as well as a substitute for non-biodegradable synthetic fibres.

References

Abas, N.F., Abd-Rased, A.N.N.W., 2016. The thermal performance of manufactured concrete roof tile composite using clay and rice straw fibers on a concrete mixture. Jurnal Teknologi 78 (5). https://doi.org/10.11113/jt.v78.8352.

Alvarez, J.A., Becares, E., 2006. Seasonal decomposition of Typha latifolia in a free-water surface constructed wetland. Ecological Engineering 28 (2), 99—105. https://doi.org/10.1016/j.ecoleng.2006.05.001.

Baron, E.M.D., Tomawis, A.S., Briones, M.M., Ibaos, D.A., Sabando, M.C., 2018. Lead absorption abilities of bougainvillea (bougainvillea spectabilis willd.) and money plant (Epipremnum aureum GS bunting) in lead-based paint coated compartments. Journal of Engineering, Environment and Agriculture Research 1. https://doi.org/10.34002/jeear.v1i0.18, 6—6.

Biagiotti, J.D., Puglia, D., Kenny, J.M., 2004. A review on natural fiber-based composites-Part I. Journal of Natural Fibers 1 (2), 37—68. https://doi.org/10.1300/J395v01n02_04.

Brinkmann, K., Blaschke, L., Polle, A., 2002. Comparison of different methods for lignin determination as a basis for calibration of near-infrared reflectance spectroscopy and implications of lignoproteins. Journal of Chemical Ecology 28, 2483−2501. https://doi.org/10.1023/A:1021484002582.

Chehrenegar, B., Hu, J., Ong, S.L., 2016. Active removal of ibuprofen by Money plant enhanced by ferrous ions. Chemosphere 144, 91−96. https://doi.org/10.1016/j.chemosphere.2015.08.060.

Chonsakorn, S., Srivorradatpaisan, S., Mongkholrattanasit, R., 2019. Effects of different extraction methods on some properties of water hyacinth fiber. Journal of Natural Fibers 16 (7), 1015−1025. https://doi.org/10.1080/15440478.2018.1448316.

Dubey, A., Philip, D., Das, M., 2017. A sustainable, eco-friendly charge storage device from bio-charred jute: an innovative strategy to empower the jute farmers of India, smart innovation systems and technologies. Springer 66, 397−407. https://doi.org/10.1007/978-981-10-3521-0_34.

El-Aimer, Y.A., 2013. Spatial distribution and nutritive value of two Typha species in Egypt. Egyptian Journal of Botany 53 (1), 91−113.

Elhaak, M.A., Mohsen, A.A., Hamada, E.S.A., El-Gebaly, F.E., 2015. Biofuel production from Phragmites australis (cav.) and Typha domingensis (pers.) plants of Burullus Lake. Egyptian Journal of Experimental Biology 11 (2), 237−243. http://my.ejmanger.com/ejeb/.

Haider, S.Z., 1989. Recent work in Bangladesh on the utilization of water hyacinth. Commonwealth Science Council 278, 37. https://www.cabdirect.org/cabdirect/abstract/19922319442.

Hassan, A., Qibing, C., Yinggao, L., Tao, J., Li, G., Jiang, M., Nian, L., Bing-Yang, L., 2019. Psychological and physiological effects of viewing a money plant by older adults. Brain Behav 9 (8), 1359. https://doi.org/10.1002/brb3.1359.

He, M.J., Zhang, J., Li, Z., Li, M.L., 2016. Production and mechanical performance of scrimber composite manufactured from poplar wood for structural applications. Journal of Wood Science 62 (5), 429−440. https://doi.org/10.1007/s10086-016-1568-1.

Indian Creates Vegan Wool from a Wasteland Plant, 2020. https://www.veganfirst.com/article/indian-creates-vegan-wool-from-a-wasteland-plant.

Jangir, H., Pandey, M., Jha, R., Dubey, A., Verma, S., Philip, D., Sarkar, S., Das, M., 2018. Sequential entrapping of Li and S in a conductivity cage of N-doped reduced graphene oxide supercapacitor derived from silk cocoon: a hybrid Li−S-silk supercapacitor. Applied Nanoscience 8 (3), 379−393. https://doi.org/10.1007/s13204-018-0641-z.

Karthik, T., Murugan, R., 2013. Characterization and analysis of ligno-cellulosic seed fiber from Pergularia daemia plant for textile applications. Fibers and Polymers 14 (3), 465−472. https://doi.org/10.1007/s12221-013-0465-0.

Khetan, S.K., Collins, T.J., 2007. Human pharmaceuticals in the aquatic environment: a challenge to green chemistry. Chemical Reviews 107 (6), 2319−2364. https://doi.org/10.1021/cr020441w.

Kim, H., Lim, J., 2009. Effect of surface treatment on the mechanical properties of rice straw fiber. International Journal of Materials and Product Technology 36 (1−4), 125−133. https://doi.org/10.1504/IJMPT.2009.027825.

Köbbing, J.F., Thevs, N., Zerbe, S., 2013. The utilisation of reed (Phragmites australis): a review. Mires & Peat 13, 1−14. http://www.mires-and-peat.net/.

Kozlowski, R., 2009. Inventory of Natural Fibers and Their Potential in Diversified Applications. http://www.fao.org/DOCREP/004/Y1873E/y1873eOb.htm.

Kumar, A., Jash, A., Dubey, A., Bajpai, A., Philip, D., Bhargava, K., Singh, S.K., Das, M., Banerjee, S.S., 2018. Water mediated dielectric polarizability and electron charge transport properties of high resistance natural fibers. Science Rep-UK 8 (1), 2726. https://doi.org/10.1038/s41598-018-20313-4.

Li, F., Fu, H., 2015. Effect of alkaline degumming on structure and properties of lotus fibers at different growth period. Journal of Engineered Fibers and Fabrics 10 (1). https://doi.org/10.1177/2F155892501501000114, 155892501501000130.

Lichvar, R., Melvin, N.C., Butterwick, M.L., Kirchner, W.N., 2012. National Wetland Plant List Indicator Rating Definitions. Cold Regions Research and Engineering Laboratory (U.S.) Engineer Research and Development Center (U.S.). https://hdl.handle.net/11681/2616.

Lichvar, R.W., Banks, D.L., Kirchner, W.N., Melvin, N.C., 2016. The national wetland plant list: 2016 wetland ratings. Phyton 30, 1−17.

Liu, Y., Wang, Y., Yuan, X.H., Prasad, G., 2020. The function of water absorption and purification of Lotus fiber. Materials Science Forum 980, 162−167. https://doi.org/10.4028/www.scientific.net/MSF.980.162.

Maheshwaran, M.V., Hyness, N.R.J., Senthamaraikannan, P., Saravanakumar, S.S., Sanjay, M.R., 2018. Characterization of natural cellulosic fiber from Epipremnum aureum stem. Journal of Natural Fibers 15 (6), 789−798. https://doi.org/10.1080/15440478.2017.1364205.

Mathura, N., Cree, D., 2016. Characterization and mechanical property of Trinidad coir fibers. Journal of Applied Polymer Science 133 (29). https://doi.org/10.1002/app.43692.

Mathuriya, A.S., Bajpai, S.S., Giri, S., 2015. Epipremnum aureum (money plant) as cathode candidate in microbial fuel cell treating domestic wastewater. Journal of Biochemical Technology 6 (3), 1025−1029.

Mishra, S., Pandey, R., Singh, M.K., 2016. Development of sanitary napkin by flax carding waste as absorbent core with herbal and antimicrobial efficiency. International Journal of Science Environment and Technology 5 (2), 404−411.

Pandey, R., 2016. Fiber extraction from dual-purpose flax. Journal of Natural Fibers 13 (5), 565−577. https://doi.org/10.1080/15440478.2015.1083926.

Pandey, R., Pandit, P., Pandey, S., Mishra, S., 2020a. Solutions for Sustainable Fashion and Textile Industry, Recycling from Waste in Fashion and Textiles: A Sustainable and Circular Economic Approach. Scrivener Publishing LLC, pp. 33−72. https://doi.org/10.1002/9781119620532.ch1.

Pandey, R., Sinha, M.K., Dubey, A., 2020b. Cellulosic fibers from Lotus (*Nelumbo nucifera*) peduncle. Journal of Natural Fibers 17 (2), 298−309. https://doi.org/10.1080/15440478.2018.1492486.

Pandey, R., Jose, S., Sinha, M.K., 2020c. Fiber extraction and characterization from Typha domingensis. Journal of Natural Fibers 1−12. https://doi.org/10.1080/15440478.2020.1821285.

Pandey, R., Jose, S., Basu, G., Sinha, M.K., 2021. Novel methods of degumming and bleaching of Indian flax variety Tiara. Journal of Natural Fibers 18 (8), 1140−1150. https://doi.org/10.1080/15440478.2019.1687067.

Pandey, V.C., Bajpai, O., Pandey, D.N., Singh, N., 2015. Saccharum spontaneum: an underutilized tall grass for revegetation and restoration programs. Genetic Resources and Crop Evolution 62 (3), 443−450. https://doi.org/10.1007/s10722-014-0208-0.

Pandiarajan, P., Kathiresan, M., 2018. Physicochemical and mechanical properties of a novel fiber extracted from the stem of common reed plant. International Journal of Polymer Analysis and Characterization 23 (5), 442−449. https://doi.org/10.1080/1023666X.2018.1474327.

Ramamoorthy, S.K., Skrifvars, M., Persson, A., 2015. A review of natural fibers used in biocomposites: plant, animal and regenerated cellulose fibers. Polymer Reviews 55 (1), 107−162. https://doi.org/10.1080/15583724.2014.971124.

Rawal, S., Joshi, B., Kumar, Y., 2018. Synthesis and characterization of activated carbon from the biomass of Saccharum bengalense for electrochemical supercapacitors. Journal of Energy Storage 20, 418−426. https://doi.org/10.1016/j.est.2018.10.009.

Rawlatt, U., Morshead, H., 1992. Architecture of the leaf of the greater reed mace, Typha latifolia L. Botanical Journal of the Linnean Society 110 (2), 161−170. https://doi.org/10.1111/j.1095-8339.1992.tb00289.x.

Reddy, N., Yang, Y., 2006. Properties of high-quality long natural cellulose fibers from rice straw. Journal of Agricultural and Food Chemistry 54 (21), 8077−8081. https://doi.org/10.1021/jf0617723.

Roy, M., Dubey, A., Singh, S.K., Bhargava, K., Sethy, N.K., Philip, D., Sarkar, S., Bajpai, A., Das, M., 2016. Soft magnetic memory of silk cocoon membrane. Sci. Rep-UK 6 (1), 29214. https://doi.org/10.1038/srep29214.

Sahu, R., Chandravanshi, S.S., 2018. Lotus cultivation under wetland: a case study of farmers innovation in Chhattisgarh, India. International Journal of Current Microbiology and Applied Sciences 7, 2319−7706.

Satya, P., Maiti, R., 2013. Bast and Leaf Fiber Crops, International, Biofuel Crops: Production Physiology and Genetics. CABI.

Schneider, E.L., Carlquist, S., Hellquist, C.B., 2009. Microstructure of tracheids of nymphaea. International Journal of Plant Sciences 170 (4), 457−466. https://doi.org/10.1086/597783.

Singh, G.P., Madiwale, P.V., Adivarekar, R.V., 2017. Investigation of properties of cellulosic fibers extracted from Saccharum Munja grass and its application potential. International Journal of Fiber and Textile Research 7 (1), 30−37.

Szijártó, N., Kádár, Z., Varga, E., Thomsen, A.B., Costa-Ferreira, M., Réczey, K., 2009. Pretreatment of reed by wet oxidation and subsequent utilization of the pretreated fibers for ethanol production. Applied Biochemistry and Biotechnology 155 (1), 83−93. https://doi.org/10.1007/s12010-009-8549-4.

Uddin, M.B., Mukul, S.A., Khan, M.A.S.A., Chowdhury, M.S.H., Uddin, M.S., Fujikawa, S., 2006. Indigenous management practices of Hogla (Typha elephantinaRoxb.) in local plantations of floodplain areas of Bangladesh. Journal of Subtropical Agriculture Research Development 4 (3), 114−119.

Vasudevan, P., Gujral, G.S., Madan, M., 1984. Saccharum munja Roxb., an underexploited weed. Biomass 4 (2), 143−149. https://doi.org/10.1016/0144-4565(84)90062-3.

Vijay, R., Singaravelu, D.L., Vinod, A., Raj, I.F.P., Sanjay, M.R., Siengchin, S., 2019. Characterization of novel natural fiber from Saccharum bengalense grass (Sarkanda). Journal of Natural Fibers 17 (12), 1739−1747. https://doi.org/10.1080/15440478.2019.1598914.

Zhang, H.W., Li, Y.L., 2011. Lotus fiber production and its property, Advanced Materials Research. Trans Tech Publications Ltd 146, 93−96. https://doi.org/10.4028/www.scientific.net/AMR.146-147.93.

Mushroom and corn fibre—the green alternatives to unsustainable raw materials

7

Yamini Jhanji

Fashion and Apparel Engineering Department, The Technological Institute of Textile and Sciences, Bhiwani, Haryana, India

7.1 Detrimental impact of textile and fashion supply on environment

The textile and fashion industry in spite of being one of the major revenue-generating and employment offering industries is also denigrated and presented in bad light owing to deleterious environmental impacts throughout the textile and fashion supply chain. The industry is regarded among the world's most wasteful, polluting and energy-intensive industry, owing to widespread exploitation of natural resources and imprudent consumption of pesticides, fertilizers and other toxic chemicals that are utilized at different stages of manufacturing and processing. Additionally, improper disposal of waste water, industrial effluents and warehouse and sampling waste ending as landfills do no good to environment except for contributing to massive soil degradation, emission of greenhouse gases and increase in carbon footprint and water pollution. The apparel and accessory industry primarily rely on cotton as the major raw material owing to its abundance and conducive properties for varied apparel applications. Needless to mention, while manifesting all the positive attributes of the fibres, the manufacturers contravene the fact that cotton production is one of the most water, pesticide and fertilizer intensive processes. The massive utilization of chemicals to increase the yield not just pollutes the groundwater and air but also results in reduced soil fertility. The inability of synthetic fibres to decompose and release of methane by wool during landfill decomposition further poses environmental concerns. Apart from procurement stage, dyeing, printing and bleaching are other energy and chemical intensive stages involved in manufacturing line. Additionally, production nonconformities lead to large quantities of textiles ending up as waste annually.

7.1.1 Fast fashion—a big barrier to sustainable approach

There is no denying the fact that the fashion industry, as a result of its intrinsic characteristics, is violating environmental and social norms that are fundamental to sustainable concepts owing to the full exploitation of natural resources, the usage of toxic, chemical products, poor labour conditions and the trend of outsourcing manufacturing and job work to low labour cost countries.

Sustainable Fibres for Fashion and Textile Manufacturing. https://doi.org/10.1016/B978-0-12-824052-6.00012-3
Copyright © 2023 Elsevier Ltd. All rights reserved.

Furthermore, the fashion industry follows an ephemeral life cycle owing to its deep-rooted dependence on trend, season and consumer eccentricity as far as their preferences and buying behaviours are concerned. The manufacturers and retailers in the quest to renew their merchandise every season and stay ahead of their competitors have tuned out to be ardent followers of fast fashion aiming at improved productivity and reduced lead times thereby burdening the workforce and completely snubbing the well-being and safety concerns of blue collared employees and all stakeholders.

The emergence and rapid progression of fast fashion is undoubtedly violating the sustainable approaches and ethical practises in textile and fashion industry. The business model of fast fashion emphasizes on reducing the processes involved in the buying cycle and shortening the lead times to keep the stores replenished with new merchandise and to meet the seasonal demands of consumers. The adverse impacts of fast fashion, wrong procurement decisions and production processes in the textile and fashion industry has hit the industry hit deviating it from sustainable principles.

The negative impacts of micro-seasons and fast fashion on the environment cannot be undermined with the manufacturers and retailers paying little heed to the unsustainable characteristics associated with procurement, manufacturing, dyeing, printing, finishing, processing, distribution and packaging of their merchandise, and end up messing with the ecological balance just for the sake of surviving in the competitive and dynamic industry (Jhanji, 2021; Nayak, 2020).

7.1.2 Animal leather for fashion ensembles—nonconformity of sustainable principles

Leather undoubtedly holds lion's share of market as far as apparel and accessory manufacturing is concerned. The millennials and generation Z like to flaunt dapper look by embracing leather biker jacket, leather duffle bag or quirky pair of leather boots (Fig. 7.1) without paying any heed to the detrimental effect it poses on the planet.

The animal groups that are mainly reared for their hides and skin to be ultimately processed into leather are cattle, cow, bulls, buffalo, sheep, lambs, goat, equine horse, zebra, pig, reindeer, kangaroo and ostrich, etc. A range of amphibians, aquatic as well as land animals are exploited for exotic leather. Frog, seal, shark, walrus and turtles are the aquatic of the group, camel, ostrich, elephant and pangolin are the land classification, and alligator, crocodile, lizard, and snake represent the reptiles which are used for leather manufacturing. Each has distinctive characteristics and markings that make them appropriate for unusual apparel and accessories. Although, a variety of land and aquatic animals serve as leather source however, the most common source of leather—bovine leather, by far is from cattle industry. Apart from cruelty and exploitation to animals due to their rearing and slaughtering for the raw skin, the animals rearing are also associated with deforestation and gas emissions. Furthermore, the processes involved in treating raw hide and conversion to wearable leather possessing textile like properties further add to the woes of environment.

Leather in order to be used for next to skin applications and for design and development of range of apparels and accessories should be imparted textile like properties

Figure 7.1 Leather apparels and accessories.

like moisture absorption, softness and comfort next to skin. The hide obtained in raw form from animals is not suitable for next to skin and other textile applications. Consequently, the hides or skins are subjected to a series of chemical intensive operations to convert it into useable and wearable leather.

The processes involved in treating raw hide for conversion to useable, wearable form themselves are very chemical and energy intensive. A range of toxic and hazardous pollutants such as sodium chloride, sulphate, sulphide, chromate, azo dyes, cobalt, mercury, copper and formaldehyde resins are utilized at different stages of leather manufacturing process (Jhanji, 2021; Nayak, 2020). Imitation or simulated leather has been doing rounds and gaining popularity among fashion buffs. Leatherette, vinyl, pleather, liquid leather, super-suede, facile and ultrasuede are the varieties of simulated leather that are specially treated to acquire washability and durability matching that of real leather.

Cheaper substitutes of leather aimed at cost cutting prompts the creation of simulated, imitated leather by shredding and reconstitution of leather scraps, however it cannot match the durability associated with patent leather. Leather both patent, virgin as well as imitated version like suede, bonded leather find application in myriad of accessories like footwear, belts, wristbands, watch straps, gloves, luggage, handbags, wallets, etc. Nevertheless, the harm brought about by imitated leather versions is no less owing to varied dyeing and finishing treatments they receive to simulate the texture and appearance of virgin leather.

The processed raw hides converted to exotic leather varieties as well as the ones treated to obtain simulated leather are equally contributing to solid waste that comprises of the by-products of leather industries like finished leather fragments, chrome shavings, keratin wastes, hides trimmings and buffing dust. The discarded and damaged leather ensembles do not contribute as much to waste as the industries involved in processing and production of leather. The waste leather disposed off during footwear production amounts up to 20%–30%. Moreover, the manufacturing process results in conversion of just 150 kgs of leather from 1000 kg of raw hide while remaining amount ending up as solid waste (Hashmi et al., 2017).

Tanning, the process of treating raw hide to prevent its putrefaction and imparting textile like properties consume substantial quantities of toxic chemicals like chrome salts, tannins, lime, ammonium sulphate, hydrogen sulphide, sodium chloride, etc. Chromium-based tanning agents extensively utilized in leather processing are the major culprits of environmental pollution (Hashmi et al., 2017; Song et al., 2000). Chromium tanning salts, vegetable tannins and glutaraldehyde are some of the tanning agents used for tanning. Chrome tanned leathers tanned with mineral-based tanning agents like salts of chromium are comparatively less desirable than leather tanned with other agents because of the mineral content of the former.

Tawing, another chemical intensive process is accomplished by subjecting animal hide to a mixture of alum (aluminium sulphate) and saline producing white leather. Aldehyde based tanning agents used to obtain soft, dry cleanable and washable leather poses issues related to effluent treatment and higher energy consumption. Furthermore, the process of neutralizing leather post-tanning to remove any leftover acids from tanning process also utilizes abundant toxic and hazardous chemicals.

Although the extensive utilization of leather in fashion industry is a great boast for scaling up leather apparel and accessory business and the promotion on global arena, the downside is the leather's deleterious impacts on environment accounted to excessive usage of hazardous chemicals during the leather manufacturing processes. The industries do not adopt any standard operating procedures as far as discharge of toxic liquid wastes are concerned with most of the waste released into water bodies, soil and air. Additionally, toxic tannery waste is hazardous to human beings as well resulting in several chronic diseases and health ailments like bladder, pancreatic, lung cancer and eye irritation (Song et al., 2000; Senthil et al., 2000; Kanagaraj et al., 2000).

The leather apparel and accessory manufacturing units further contribute to waste owing to scrap and left-over leather pieces from sampling, cutting and sewing room ending up as landfills. Additionally, leather postconsumer waste includes the discarded and disposed off leather ensembles by consumers (Kanagaraj et al., 2000). The need of the hour is, therefore, to revamp the textile and fashion industry by switching over to sustainable and ethical practises in order to relinquish the adversities offered by the emerging fast fashion industry. The adverse impacts of leather processing and manufacturing leather goods on social and environmental well-being have been discussed elaborately in the previous section. It is thus imperative to prudently switch over to sustainable substitutes to subvert the deleterious impacts inflicted by leather on flora, fauna and mankind.

The next section of chapter will highlight measures adopted by fashion industry to transform towards sustainable supply chain. The manufacturers and retailers in the pursuit to adopt sustainable approaches have been experimenting with eco leather and other sustainable options like mushroom leather, corn leather, etc. that are promising and less hazardous compared to leather obtained from various animal sources.

7.2 Eco leather/environmentally preferred leather

The environmental hazards associated with leather manufacturing cannot be undermined which calls for stringent actions on the part of leather manufacturers, tanners, leather apparel and accessory manufacturers to comply with eco-friendly norms so as to minimize any environmental issues. The compliance of eco-friendly practises in leather processing is gauged by the absence of certain restricted chemicals like banned azo dyes, PCP, chrome VI, formaldehyde, etc. Therefore, eco labels, high street retailers or retailers who want to have cutting edge over their competitors through product positioning have been adopting sustainable practices during leather manufacturing and processing. The key elements determining an eco-leather/environmentally preferred leather include

- Control of manufacturing method of leather and inputs required for its manufacturing
- Clean technology and non-toxic chemical selection during processing
- Effective management of restricted substances
- A measure of the end-of-life impact

Cleaner production applications include green hide and skin processing (supply of raw material from slaughterhouses without preservation, e.g., salting), water management (utilization of minimum volume of process water), recycling (e.g., in liming) and chromium recovery (after tanning), hair saving (for reduction of dissolved solids in effluent) and application of environmentally friendly chemicals (e.g., enzymes). The practices followed by tanners coupled with chemical selection can go a long way in overcoming the adverse environmental impacts wreaked during leather manufacturing. Accordingly, following aspects need special consideration for hazzle-free leather processing and manufacturing:

- Energy consumption
- Air emissions
- Waste management (hazardous and non-hazardous)
- Environmental management systems
- Water consumption
- Control of manufacturing processes
- Effluent treatment
- Chrome management
- Traceability of material

The selection of inputs for manufacturing processes thus demand usage of materials that enables improvement in ecoprofile to leather. Accordingly, procurement of sustainable and eco-friendly raw materials as a substitute to leather in fashion industry has been explored and experimented by apparel manufacturers for adherence to sustainable norms in their manufacturing processes. The bio materials that have been gaining traction among designers, brands and gradually marching towards mainstream market include 3D printed biomaterials, pineapple, bamboo, hemp, corn besides mushroom as an eco-friendly alternative to synthetic fabrics, plastic and animal leather. Mushroom leather is considered to be one of the most promising and eco-friendly option that has the potential to completely replace leather from fashion arena (ecowarriorprinces; textilevaluechain). The unique characteristics and properties of mushroom fibres, production process of mushroom leather and merits of former over conventional leather will be discussed in the following section of chapter in details.

7.3 Mycelium and mushroom leather

Moulds, yeasts and mushrooms are classified as eukaryotic organisms. Mushroom belongs to an underground honey fungus family called mycelium that features tiny white threads and enormous underground network. The mycelia from various genetically identical individual honey fungus combine together forming a large fungal body with subsequent development of single individual by blending of extensive networks of fungal clones. The vegetative stage of mushroom also referred to as pre-mushroom stage is the mycelium growth. Mycelium also referred to as wood wide web is the underground network of thread-like branches that grow below mushrooms and fungi, connecting living flaura, enabling nutrient exchange, breaking down of decaying stock, earth regeneration and sequestering carbon.

The cultivation of mycelium apart from natural growth is also possible from agricultural waste like sawdust and pistachio shells. Fine Mycelium grown from fungi very closely simulates the appearance and texture of leather even outclassing the strength and durability of latter. Mycelium is a living, natural fibre that features its own distinct grain and performance with feel resembling leather but as adaptable as plastic. The most promising aspect of mushroom leather obtained from mycelium is its ability to be grown into desirable shape, size and in multiple dimensions as per the selected substrate in the in vitro conditions thereby eliminating cutting room waste. The mushroom leather from mycelium finds application in high street fashion, car upholstery and high end, exotic accessory collection. The next section of chapter will discuss the raw material, production process and properties of mushroom leather over animal leather.

7.3.1 Raw materials for mushroom leather

The commonly used mushroom variety for developing mushroom leather is commercial Oyster mushrooms; however, specific mushroom types may be selected depending on the intended leather attributes in the sustainably grown mushroom leather. A gigantic parasitic fungus, namely, Phellinus ellipsoideus generally found in subtropical forests, is utilized to obtain an innovative material—MuSkin.

Bolt thread, pioneer in mushroom leather manufacturing has developed an innovative and the most suitable alternative to leather—Mylo.

The raw material for mushroom leather is extracted from a skin friendly fungi known as Ganoderma lucidum or reishi mushroom grown in Asia. Majority of manufacturers rely either on Phellinus ellipsoideus, a huge-growing species or the vegetative part of mushroom, the mycelium of reishi, pearl oyster mushrooms as a source of mushroom leather. Some big players experimenting with mycelium include Mycoworks, Mogu, Mylo by Bolt Threads, while MuSkin is utilizing Phellinus ellipsoideus for developing mushroom leather.

Myco Works uses a wood eating fungi species known as Ganoderma or Reisi in developing their mushroom leather—Reisi that is claimed to be viable fashion raw material owing to its strength, durability and appearance comparable to animal leather. Ganoderma is fed on agricultural waste such as sawdust to provide conducive conditions for growth of mycelium cells into a dense and intertwined structure. The sheet of fibres from mycelium can be obtained by stringent control on ambient conditions like temperature, humidity in situ.

7.3.2 Production of mushroom leather

The environmental adversities inflicted by animal leather have already been discussed in previous section. The aftermath of animal leather on environmental and social wellbeing can be subverted by switching over to vegan and eco-friendly options like mushroom leather. The vegetative component of fungus—mycelium is the main constituent of mushroom leather. The versatility of mushroom leather makes it a preferred choice for a myriad of applications like homeware products, batteries, spaceships and fashion ensembles.

Fig. 7.2 illustrates the process flow for mushroom leather production. The process of mushroom leather making commences with selection and moisturization of the appropriate substrate (organic waste). Substrates are the materials that provide nutrition for the growth of mushroom. The commonly used substrates for mushroom growth include wood chips, straw, corn, and such substance on which mushroom can adhere and grow (Fig. 7.3).

The affixation and development of mushroom gets accelerated as the substrate is wettened during the process. Composites of varied sizes, shapes and performance can be achieved by variation in growth conditions such as amount of light, humidity, gaseous exchange, temperature, and nutrients fed to mushroom. Hence, the growth conditions are of paramount importance in determining the characteristics and texture of mushroom leather thereby achieving material as hard as enamel, shell-like or soft and porous like sponge.

The modification of substrate and variation in growing conditions is feasible in the premature stage of process. The treatment methods, namely, autoclaving, composting, pasteurization to induce mycelium growth and cultivation method (to deposit the material in bag, bottle, tray, shelves) vary as per the chosen substrate as shown in Fig. 7.3. Irrespective of the chosen substrate, the process of substrate dampening is carried out, placed in a bag or tray and is subjected to pasteurization thereby ensuring convenient and rapid mycelium growth devoid of any microbes. Closed tray system ensures

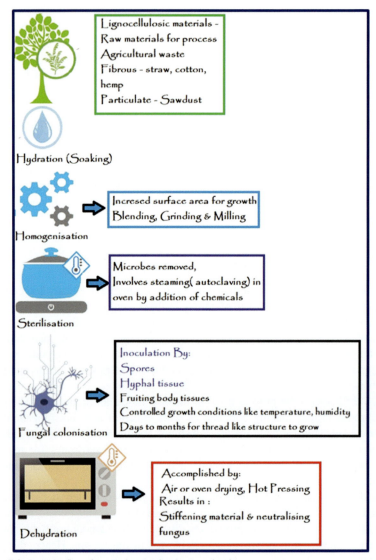

Figure 7.2 Process flow for mushroom leather production.

complete control on growth environment. The material is preferably grown in vertically stacked layers for better space utilization.

Development of mycelium leather demands attention and time. The mushroom takes anywhere between two to 3 weeks to grow invitro conditions depending on type of mushroom and substrate, ambient conditions like sunlight, moisture and air. Extraction and compression of mushroom is accomplished once the mycelium has grown to the desired size and shape. Each sheet of fine mycelium can be optimized

Mushroom and corn fibre—the green alternatives to unsustainable raw materials

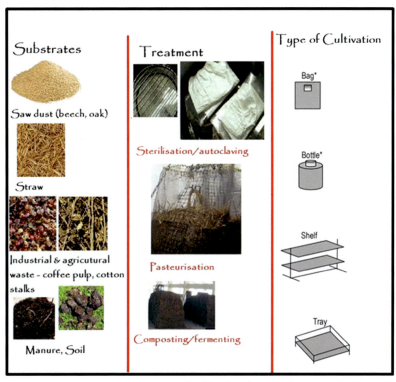

Figure 7.3 Substrates, treatment and type of cultivation for Mycelium growth.

as it grows, and exact specifications can be achieved when bulk production of garments has to be carried out.

The compression process is quite crucial as it can affect the texture and colour of the obtained material. Cow, alligator and python leather can be easily imitated at this stage by dyeing or varying the pressing form. The compression is followed by drying of mushroom leather. The processing of mycelium-based leather is eco-friendly process with no hazardous chemicals involved during processing unlike processing of PVC leather that is obtained by chemical alteration of vinyl polymer (ecowarriorprinces; textilevaluechain).

7.3.3 Properties and benefits of mushroom leather over animal leather

The raw materials and production process of mushroom leather have been discussed in detail in previous section. It has already been stated that cultivation conditions of mushroom can be watchfully controlled to obtain an exact leather replica featuring distinct textures and colours. Apart from closely resemblance to animal leather in terms of appearance and durability, mushroom leather exhibits a myriad of remarkable

138   Sustainable Fibres for Fashion and Textile Manufacturing

properties such as non-toxicity, breathability, waterproofing, antimicrobial property, flame and abrasion resistance thereby making it a preferred vegan option for a range of fashion ensembles. The flexibility of material can be tailored as per the end use such that mushroom leather can be crafted as thin as paper to be used for attires and lamp shades or it can be made into thick sheets for heavy-duty items while still retaining its flexibility and strength.

7.3.3.1 Mushroom leather outshines animal leather

The mushroom leather is being developed not just to be aesthetically identical to animal leather but former is comparable to latter as far as physical and mechanical properties are concerned. Table 7.1 compares the mushroom leather with conventional leather in terms of strength, durability and appearance. Reisi leather is comparable to cowhide leather and appears to be a promising alternative to animal leather (madewithreishi).

The production of animal leather is an environment polluting process. Leather in its final finished form ready for a myriad of textile and fashion applications has been subjected to several chemical intensive and deleterious processes like tanning, dyeing and finishing as discussed in section 1.2. The hardening agent, namely, chromium utilized during leather production is the main culprit for chromium contamination with water and solid waste from tanneries containing chromium content.

Subsequently, the metal ends up in water bodies, air and soil rendering water unsuitable for drinking ultimately contaminating the aquatic food chain and major cause of prolonged illness and other health ailments like liver failure, kidney damage, lung cancer and premature dementia. Moreover, the toxic discharge from tanneries leads to soil infertility in regions of close proximity to tanneries and leather processing units.

The practise of raising animals to be subsequently slaughtered for their hides and skin in the process of leather making is doing no good to eco system. In fact, the exploitation of animals perturbs the ecological balance and poses serious threat to flaura, fauna and disrupted food chain. The leather manufacturing process not just utilizes toxic and hazardous chemicals and auxiliaries but the industry is also credited to wasteful utilization of resources since cattle raising demands approximately 30% of the world's ice-free land while one-third of the world's fresh water is utilized to support the industry unlike the sustainable mushroom leather that does not require such capacious quantities of natural resources and hazardous chemicals (Fig. 7.4).

Mushroom leather making process demands minimal utilization of resources, water and energy. The conventional mode of raising livestock for their hides, associated

Table 7.1 Comparison of mushroom leather with animal leather.

Leather type	Durability	Strength (Mpa)	Appearance (rating 1−5)
Reisi leather	100,000+ cycles	10	4.5
Cowhide leather	100,000+ cycles	8	4.5

Mushroom and corn fibre—the green alternatives to unsustainable raw materials

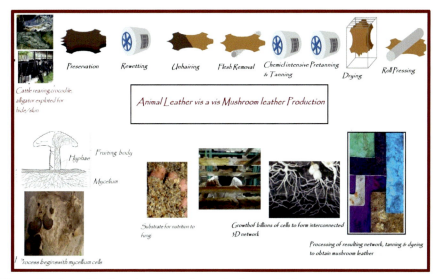

Figure 7.4 Animal leather vis a vis mushroom leather production process.

greenhouse gases and material waste is eliminated as mushroom leather replaces animal leather at massive scale and hits the main stream market. Furthermore, mycelium obtained from vegetative component of mushroom belongs to genre of sustainable natural fabrics in contrast to fossil fuel-based synthetic leather like polyurethane and PVC which involve plentiful wasteful and hazardous processes and chemicals during their production and processing.

The process of mushroom leather making is comparatively much quicker involving fewer processes (Fig. 7.4) with just couple of weeks for the vegetative root to grow and transform into leather like material as compared to animal leather which involves an exhaustive process of as long as 3 years to raise cattle to the size sufficient to obtain a workable piece of their hide or skin. Mushroom leather unlike conventional leather offers recycling and reusability options owing to the repeated usage of products even after end-of-life cycle as post-consumer waste.

The mushroom leather making process offers manufacturing agility as the surface can be transformed and tailored as per design requirements either into flat sheets or multidimensional 3d structures. Although mushroom leather offers fragile look however its strength is comparable to deerskin. The designers working with mushroom leather explore myriad of patterns, colours and textures that remain unachievable with conventional leather.

The cost of mushroom leather and exotic animal leather is comparable owing to less production of the former however with more and more brands and apparel manufacturers switching over to eco-friendly approaches, a day is not far away when productions volumes of mushroom leather will increase considerably thus making it more affordable compared to animal leather.

7.3.4 Sustainable and beneficial attributes of mushroom

The in vitro cultivation of mushroom leather completely conforms to the principles of sustainability and circular economy. Moreover, the fibre is biodegradable as it reaches end of its life cycle. Mushroom leather cannot just minimize the need for industrial animal agriculture but also has positive environmental impact by offering a plausible and better substitute for leather, and further a remedy for hazards associated with plastic consumption.

The production of mushroom leather follows a closed-loop process (Figs 7.5 and 7.6) with post-consumer waste namely corn cobs, wood chips, and straw subjected to recycling, repurposing and transforming into eco-friendly end products. These discarded materials are mixed in with mushroom spawn to create mycelium that later on is used to make vegan-friendly leather. Moreover, the waste resulting from the making of mushroom leather can be reused as a smoking product in beekeeping or as organic crop fertiliser. The various benefits that mushroom bestows are enlisted below:

- Mushroom is a natural living organism that is biodegradable, non-toxic, gluten and chemical free and does not require toxic and hazardous chemicals, synthetic fertilizers, pesticides, etc. during its production.
- The mushroom leather making process can considerably bring down the water consumption involved in growing and dyeing the material by up to 99% when compared to traditional textile raw materials.
- The production process of mushroom ensures effective waste management with post-consumer like corn cobs, wood chips and straw utilized in mushroom production and by

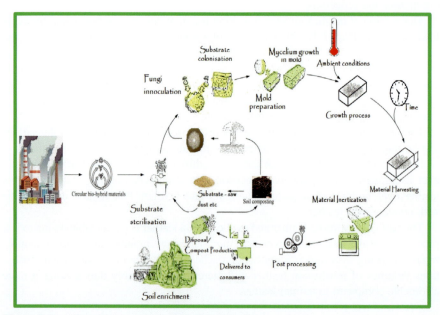

Figure 7.5 — Closed loop system of mushroom leather.

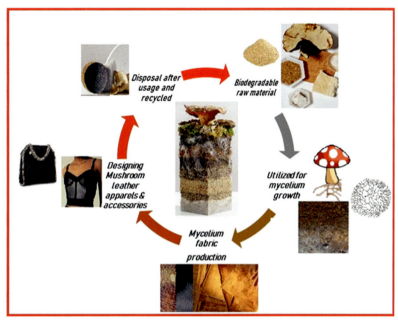

Figure 7.6 Recycling and reusability of mushroom-based end products.

reusing the by-products of mushroom leather making process as organic crop fertilizer and smoking product in beekeeping.
- The adaptability and ability of fibre to grow into any required size, shape eliminates the cutting and sewing room operations thereby eliminating the cutting room waste as with conventional garments (yarnsandfibers).
- Apart from biodegradability of mushroom, it is also capable of composting as its life cycle ends thereby exhibiting negative carbon footprints and following closed-loop model in its production process.
- The ability of fungi mycelia to decompose and recycle has paved the way for optimization of textile wastes, by employing an already existing technique of producing synthetic fibre fabrics at their end use phase. The method delivers a short fibre to fabric chain of production and is derived from free-form machine embroidery techniques. Bespoke method of sustainable textile production is thus made feasible by exploring the intricate structural characteristics of the mushrooms to develop abstracted 2D and 3D designs.
- The garment composed of mushroom leather does not end up as landfill as with all other garments composed of synthetic fabrics and animal leather. As the mushroom leather garment reaches its obsolescence phase, the biodegradability of fibre ensures that it follows a perfect circle of recycling from cradle to grave and then back to cradle with the end-product getting converted back to the fungi that created it (Fig. 7.6).
- Mycelium-based biocomposites utilizing mushroom, chicken feathers, agricultural waste and textile waste are anticipated to be cost effective strategies for solid waste management and a potential source of revenue for local mushroom farmers. The fashion and life style products composed of mycelium biobased composites will be biodegradable enabling their

decomposition thereby eliminating the risk of disposed off articles ending as landfills like conventional fashion ensembles.
- Mushroom exhibits skin curing properties and has been successfully utilized by some civilizations for developing skin medicines from mushroom. The skin friendly and antimicrobial property of mushroom makes it suitable for intimate wear.
- Mushrooms is also known for its anti-ageing properties as it aids in slowing down the ageing process and has non-carcinogenic properties
- The fabrics composed of mushroom are thin, flexible, water repellent, breathable and comfortable even for next to skin applications thus suitable for designing athletic wear and intimate wear, etc.
- Apart from functional brilliance, mushroom fabric is aesthetically appealing as well with possible customization of colour, textures and patterns to meet the consumers aspirations.

7.3.5 Mushroom leather—the preferred choice of sustainable fashion brands

The negative carbon footprints, production agility, easy dyeability, biodegradability, plastic free and ability to be tailored to required shapes, sizes, colours and textures make mushroom leather a preferred choice in a variety of industrial establishments besides textiles. The light weight and flexibility of mushroom leather makes it a preferred option for myriad of apparel and accessory categories. Mushroom is not just being explored as an eco-friendly and vegan leather substitute for designing haute couture and fashion ensembles intended for runways but it has also served as a source of design inspiration and aesthetic appreciation for designers like Rahul Mishra. Mishra in his Spring 2021 collection glorified the magnificence of mushroom fibre by designing minidresses featuring hand-embroidered mushroom motifs and flowing gowns layered to create the illusion of fungi sprouting from trees (Fig. 7.7). Likewise, the

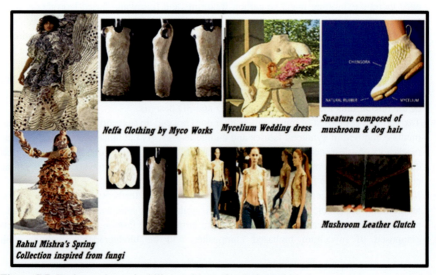

Figure 7.7 Design and sustainability as drivers for designers.

exceptionally brilliant aesthetic and functional properties of mycelium apart from its inherent abundance in nature, rapid growth and compostable attributes prompts several start-ups, brands, designers and stakeholders to use fungi as a vegan substitute for animal leather and explore the versatility and potential of mushroom leather in a gamut of applications.

Hermes and Stella McCartney are some high-end clothing and accessory brands that are utilizing mushroom leather in their exotic design collections. A line of mushroom decorated t shirt by Bella Hadid, amadou mushroom hat by Daniel Del and fungi inspired spring collection by national and international designers like Rahul Mishra and Iris Van Herpen are some examples to narrate the popularity and magnificence of this biodegradable and eco-friendly material. Additionally, mushroom inspired fashion has captivated the attention of vegan Indian brands such as Paio, Aulive and Escaro as well. Lululemon, Adidas, Stella McCartney and Gucci are joining hands with Bolt Threads for their patented mushroom leather—Mylo to be transformed into eco-friendly and biodegradable fashion apparels and accessories.

The designers and material engineers at Bolt Threads firmly believe that mushroom leather is a plausible solution to deleterious environmental impacts and a viable replacement for plastic, animal and synthetic leather. Mylo takes less than 2 weeks to fully grow into thick sheets on a bed of sawdust and other organic materials by stringent control on all growth parameters. The processing, tanning, dyeing and embossing ultimately produces a perfect replica of animal leather in terms of form and function.

The future belongs to this wonder material which does not require wasteful utilization of depleting natural resources with just half the volume of water compared to cotton cultivation is required in production of Mylo. The suppleness and warmth associated with Mylo is missing in synthetic leather that may at first instance give sensation of cold plastic. Thus, apart from appearance, the superior tactile properties of Mylo also make it popular among fashion aficionados.

The traditional leather brands have ruled the fashion arena for ages but still some designers like Stella McCartney have adopted a vegan route and refrain the usage of animal hides or fur to design her clothing and accessory line. Cartney can be credited as the first designer to recognize the potential of mushroom leather and creating Mylo garment like bustier, leggings and her signature Falabella bag for exhibiting in Victoria and Albert Museum (Fig. 7.8). She has been working in close association with scientists at Bolt Threads to attain perfection as far as drape, texture and weight of Mylo is concerned.

Ecovative Design has partnered with Bolt threads. The association aims at bringing the sustainable material into mainstream textiles and making it accessible to the fashion world owing to its outstanding properties like suppleness, durability, biodegradability, reduced environmental impact and a promising replacement for real and synthetic leather. Furthermore, Mylo can be produced in a couple of days unlike the conventional leather that involves cumbersome and exhaustive processes right from animal rearing to tanning and finishing treatments.

Lululemon, a Vancouver-based athletic wear manufacturers are pioneer in developing state of the art yoga accessories such as yoga mats and bags utilizing Mylo. The designers have experimented with varying patterns in the woven Mylo yoga

Figure 7.8 Exotic fashion accessories composed of mushroom leather.

mat to offer comfortable experience to wearer by better placement of hands and feet when practising yoga. Another innovative and premium accessory designed by Lululemon utilizing mushroom leather, Mylo is the Barrel Duffel Bag which the gym goers and fitness enthusiasts can embrace for their gaming accessories.

Furthermore, Lululemon has collaborated with brands like Stella McCartney and Allbirds who are known for extensive usage of mushroom leather for clothing and sneakers respectively (yarnsandfibers; fashionnetwork; vogue). Allbirds and Adidas in the pursuit to meet their sustainable goals and develop a footwear with low carbon footprint have collaborated for Futurecraft. Footprint Shoe that has been designed to reduce the carbon footprints to 2.94 kgs in contrast to7.85 kgs for running shoes designed conventionally. The launch of no carbon footprint performance sneakers is a landmark for both the brands with footwear designed with watchful selection of sustainable raw materials for different footwear components.

The midsole of the sneakers uses all bird's sugarcane-based SweetFoam that offers a low-carbon natural component while the shoe upper comprises of 77% recycled polyester and 23% Tencel. An innovative tangram principle employed to fit each individual component like a tangram puzzle for construction of upper and outer footwear components ensure that each part contributes to minimal scrap in production thereby resulting in considerable waste reduction (fashionnetwork; vogue; bustle; neffa).

Apart from Mylo, Muskin by Zero Grado Espace is another eco-friendly leather substitute preferred by designers and brands in their design collections. Muskin obtained from parasitic fungi Phellinus ellipsoideus is a virgin vegetable layer that can replace animal leather. The bacterial proliferation in clothing comprising Muskin is restricted owing to presence of natural penicillins in the fungi. The material after its extraction is subjected to similar treatments like animal leather however eco-friendly substances like eco-wax are utilized to render leather like characteristics to Muskin. Muskin can be tailored to exhibit suede like touch or soft to cork-like texture and is deemed suitable for limited edition collections owing to pint-sized monthly production of 40–50 m^2 of Muskin.

MycoTEX is pioneer in utilizing sustainable and vegan raw materials like mycelium for developing customized end products with perfect fit and zero waste production. The seamless production technology employed by MycoTex ensures perfect fit of designed garments eliminating the necessity of cut and sew operations. Furthermore, offcuts associated with unnecessary textile waste is also avoided due to exclusion of usual measuring and trimming.

Myco Tex attempts to explore the consumptive mannerism of nature with trees shedding leaves to be soon replenished with new ones. The team is utilizing the same principle in designing garments which are capable of getting composted after being discarded and then being converted back to the raw material just like trees with their leaves. NEFFA, a state-of-the-art jacket (Fig. 7.7) is designed by Myco Tex taking inspiration from biological life cycle. Other clothing line developed by the brand also perfectly adheres to sustainable principles with the entire process chemical and pesticide free, replacing plastics and leather with home compostable stuff (bustle; neffa; gzespace; lifegate).

Dutch textile designer Aniela Hoitink has used disc shaped mushroom mycelium pieces to develop a Neffa dress with an intent to utilize living organism in a wearable garment. The dress was designed with combination of mycelium and textile elements thereby developing a flexible composite product referred to as MycoTex. Subsequently, a novel production method eliminating the need of any additional fibres which allowed the material to retain its shape and flexibility was introduced by the designer. The circular pieces were moulded around dress to give them the look of the dress (Fig. 7.7).

Mycelium patterns can be created for adjusting garment's length or for adding some elements. Further, the dress can be composted after end-of-life cycle owing to biodegradability of raw materials comprising the dress (leather; lifestyleasia; sciencedaily; dezeen).

German designer Emilie Burfeind is pioneer in designing sock shoe known as Sneature (Fig. 7.7), the term coined by combination of sneakers and future. The salient feature of the so-called future footwear is unconventional and sustainable raw materials such as dog hair and mushroom used to develop the footwear. The innovative shoe by Emilie is designed as a seamless sock thereby reducing the number of components unlike the traditional trainers which may require as many as 12 different components to be assembled together. The disassembling and recycling of traditional trainers stances a challenge owing to large number of components that constitute the finished footwear. The footwear uses specialized 3D knitting technology in its construction enabling each shoe to be created in one print thereby eliminating any cutting waste. The sole of footwear is constructed from fungi mycelium which can be shredded and reused rendering biodegradability to the footwear (lifestyleasia; sciencedaily; dezeen; springwise).

MycoWorks is pioneer in creating a patented process for transforming mycelium into amber hued vegan leather referred to as Sylvania that closely resembles conventional leather. Hermes, an exotic French label uses vegan leather, Sylvania (Myco Works) for their range of luxury and life style products. Accordingly, the label is launching its popular Victoria travel bag composed of mushroom leather, canvas and elements of calfskin. Reishi, a new generation sustainable mushroom leather is another feather in MycoWorks' crown as the vegan leather is suitable for multitude textile applications owing to its functional and aesthetic properties at par with animal leather. Furthermore, the innovative material made it to runways with apparels comprising Reishi unveiled at New York Fashion Week.

An Indonesian footwear brand, Brodo has associated with a bio-material startup MYCL (Mycotech Lab) and plans to launch mushroom footwear composed of MYLEA. The mushroom leather by MYCL is composed of agro forestry waste and its processing does not utilize hazardous, toxic chemicals conventionally used in leather processing. Erin Smith, Microsoft's Artist-in-Residence utilized mycelium and tree mulch to design her wedding dress. The wonder fibre finds application in home decor as well with lighting designer Danielle Trofe launching biodegradable light fixtures composed of mycelium (Fig. 7.7).

Nina Fabert, a Berlin-based designer, has conceptualized and designed vegan material for high end fungus footwear that claims to be eco-friendly, organic, vegan, gluten and chemical free. The footwear component uses parasitic and decomposable tinder fungus and a combination of cork, rubber, eco-cotton terry cloth and microfiber suede from recycled bottles rendering the footwear vintage look and a soft texture (Fig. 7.8).

Zvnder, a German-based mushroom leather accessory company, pay due consideration to comfort of mushroom leather in intimate wear and next to skin applications. Mushroom leather sports footwear developed by Zvnder exhibits excellent moisture absorption and improved foot conditions for athletes Furthermore, the utilization of the vegan material eliminates the need of hazardous chemical sprays for achieving anti-odour properties in footwear. Additionally, the mushroom leather footwear and other accessories exhibit brilliant insulation properties and find application as winter wear. The watch straps comprising mushroom leather designed by the company is claimed to be effective against skin irritation and suitable for consumers suffering from eczema (dezeen; springwise; nytimes; theguardian).

Researchers from University of Delaware crafted a biodegradable, nontoxic footwear proto type comprising of mushrooms, agriculture waste and fabric scraps offering a perfect solution to reduce the impacts of textile waste, toxic inputs and utilization of renewable resources. A variety of mushrooms like reishi, oyster, king oyster and yellow oyster varieties are preferred for footwear owing to their impeccable aesthetics and strength. Mushroom was grown exactly in footwear shape by utilizing a shoe sole mold which was filled by mycelium in less than a week. As the mold got filled with mycelium, baking was accomplished to halt the growth and prevention of mushrooms from fruiting on the surface (Fig. 7.7).

Researchers at University of Pittsburgh have taken inspiration from Enoki mushroom to develop a water proof polyethylene terephthalate (PET) flexible coating material suitable for flexible applications like apparels and home textiles.The coating features nanostructures resembling the shape of enoki mushrooms. The shape with long, thin stalks and large caps holds liquid on top of the nanostructure, making the coating transparent and thus suitable for flexible solar cells (smithsonianmag; fibre2-fashion). The innovative utilization of mushroom leather by high end designers and

Mushroom and corn fibre—the green alternatives to unsustainable raw materials   147

brands in their sustainable apparel and accessory line has been discussed in the previous section.

Apart from clothing, footwear, watch straps and handbags, designers and brands are exploring the versatility of material in packaging industry, plant-based meat, home decor, durable furniture, lamp shades and building bricks as well (Fig. 7.9A and B. Furthermore, the potential of mycelium is also explored in medical arena with the material serving as scaffolds for growth of new organs.

7.3.6 Challenges for mass adoption of mushroom leather

The ardent leather lovers and proponents are of the view that traditional industry serve as waste stream for skins and hides from beef production which would otherwise end up as landfills, however, they tend to undermine the environmental hazards caused by

Figure 7.9 (A) Application of mushroom leather in fashion and life style product mix. (B) Application of mushroom leather in home decor and packaging.

toxic chemicals, auxiliaries and metal compounds involved in leather processing and finishing. Although mushroom leather is sustainable and most promising animal leather substitute however it would take some time for the material to hit the mainstream textile and fashion segment and enjoy mass acceptance and consumption.

The biggest barrier to adoption of mushroom leather by consumers is exorbitant prices of mushroom leather owing to less volumes being produced. Moreover, the consumers tend to be negligent to eco-friendly options and bear a preconceived notion of superiority and penchant for animal leather.

The designers and brands still need to address certain challenges before mycelium can completely substitute the unsustainable raw materials and hit the mainstream. The growth of mycelium requires controlled ambient conditions to be maintained via installation of humidifiers and grow lights that demands considerable energy, space and initial capital investment. It thus becomes imperative for start-ups to figure out some effective measures to cut down the energy consumption during mycelium growth to fully harness the benefits of the fungi. Additionally, it is not just about the end product but the sustainability has to be incorporated at the processing, manufacturing and growing stage of mycelium (theguardian; smithsonianmag).

The brands can commit to sustainable principles by switching over to mushroom leather replacing traditional animal leather however caution is also required as far as biodegradability of the finished product is concerned by ensuring that any hardware, trims, notions and fastening elements used in the apparel and accessories comprising of mushroom leather are sustainable too. Accordingly, designers and brands are exploring a myriad of sustainable options in their design collections apart from mushroom such as pineapple, apple and corn leather, hemp, cactus, etc. The next section of chapter shall discuss the corn fibre in details with emphasis on salient features, production process and application areas of the fibre.

7.4 Introduction to corn fibre

Corn fibre also referred to as ingeo fibre belongs to genre of man-made fibres derived from renewable resources and is the by-product of corn wet milling industry. The main component of corn fibre is pericarp consisting of 35% hemicellulose, 18% cellulose and 20% remaining starch (protein, fibre oil and lignin). The fibre serves best of both worlds by offering advantages of synthetic fibres with complimenting properties of natural fibres like cotton and wool. The fibre is available in both spun and filament forms in varying count range as per required fabric aerial density ranging from microdenier for light weight fabrics to high counts for heavy duty or dense fabrics.

Apart from serving as animal feed, the fibre is also used for production of corn leather, bio plastic and ethanol. The next sections of chapter will cover various aspects pertaining to corn leather and corn plastic. However, first let us discuss some salient features and production process of corn fibre (smithsonianmag; Aggarwal et al., 2009).

7.4.1 Salient features of corn fibre

- Corn fibre fabrics exhibit strength, resilience, softness, drape and comfort characteristics thus making them ideal choice for a gamut of textile applications
- Eliminate the need of chemical additives and surface treatments owing to its inherent functional attributes like flame retardance and softness.
- Strength retention, colour fastness and UV resistance of corn fibre are far superior compared to synthetic fibres.
- A viable, eco-friendly substitute for polyester and nylon in development of outerwear and padded garments.
- Outstanding moisture management and anti-odour properties thus opening the avenues for fibre in sportswear and intimate wear development.
- Corn fibre filament exhibit subtle luster and drapeability rendering it suitable for evening wear.
- Corn fibre fabrics and garments designed thereof exhibit good soil resistance, dry quickly and demonstrate excellent after wash appearance.
- Ease of washing and after care
- Since corns are derived from plant sugars, the corn-based end products after end-of-life cycle are compostable
- The complete life cycle of production, consumption, disposal and reuse of corn fibre follows a closed system owing to fibres's ability of composability and chemical recyclability (Fig. 7.10).

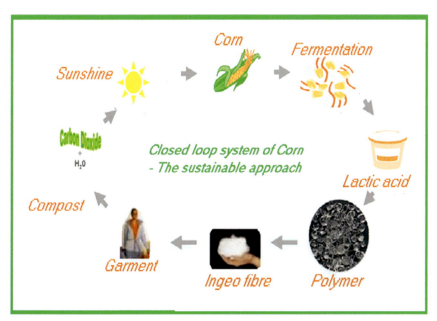

Figure 7.10 Closed loop system of corn production, usage and disposal.

7.4.2 Production process of corn fibre

Corn or ingeo fibre is an agricultural product and is utilized for starch extraction by breaking down into sugars followed by fermentation and separation into polymers. The growth of crop requires conducive environmental conditions like soil, water, sunlight, etc. The harvested crop is soaked to separate the endosperm from fibres and gluten.

Wet milling is a mechanical process of separation of starch and other components from corn kernel. Corn wet milling involves separation of kernel into constituent chemical components. Wet milling processing commences with steeping kernel corn in aqueous solution of sulphur di oxide and lactic acid (produced by microorganisms) at 50°C for 24–48 h followed by coarse grinding of corn (Fig. 7.11).

Thereafter, separation of lipids containing germ and fibrous hull portions is accomplished. The remaining components are finely ground and separation of starch and protein is carried out by hydro clones, that is, continuous centrifuges. Germ can be processed into oil while protein, fibre component is blended to serve as animal feeds.

Wet milling is followed by heating the starch in the presence of acid and enzymes to hydrolyse starch to dextrose. Dextrose can be isolated by crystallization or can be used as liquid concentrate. Conventional wet milling employed for recovery and purification of starch and co-products has been modified as enzymatic corn wet milling/E milling. The process involves the usage of proteases for breaking corn into constituents thereby eliminating the need of sulfites as with wet milling that uses pre-treatment of corn in sulfonic acid (SO_2 in water) for physical separation of co-products and starch. The process is not just energy and time consuming, but also has negative environmental impacts owing to usage of sulphur di oxide which is associated with chronic health ailments like respiratory disorders adversely affecting asthmatic patients.

Thereafter, fermentation of dextrose (D glucose) to lactic acid is accomplished by a process called fermentation. Fermentation involves biochemical extraction of energy in form of ATP (Adenosine Triphosphate) from carbohydrates like glucose in anaerobic conditions and akin to yoghurt making process. Fermentation of glucose results in production of L-lactic acid (long chain polymer of lactide).

The fermentation is followed by transformation into polylactide. Polylactide, a high-performance polymer can either be spun or processed into corn fibre (Figs 7.11 and 7.12). At this stage, corn fibre features a paste like appearance which is then extruded into delicate strands to be cut, carded, combed and spun into yarn (Fig. 7.13). The production and usage of corn fibre is an eco-friendly process with less greenhouse gas emission to environment (smithsonianmag; Aggarwal et al., 2009; fibre2fashion; unnatisilks).

7.4.3 Corn leather

Corn eco-leather or Bioskin obtained from maize processing waste consists of 60% bio-based content and is a promising bio-based alternative to animal leather. The environmental impact of corn production can by considerably reduced by transforming it to corn leather since corn waste left over of food industry is utilized in the process. Corn

Mushroom and corn fibre—the green alternatives to unsustainable raw materials 151

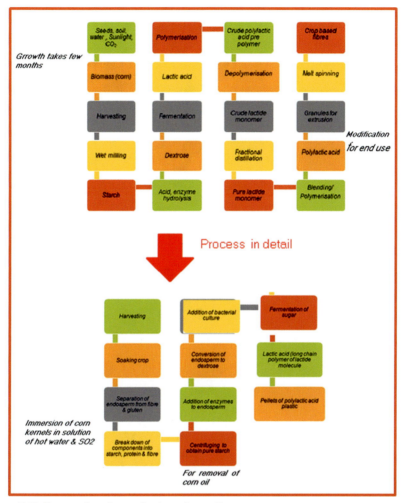

Figure 7.11 Process flow for production of corn fibre.

leather exhibits elasticity, toughness and tactile properties comparable to animal leather although offering an eco-friendly and easy handling, after use of the vegan leather.

The salient feature of corn leather includes the usage of resin from corn waste industry for coating a waxed canvas intended for footwear development. 50% corn waste and polyurethane comprise the coated canvas. The brands in pursuit to achieve sustainability have been experimenting with varying corn/polyurethane blends with 47% PU and 82% corn waste. The succeeding section of chapter will discuss some sustainable brands that are utilizing corn leather in assortment of fashion products.

Corn leather owing to blending with PU; is moderately sustainable exhibiting up to 63% biodegradability. Although the production of corn leather does not involve

Figure 7.12 The building blocks of corn fibre.

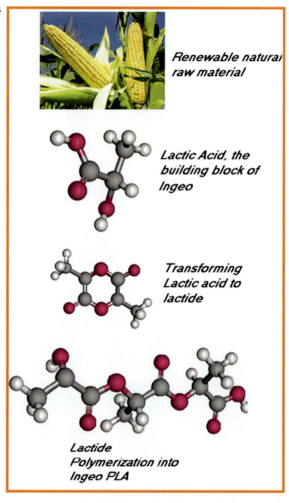

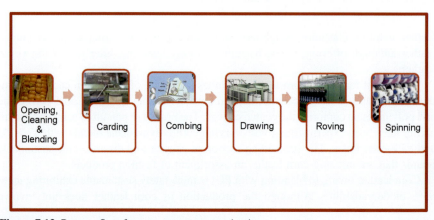

Figure 7.13 Process flow for corn spun yarn production.

animal rearing and slaughtering however, if genetically modified corn crops are utilized, it can be harmful to wildlife and environment due to spraying of crops with chemicals and pesticides. The need of the hour is thus to explore avenues for PU free carbon leather to make the material completely bio degradable.

7.4.4 Corn plastic

Another innovative application of corn fibre includes the production of corn plastic which is claimed to be an ideal replacement for plastic materials conventionally being used since ages. The traditional plastic wreaks deleterious environmental impacts owing to its petro chemical-based origin while endorsement of corn plastic can be a judicious move towards reduced footprints. Accordingly, several industries are looking up to corn starch plastic as an alternative to petro-based plastic (Line, 2021; healabel; seventhvegan). Corn plastic or bioplastic is composed of biodegradable and renewable corn starch polymer—polylactic acid obtained from fermented plant starch.

The endosperm of kernel is used to obtain corn starch polymer and find application in production of corn starch plastic and airbags. Bioplastic is increasingly used in packaging as a promising replacement for traditional plastics owing to its renewable biomass constituents, that is, corn. Some of the advantages of corn plastic are enlisted below (climate; Elizabeth, 2006; McInnes, 2008):

- Corn plastic does not emit any toxic fumes and gases on being incinerated.
- Less greenhouse gas emission with 68% reduction in emissions as against conventional fossil fuel-based plastics.
- Energy consumption in production of corn starch plastic is 65% less compared to traditional non-renewable plastics
- Offer cost competitiveness compared to conventional ones
- Safer to work and handle
- Offer less static charge generation compared to synthetic packaging materials
- Polylactic acid plastic in film or fibre is suitable as packaging materials offering low flammability, UV resistance and recycling by regrinding.
- Find application in various segments like packaging, automotive textiles, agriculture, horticulture and last but not the least medical implants—sutures, ligatures, meshes

7.4.5 Application areas of corn

- The outstanding properties of corn fibre, namely, strength, resilience, drape, moisture management, anti-odour, soil release and easy-care properties make the fibre suitable choice for a range of apparels like athleisure, active wear, duvet jackets, intimate wear etc.
- Furthermore, corn fibres find application in myriad of home textile products like bedding fibrefill—pillow, duvets, quilts, mattresses, blankets, carpets, draperies, wall panelling, upholstery fibrefill, etc.
- Corn fibre also finds application in non-woven segment with extensive usage in cosmetics, diapers, wipes and female hygiene products owing to reusability of fibres after end of its life cycle.

- The technical textile segment, namely, geotextiles, agrotextiles and filtration is also exploring the potential of the sustainable fibre.
- Medical applications of corn fibres include post operation seams, bandages, gauze, absorbent cotton. Corn fibre is biodegradable, compostable, burn without producing toxic fumes and is recyclable.
- Vegan leather and bioplastic produced from corn are extensively utilized for fashion and life-style products and packaging, etc.
- As far as non-textile application is concerned, corn in mainly used for ethanol production to be subsequently used as automotive fuel and as animal feed.

7.4.6 Corn—the prime choice of sustainable fashion brands

A gamut of apparel and accessory brands is switching to sustainable alternative replacing fossil fuel-based materials. The major footwear brands are fostered to incorporate sustainable principles in their supply chain by sustainable raw material procurement owing to consumer mindfulness and inquisitiveness in components of merchandise they purchase. Moreover, there has been rising world-wide clamour of deleterious environmental impacts associated with traditional footwear making process that primarily utilizes oil-based plastics, being non-biodegradable and will ultimately end up as landfills as the footwear reach the grave stage in their life cycle. Accordingly, footwear designers are prompted to reconnoiter sustainable alternatives for footwear making and are thus exploring corn-based products such as corn leather along with other vegan options. The footwear and accessories designed from corn leather results in lower environmental impact compared to synthetic polyurethane leather owing to carbon neutrality and renewable origin of corn fibre.

The sustainable brands that are designing exclusive range of corn apparels and accessories are being discussed in the succeeding section. Veja and *ACBC* are pioneering in utilizing vegan leather in assortment of their product mix. *Veja*, French sustainable footwear brand uses vegan leather derived from corn in their exclusive pair of sneakers.

ACBC (Anything can be changed) is another footwear brand incorporating sustainable raw materials in their product line (Fig. 7.13). The exclusive, high end unisex Bio-milan corn footwear by ACBC is a sustainable footwear that is a classic example of blending fashion and eco-friendly concepts. Corn Eco-leather, Bioskin comprises the footwear upper. The insole and lining of footwear are biodegradable with 30% corn waste and recycled form used in insole while footwear lining is produced with rebotilia obtained from recycled bottles.

Louis Vuiton, a fashion luxury brand has incorporated an innovative and sustainable corn-based plastic, Biolpolioli in their Charlie sneakers (Fig. 7.13). The corn-based footwear is exclusively designed to be recyclable with sole comprising of 94% recycled rubber and 100% recycled fibre used in footwear laces. The sneakers are believed to be landmark in genre of sustainable luxury shoe line considerably reducing the environmental impact.

Reebok in an endeavour to promote their 'Corn + Cotton Initiative' has launched a pair of sneakers comprising corn. The sole of sneakers features eco-friendly, non-toxic

Mushroom and corn fibre—the green alternatives to unsustainable raw materials 155

Figure 7.14 Corn-based exotic fashion accessories.

industrial grown corn while shoe body is composed of 100% organic cotton. The compost ability of the footwear's constituents enables recycling of sneakers to obtain the raw material for a fresh pair of footwear.

Hugo Boss, Native Shoes are abiding the sustainable principles by introducing an innovative, sustainable raw material fordeveloping their footwear line. The brand utilizes Pinatex composed of pineapple leave fibres mixed with polylactic acid (corn-based bio plastic) to form non-woven mesh for footwear.

LUXTRA London, an Australian luxury brand focuses on cruelty free and sustainable fashion. The latest addition to the brand's collection of eco-friendly and vegan raw materials is bio-based E-ULTRA. The material obtained from non-edible part of corn is claimed to be a promising leather substitute owing to its softness, appeal, smoothness and recyclability with a high bio-content of 69%.

Alexandra K, an Italian clothing brand has used corn leather in their spring summer collection (vkind; livekindly; vegnews; lifestyleasia).

7.4.7 Challenges associated with corn fibre production and usage

- Corn production is considered to be relatively destructive process due to nitrogen and nutrient depletion in soil.

- Another problem associated with corn is genetically modification of crops for increasing yield and enhanced resistance to insects, diseases, chemicals etc. Moreover, corn production involves monocropping that leads to poor soil quality thus necessitating the usage of chemicals, fertilizers and pesticides that are major contributors to environmental pollution.
- Corn leather owing to blending with PU is moderately sustainable exhibiting up to 63% biodegradability. Although the production of corn leather does not involve animal rearing and slaughtering, if genetically modified corn crops are utilized, it can be harmful to wildlife and environment due to spraying of crops with chemicals and pesticides.
- Bioplastic production involves the utilization of Corn-2 Yellow Dent variety that serves as animal fodder thus creating inadequacy in food chain for fauna.
- Corn plastic requires hot and humid environment of commercial composting facility to break down and decompose. Therefore, the dearth of commercial composting facilities poses a major challenge in composting corn starch plastic presently. Inadequate infrastructure for composting PLA leads to bioplastic ending up as landfills. Since PLA is plant based, it can be disposed only be composting which thus leads to increased acidity of regular compost, greenhouse gas and methane production during their composting.
- Contamination of recycling stream occurs if corn starch plastic is attempted to be recycled rather than composted. Also, bio degradation and recycling of corn plastic is a slow process with breaking down of PLA in landfills as long and exhaustive as with traditional plastic (seventhvegan; climate; Elizabeth, 2006; McInnes, 2008).

7.5 Conclusions

The adverse impacts of fast fashion, wrong procurement decisions and production processes in the textile and fashion industry have deviated the supply chain from sustainable principles. The need of the hour is, therefore, to revamp the textile and fashion industry by switching over to sustainable and ethical practises in order to relinquish the adversities offered by the emerging fast fashion industry and massive consumption of non-renewable and unethically produced raw materials like animal leather and plastic. The designers and brands in the pursuit to compete, scale up their businesses, follow the sustainable revolution and satisfy the consumers 'aspirations are open to exploration and experimentation with a range of unconventional and sustainable materials like mushroom, corn, etc.

Mycelium undoubtedly is a promising candidate in bringing down the fashion industry's carbon footprint and serves as green alternatives to plastic, leather and synthetic fabrics. Although mushroom leather is sustainable and most promising leather substitute, it would take some time for the material to hit the main stream textile and fashion segment and enjoy mass acceptance and consumption.

The biggest barrier to adoption of mushroom leather by consumers is exorbitant prices of mushroom leather owing to less volumes being produced. Moreover, the consumers tend to be negligent to eco-friendly options and bear a preconceived notion of superiority and penchant for animal leather.

Apart from mushroom leather, designers are also exploring other eco-friendly options such as corn. Corn belongs to genre of man-made fibres derived from renewable resources and is the by-product of corn wet milling industry. Corn fibre follows a closed loop system and can be credited to production of several other useful end product like corn leather and corn starch or bioplastic. Although corn leather and corn starch plastic are potential candidates to replace animal leather and petro-chemical-based plastic respectively, there are some challenges pertaining to their sustainability and composting (use of genetically modified corn crop) which need to be addressed for massive usage of corn-based products. Nevertheless, both the fibres are extremely versatile and can prove to be a promising ultimate green material of the future supporting the circular economy.

References

Aggarwal, R., Singh, R., Shakiya, M.P., 2009. Corn Fibre: A New Fibre on Horizon. https://www.fibre2fashion.com/industry-article/4455/corn-fiber-a-new-fiber-on-horizon.

https://www.bustle.com/p/clothing-made-of-mushrooms-might-just-be-the-future-its-actually-pretty-cool-8018663.

https://news.climate.columbia.edu/2017/12/13/the-truth-about-bioplastics.

https://www.dezeen.com/2016/04/01/aniela-hoitink-neffa-dress-mushroom-mycelium-textile-materials-fashion.

https://ecowarriorprincess.net/2017/09/wearable-mushrooms-fungus-future-of-fashion/.

Elizabeth, R., 2006. Corn Plastic to the Rescue. https://www.smithsonianmag.com/science-nature/corn-plastic-to-the-rescue-126404720/.

https://us.fashionnetwork.com/news/Allbirds-and-adidas-release-their-first-low-carbon-performance-running-shoe,1361776.html.

https://www.fibre2fashion.com/industry-article/7249/corn-fabric-superior-in-use-and-fine-in-comfort.

http://www.gzespace.com/research-muskin.html.

Hashmi, G.J., Dastageer, G., Sajid, M.S., Ali, Z., Malik, M.F., Liqat, 2017. Leathe industry and environment: Pakistan scenario. International Journal of Applied Biology & Forensics 1 (2), 20−25.

https://healabel.com/c-fabrics-materials-textiles/corn-leather.

Jhanji, Y., 2021. Problems & hazards of textile and fashion waste, Waste Management in the fashion& Textile Industries. In: Nayak, R. (Ed.), 13 the Textile Institute Book Series, 253. eBook, ISBN 9780128187593.

J. Kanagaraj, V.kandukalpatti chinnaraj, N. K. Chandra Babu, S. Sayeed, Solid Wastes Generation in the Leather Industry and its Utilization for Cleaner Environment, DOI: 10.1002/chin.200649273.

https://www.leather-dictionary.com/index.php/Muskin_-_MuSkin.

https://www.lifegate.com/muskin-leather-mushrooms.

https://www.lifestyleasia.com/ind/style/fashion/lululemon-is-making-yoga-accessories-using-mylo-from-mushrooms/.

https://www.lifestyleasia.com/sg/style/fashion/louis-vuitton-vegan-charlie-sneakers/.

Line, C., 2021. What Is Fabric Made with Corn. Eco World. https://ecoworldonline.com/what-is-fabric-made-with-corn/.

https://www.livekindly.co/vejas-new-vegan-sneakers-made-with-corn-leather/.

https://www.madewithreishi.com/stories/performance-results-q120).

McInnes, L., 2008. The Environmental Impact of Corn Based Plastic. https://www.scientificamerican.com/article/environmental-impact-of-corn-based-plastics/.

Nayak, R., 2020. Supply Chain Management and Logistics in the Global Fashion Sector—the Sustainability Challenge. In: Nayak, R. (Ed.). ISBN13: 9781000206814.

https://neffa.nl/mycotex/.

https://www.nytimes.com/2020/10/02/fashion/mylo-mushroom-leather-adidas-stella-mccartney.html.

https://www.sciencedaily.com/releases/2018/04/180411174157.htm.

Senthil A., Balasubramanian A., Palsamy J., Gurusamy T., Diana J., Ravindran Y. & Balakrishnan K., Perception and prevalence of work-related health hazards among health care workers in public health facilities in southern India, International Journal of Occupational and Environmental Health, doi: 10.1179/2049396714Y.0000000096. Epub 2014 Dec 8.

https://www.seventhvegan.com.

https://www.smithsonianmag.com/science-nature/are-baked-mushroom-shoes-future-fashion-180969152/.

Song, Z., WilliamsC, Robert, G.J., 2000. Sedimentation of tannery wastewater. Water Research 34 (7), 2171−2176. https://doi.org/10.1016/S0043-1354(99)00358-9.

https://www.springwise.com/sustainability-innovation/fashion-beauty/emilie-burfeind-sneature-trainer-dog-hair.

https://textilevaluechain.in/in-depth-analysis/an-environment-friendly-mushroom-leather.

https://www.theguardian.com/science/2021/dec/02/californian-firm-touts-mushroom-leather-as-sustainability-gamec.

https://www.unnatisilks.com/blog/corn-fiber-an-exciting-addition-to-the-world-of-fabrics/.

https://vegnews.com/2021/11/louis-vuitton-sustainable-vegan-shoe.

https://www.vkind.com/louis-vuitton-vegan-corn-leather-shoes/.

https://www.vogue.com/article/fungi-mushrooms-fashion-inspiration-mycelium.

https://www.yarnsandfibers.com/textile-resources/other-sustainable-fibers/natural-fibers.

Wool and silk fibres from sustainable standpoint

Vinod Kadam and N. Shanmugam
Textile Manufacturing and Textile Chemistry Division, ICAR-Central Sheep and Wool Research Institute, Avikanagar, Rajasthan, India

8.1 Introduction

The literate meaning of sustainability varies with the context in which it is discussed. In biology, sustainability means living to survive, reproduce, and avoid extinction. A biological system that can remain diverse and productive indefinitely is sustainability (Brundtland et al., 1987). In the economy, it means avoiding financial crises and avoiding major disruptions. In the modern world, sustainability is a concept that describes the survival or persistence of a 'system' (Costanza and Patten, 1995) over the interrelated scale of time and space (Fig. 8.1). The word 'system' refers to one of the following—a particular species, biodiversity, culture, custom, biological process/product, non-biological process/product, business unit, industry and eco-system. To simplify, sustainability is the endurance of a system, process, and product. Overall, sustainability is a socio-ecological process characterized by the pursuit of a common ideal (Karthik and Rathinamoorthy, 2017). Ecology, economics, politics, and culture are interconnected domains of sustainability. However, it is often a trade-off between economic productivity and environmental impact. Sustainable development should integrate social, economical and environmental sustainability and should use these three to make new development sustainable (Goodland, 1995).

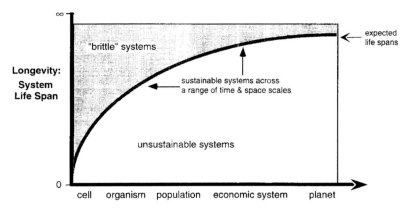

Figure 8.1 Space and time scale dependant sustainability concept (Costanza and Patten, 1995). with permission from Elsevier.

Sustainable Fibres for Fashion and Textile Manufacturing. https://doi.org/10.1016/B978-0-12-824052-6.00007-X
Copyright © 2023 Elsevier Ltd. All rights reserved.

In the textile system, sustainability is mostly assessed using life cycle assessment (LCA) beside eco-efficiency and Higg index. LCA is a cradle to grave approach wherein the environmental impact of each process in terms of CO_2 is evaluated (Costanza and Patten, 1995). Other greenhouse gases (GHGs) like CH_4 and N_2O are converted to equivalent CO_2 emissions. Eco-efficiency is measured as the ratio of the economic value of a product to the environmental impact created due to its manufacture. The Higg index is a scoring tool used by major textile brands-it contains qualitative and quantitative scoring. Forced labour, child labour, and social factors are also included in arriving at the Higg index.

Wool and silk are known to human mankind since early civilization. These fibres have been used primarily for apparel fabrics. Wool and silk fabrics are always been symbols of richness and luxury. The scientific interventions over time increased wool and silk production. However, since the inception and sudden rise of synthetic fibres in the last century, the importance of wool and silk fibre is diminished. It is worth noting that despite so much advancement in synthetic polymers, no fibre could match the properties of wool and silk. It infers wool and silk fibres science could not be fully understood yet. The inherent biodegradable nature of these protein polymers is now getting attention to develop advanced materials such as tissue engineering. In such a scenario, it is important to understand the sustainability standpoint of wool and silk fibres.

This chapter briefly introduced wool and silk and discussed the sustainability aspects of both fibres separately. The discussion is focused on three aspects: LCA, sustainable processing and prospective applications as a sustainable choice.

8.2 Wool

Wool is hair fibre obtained from the fleece of sheep. Sheep are domesticated since ancient times as a source of food and fibre. As of now, more than 1400 sheep breeds are recognized across the globe (Gowane et al., 2017). Sheep have been mainly reared for mutton due to their high commercial importance as compared to wool. Within wool fibre also, the fine (<25 μm) and long length (>45 mm) fibre fetch a high price in contrast to the medium ($25-40$ μm), coarse (>40 μm), and short length (<45 mm) fibre. The fine fibre can be spun and woven into an apparel fabric. Coarse wool is difficult to spin and hence does not attract high prices from the people in the textile trade. The majority of the sheep breeds were producing coarse wool. Selective breeding resulted in the improvement of fineness over the years. The fineness largely depends on climatic conditions. Cold regions favour fine wool production.

Wool and cotton brought the industrial revolution to Europe in the eighteenth century. Although wool is one of the oldest fibres of civilization, scientific studies on wool follicle development and fibre structure were reported from 1875 onwards (Auber, 1952). The follicle development, its interaction with sheep genetics, physiology, and environment has been initiated only after the 1950s and aimed at high wool production. At the same time, the invention of synthetic fibres with high production rates and newer properties took over and replaced natural fibres in various applications including apparel fabrics. However, no synthetic fibres to date can match the crimp, thermal insulation, and felting properties of wool. Wool has been valued for its excellent warmth, flame retardancy, water repellency, and good affinity towards natural dyes.

8.2.1 Wool production

The global wool production (2020) was 109,098 million kg produced by more than 1.163 billion sheep (IWTO, 2020). Australia, China, United States, and New Zealand are the world's leading producers of wool. The wool productivity varies from 0.nine to five kg/sheep/year depending on the climate and shearing practices.

Wool fibre has to undergo a large number of preparatory processes (fibre to yarn conversion and yarn to fabric conversion) to reach a good quality ready-to-sale fabric. Wool fibre is first sheared from the sheepskin. It is then graded and sorted based on fibre quality. Wool scouring is done to remove grease and suint in the raw wool fibre. It requires mild alkali such as sodium bicarbonate to saponify the grease while the suint is water-soluble and gets removed during the aqueous process. Scouring is a three to five bowl process. Carbonizing removes vegetable matter like a burr using acid treatment followed by hydro extraction, drying (burning), crushing, and de-dusting process. Carbonizing is an energy-intensive process.

The scoured wool fibre can be converted into yarn by three different processing setups, namely, woolen, semi-worsted, and worsted yarn. The worsted yarn is known for its better quality and requires for the fine quality of wool. The worsted yarns are mainly used for suiting and shirting. The semi-worsted yarns are relatively inferior to that of worsted yarn. Semi-worsted yarns are suitable for shawls, knitwear, and soft blankets. While the woolen yarns prepared from medium to coarse wool are used to make blankets and carpet yarns. The process flow for the three process routes is shown in Fig. 8.2.

The opening of fibre tufts makes them fluffy while blending does homogenous mixing of different lots. Oiling lubricates the fibre to avoid fibre breakage during the spinning. Carding is the heart of spinning. The objective of carding is to individualize the fibre and to remove deeply embedded dust in the fibre. The carding comprised of a set of rollers having spikes on the surface. The wire point density progressively gets dense to effectively individualize the finer fibres and align them parallel in thin roving. The gilling process is a set of combs that do align and draft the sliver. The drafting means reducing the weight per unit length of the material. Gilling is carried out in semi-worsted and worsted spinning. In worsted spinning, the number of gilling passages is more for high draft and uniformity. Combing is exclusive to worsted spinning. The objective of combing is to remove short fibres and to form a good-quality wool top. The fly frame for roving formation is also exclusive to worsted spinning. The roving is a finely drafted sliver which further refined in the spinning process to produce yarn. Twisting provides strength to the yarn. While winding converts the spun yarn into a long-length package to cater to warping and weaving.

The woolen, semi-worsted, or worsted yarn can be converted into fabric in three ways: weaving, knitting and nonwoven. The fabric is then chemically processed as per the end-use requirement. The typical chemical processing steps are scouring,

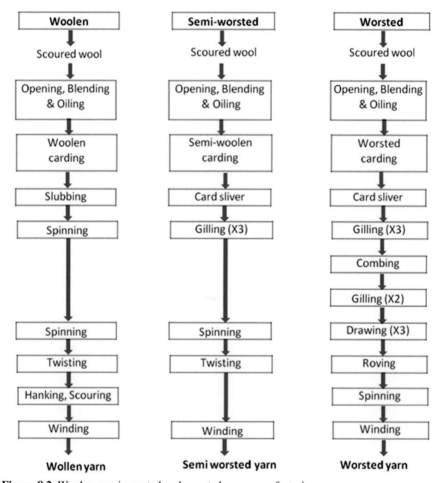

Figure 8.2 Woolen, semi-worsted and worsted yarn manufacturing processes.

dyeing, printing, milling and finishing. To sum up, wool product manufacturing starts with on-farm activities and processes through a number of the industrial process before reaching the customer.

8.2.2 Life cycle assessment of wool

Most of the LCA studies were conducted using ISO 14,040 framework with the main focus on water usage, fossil fuel consumption, energy requirement and GHG emission. The framework allows setting the scope, purpose and inventory. Table 8.1 presents a typical LCA comparison between polyester (synthetic) fibre and wool (natural) fibre. It can be seen that wool top consumes nearly 2.5 times less energy (46 MJ/kg) compared to the polyester fibre (109 MJ/kg) while, the CO_2 emission per kg of polyester (4.2

Table 8.1 Energy and CO$_2$ emission of wool and polyester fibre.

	Energy (MJ/kg)	Emission (CO$_2$/kg)
Wool top NZ	46	2.2
Polyester fibre	109	4.2

Source Russell, I.M., 2009. Sustainable wool production and processing. Sustainable Textiles 63–87; Barber, A., Pellow, G. 2006. LCA: New Zealand merino wool total energy use. In: Australian Life Cycle Assessment Society (ALCAS) Conference, Melbourne, pp. 22–24.

CO$_2$/kg) is double that of wool (2.2 CO$_2$/kg). However, synthetic fibres are cheaper may be due to industrial bulk production (Berry, 2015). Comparative wear performance of polyester t-shirt and merino wool t-shirt showed less carbon footprint (CF) of wool t-shirt and enhanced comfort (Bech et al., 2019). The frequency of replacing wool t-shirt was found less due to their odour absorbing properties. Consequently, the number of washing cycles is reduced significantly.

LCA takes into consideration the various on-farm activities of sheep husbandry such as pasture feed and nutrition supplements. The energy inputs needed at each stage are also assessed to determine the environmental impact. Fig. 8.3 shows an example of a CF assessment of a sheep farm from cradle to farm gate. Direct emissions occur on-

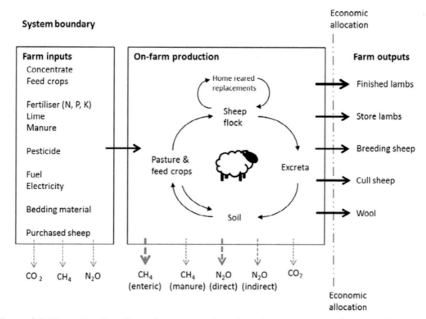

Figure 8.3 Example of cradle to farm gate carbon footprint assessment of a sheep farm using LCA approach (Jones et al., 2014).
with permission from Elsevier.

farm and indirect emissions occur elsewhere, however, attributed to the on-farm activities.

LCA study of Australian Merino wool was conducted in 2006 (Russell, 2009). The economic allocations based on overall farm receipt per sheep and wool/mutton receipt per sheep were considered as a base. The environmental impact on wool production, processing and disposal was calculated using the SimPro model. Around 2 kg of greasy wool is required to produce 1 kg of a wool garment. This 50% loss mainly accounted for weight loss of greasy wool during scouring which removes wax, dirt, sweat and salts from wool. The loss is partly due to the yarn and fabric wastage during the manufacturing processes.

Fig. 8.4 shows the carbon dioxide equivalent (CO_2-e) of the wool cradle to grave cycle. CO_2-e represents different GHGs together. It signifies the equivalent global warming impact in terms of CO_2 amount. Livestock rearing releases 18% of anthropogenic GHGs (Steinfeld et al., 2006). On-farm sheep rearing contributes nearly 50% of the CO—e from the total emissions. Biogenic methane is the largest contributor to CF followed by sheep urine which generates N_2O. The fertilizers used on-farm also produce N_2O. The other important sources that roughly contribute 25% CO_2-e are chemical processing (dyeing and finishing) and usage (washing and dry cleaning) of wool fabric. It is noteworthy that wool products do not emit CO_2-e after disposal, unlike manmade synthetic products.

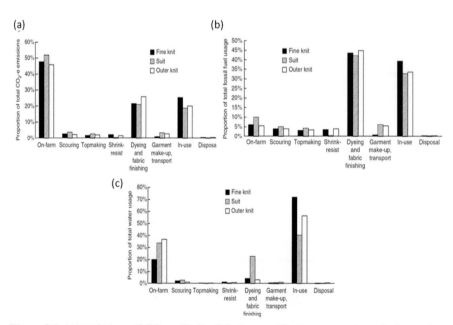

Figure 8.4 (A) emissions of CO_2-e (B) fossil fuel usage (C) water usage through the wool garment cycle (Russell, 2009).
with permission from Woodhead Publishing Ltd.

The fossil fuel used in the chemical processing of wool and during usage is in a much higher amount than the on-farm activities. The highest water consumption is due to the laundering of wool garments during use. In the case of dry cleaning, water accounts for electricity generation. While on-farm activities, water consumption is mainly attributed to fertilizer production. The dyeing process of textile fabric, especially wool suiting fabrics, is also water-intensive. Often, the suiting fabrics are dyed with very dark shades which consume a large amount of water.

In another LCA study, Australian merino wool (19 μ) emitted 24.9 kg CO_2-e/kg of greasy wool (Brock et al., 2013). The sheep manure and urine are the major emission sources of direct and indirect contribution. Direct enteric CH_4 emission from manure and urine contributed 86% of the total emissions while direct and indirect N_2O emission shared 5.2% and 1.1% contribution, respectively. The total emissions per kg of greasy wool decreased with an increase in fibre diameter, reduction in fleece weight, and selection of dual-purpose sheep (meat and wool production). Wool in the US reported comparatively lower emission (19.2 kg CO_2-e/kg greasy wool) may be because of the main difference in genetic traits and fibre quality.

It is reported (Cottle and Cowie, 2016) that the GHG emission of the sheep production system differs with climate and wool productivity. The emission of 8.6 kg CO_2-e/kg of wool was reported for wool in New South Wales and Western Australia states. The GHG emission is attributed 78% to livestock emission, 14% to transport, and 8% to pasture emission.

Studies conducted to estimate the CF of wool production to processing have revealed that the mean CF of fine-grade wool after processing was 1.20 kg CO_2-eq/kg of product (Peri et al., 2020). Wool processing made up only two to five percent of the total sheep CF. This was mainly from fuel (55% of total CF in the wool processing factories), purchased electricity (28%), and wastewater processing.

8.2.3 Climate change effect on wool

Climate change has a severe impact on sheep production affecting animal growth, milk, wool and meat (Gowane et al., 2017). Sheep rearing is vulnerable to natural calamities like extreme weather, water scarcity, drought and flood. The fodder quality and quantity, and disease epidemic have an indirect effect (Thornton et al., 2009). Altogether climate change will further increase the production cost of the sheep livestock system (Barwick et al., 2011). This is one of the biggest challenges of sheep and wool sustainability. However, it is predicted that wool and its industry can sustain the effects of climate change (Harle et al., 2007). Although small ruminants including sheep contribute to GHGs emissions (Marino et al., 2016), sheep graze, in search of food, in ranches and wastelands of Asian and African countries that enhance soil fertility and reduce GHG emissions as well (Gowane et al., 2017). The net effect is not yet studied.

8.2.4 Recent development in sustainable wool processing

Textile processing, the conversion of raw fibre to finished fabric, is highly energy-intensive, consumes a large amount of water. Textile industries also cause water, air, and noise pollution. Moreover, consumption and fashion pattern is changing very often, which leads to endless production and a lot of waste. Hence, it is important to develop environment friendly processing methods, especially in the chemical processing of textiles, to reduce the impact on the environment. Many developments have taken place in this line.

Ultrasound-assisted scouring of wool reduced the time, energy and chemicals by 25% along with increased whiteness to wool (Kadam et al., 2013). Shrinkage (poor dimensional stability) is one of the inherent limitations of the wool due to the interlocking of surface scales during washing. The conventional chlorination process of shrink resistance is very water-intensive and release hazardous absorbable organic halides (AOX) into the environment (Hassan and Carr, 2019). Newer sustainable processes such as enzymes (Kaur and Chakraborty, 2015; Kaur et al., 2016; Rani et al., 2021), biopolymers (Rani et al., 2020; Kadam et al., 2021), and click chemistry (Yu et al., 2015) have been successful in achieving good shrink resistance. Natural dyeing can greatly reduce water pollution compared to synthetic dyeing. However, bulk production and industrial scale-up are the major limitations of natural dyes. For small-scale industries, natural dyeing is sustainable.

Wool is susceptible to fungal and insect attacks. These microbes can cause allergies and diseases such as Alternaria alternate, Stachybotrys species, Acremonium species, Penicillium (biverticillata), and Fusarium species (Volf et al., 2015). Hence, wool products are often treated with a strong antimoth agent like permethrin which is a highly toxic chemical. Recently, biomaterial-based anti-moth treatments have been studied and found beneficial. The biomaterials such as neem (*Azadirachta indica*) bark extract (Kumar et al., 2016), Datura stramonium L. (Datura) (Kumar et al., 2017), pomegranate rind, and lemongrass oil (Gogoi et al., 2020) imparted antimoth properties to wool.

8.2.5 Prospective applications of wool as a sustainable choice

Fine wool is suitable for apparel and relatively expensive. Medium wool is preferred for carpets. However, the majority of the coarse wool goes underutilized and thrown as waste. The product development using coarse wool is sustainable due to a ten-fold decrease in the raw material cost. Quilt developed out of coarse wool (Shakyawar et al., 2018) will be beneficial for reducing the cost of product and decreasing the environmental pollution. Replacing polyester with coarse wool in a quilt and other products will help in reducing CFs and the process will be sustainable as the wool can be recycled again with minimal effort to prepare the quilt again. After a few years of quilt use, it can be taken back and wool can be recycled to produce new quilt, thus wool sustainability is ensured. Braiding is the process of interlacing threads to develop a continuous rope. The short-staple coarse wool can be put into the core to form a core-sheath long rope (Shanmugam et al., 2020). Various diversified products can be prepared using a wool braided rope.

Fibres from milk casein and soybean fibres were attempted to replace wool, however, all these regenerated protein fibres were having less tensile strength than that wool, and hence they did not become popular (Stenton et al., 2021).

The building uses insulation materials like polyurethane foam, polystyrene, fibre glass and mineral wool to reduce heat loss from buildings. Sustainability in a building is achieved by reducing its energy use during its operation time and also through the use of energy-efficient materials in its construction. Energy-efficient buildings with a reduction in CO_2 emissions will ensure sustainability. Cellulosic fibres like cotton and protein fibres like wool are water permeable and can regulate humidity during the operation of a building. Wool contributes to the humidity regulation of buildings. The wool absorption and desorption cycle regulate the humidity. Wool can absorb water up to 35% in weight and this property helps in balancing indoor air humidity (Parlato and Porto, 2020).

Wool was compared with polystyrene and glass wool for sound absorption and thermal insulation. Wool gave an equivalent performance and the added advantage with wool is that has low embodied energy (Rajabinejad et al., 2016). Embodied energy is the energy spent while converting the raw material into a building product and including the transport to bring into the building site. Compared to other building materials, wool construction materials are made through a nonwoven route that gives high productivity at low energy consumption. To reduce the embodied energy, lightweight materials are preferred for construction; however, it should be kept in mind that operation energy should not increase due to the use of lighter materials. Low embodied energy materials will give a low CF.

Sheep wool used as the internal wall will do micro-climate regulation and in a cold climate, it reduces condensation and absorbs odour. Another interesting application is the use of wool in cellar walls wine manufacturing plants wherein maintaining ideal temperature is important (Tinti et al., 2015). A study conducted to know the use of wool in insulation and acoustic applications (Dénes et al., 2019), it was found that wool has shown thermal conductivity in the range of 0.034–0.067 W/mK, thermal resistance values in the range of 2.5–2.6 m^2/WK, sound absorption coefficient around 0.77 Hz, which are adequate for the normal use.

Wool is flame resistant, does not burn easily which belongs to E-class (Zach et al., 2014). High nitrogen content makes it flame retardant. Wool carbonizes when exposed to heat. Wool construction materials hence will not be affected by fire easily and are scoring high in the safety point of view. Attempts made to use wool fibre as reinforcement in concrete matrix gave poor results compared to polypropylene in terms of compressive, flexural and tensile strength. Even though wool is having the highest elongation, it has the lowest modulus and tensile strength; this might have led to its poor performance in the concrete mix. However, sheep wool can be added to cement mortar to reduce cement consumption. Sheep wool reduces the cracks in mortar and a 2%–3% mix in Portland cement is found to be optimum (Maia Pederneiras et al., 2019).

Keratin is a structural protein of wool having high sulphur content. Keratin protein extracted from wool has shown promising applications in biomedical, biofertilizers, absorbents, cosmetics and animal feed (Reddy et al., 2021). In short, wool offers value-added recycling and reusing opportunities at every stage of processing. After disposal also wool improves soil fertility.

Wool is renewable and therefore sustainable. The fibre can be obtained by shearing the sheep's skin once or twice a year. The yield (0.5–6 kg) varies from breed to breed. The global average of wool fibre yield is 2.4 kg/sheep/year. Although wool is biodegradable, it takes a very long time to degrade due to the presence of hard keratin protein. If wool web (500 g/m^2) is buried under the soil, it retains its original structure even after 6 months. Disintegrated wool web can be observed after 1 year. Around 18 months later, wool degrades completely. The degradation can be accelerated using specific enzymes such as protease. After degradation also, wool is useful as a soil fertilizer. The high nitrogen content in wool (7%–8%) contributes to soil fertility. The use of waste wool in agriculture helped to retain soil moisture for a longer time (Kadam et al., 2014) and to improve nitrogen content in the soil. Consequently, the crop yield has increased. Interestingly, the unique wool fibre has importance even after degradation, unlike synthetic fibres which are posing threat to the ecosystem right from manufacturing to disposal.

8.3 Silk

Silk is the longest natural filament used in textile manufacturing for 5000 years (Xia et al., 2009). The soft feel is a natural wonder associated with silk. It is a protein fibre obtained from a silkworm, either wild or domesticated. Over 90% of the commercial silk is produced by a domesticated silkworm *Bombyx mori* (Astudillo et al., 2015). This is a monophagous insect that eats mulberry leaves. This silk is known as Mori silk or mulberry silk. Besides mulberry silk, there are several types of silk-like Tussah, Muga, Eri, Mussel and Spider silk (Karthik and Rathinamoorthy, 2017). However, their production is very limited.

8.3.1 Silk production

India and China are the leading raw silk producer and contributes 90% of the world's production(Astudillo et al., 2015). The other producers are Thailand, Vietnam, and Brazil. It takes around 5525 silkworms to produce 1 kg of raw silk. And one single cocoon can yield 900 m of silk filament.

The silk production has three steps starting with mulberry plantation which is known as moriculture followed by sericulture and reeling. Sericulture is the rearing of silkworms to the cocoon spinning stage. Reeling is unravelling and spinning long filaments from cocoons.

8.3.1.1 Moriculture

Morus alba is a fast-growing mulberry tree that grows to 10–20 m tall. The lifespan of the tree is comparable to humans. Herbicides and pesticides in specified quantities are sprayed to ensure the healthy growth of the plants. The tree starts to produce commercial quantities of leaves within 1 year of planting (Astudillo et al., 2015). One acre of mulberry trees can yield 14—16 kg of raw silk (Karthik and Rathinamoorthy, 2017). The plants, having good irrigation, can be pruned as many as six times a year. Mulberry leaves are a primary output of moriculture. The woody biomass and unsuitable leaves for silkworm rearing are by-products with commercial importance as a source of firewood/paper making and livestock fodder, respectively (Prasad et al., 2010).

8.3.1.2 Sericulture

Biovoltines (BV) (two generations/cohorts per year) is a superior silkworm variety than multivoltine (MV) (>2 cohorts/year). However, the BV variety is more susceptible to disease and environmental stress. Hence, crossbreed (CB) of BV and MV is commonly used. Silkworms can be purchased at the eggs stage or at a third stage where it is less susceptible to the environment. For maximum yield, the insects are raised indoors in well-ventilated rooms.

Fig. 8.5 shows the life cycle of mulberry silk. Moths lay eggs on specially prepared paper. The eggs are hatched. The worm eats mulberry leaves and grows old as a caterpillar in a month. The silkworms are susceptible to temperature change, noise and smell. The fully grown silkworm/caterpillar is known as a pupa. Pupa climbs a stick and spins a cocoon by a systematic head movement.

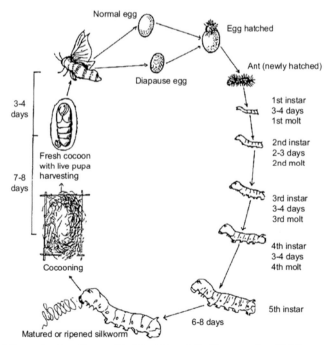

Figure 8.5 Life cycle of mulberry silk (Karthik and Rathinamoorthy, 2017). with permission from Elsevier.

The silkworm secretes two salivary glands one is fibroin and another is sericin which is a gummy substance attached to the fibroin. As the gland liquid comes in contact with air it solidifies and produces a silk filament. The silkworm can spin 1.5 km of filament inside a cocoon within 2–3 days. Once spinning is done, the newly formed moth secretes alkali-based fluid which makes a hole in the cocoon. This is the natural transmutation of the moth that ensures the breeding of the next generation. After coming out of the cocoon, the moth mates instantly and lays about 500 eggs within 4–5 days before dying. Sparing some cocoons for further breeding, rest all are heated to kill the cocoon and to obtain silk filament inside the cocoon. The heating also softens the hard sericin and assists in unravelling the filament. Approximately 10 filaments are spun together to obtain the desired strength.

Cocoons can be harvested throughout the year, however in temperate regions; these are harvested in spring and late summer. It is recommended to dry the cocoon with hot air to improve silk quality and yield. The drying ensures killing the pupae and prevents eclosion of the moth.

Sometimes killing of cocoons has not been considered eco-friendly and it can be a hindrance to the sustainability aspects. Hence researchers studied Ahimsa silk which is also known as 'peace silk' (Sannapapamma and Naik, 2015). During Ahimsa silk fibre extraction, cocoons need not be killed. Compared to the conventional growth period of 30 days, cocoons are allowed to grow for further 10 days into moth and hatch. The moth naturally secretes a liquid to dissolve the hatching hole which eventually breaks the continuity of silk fibre. The Ahimsa silk is thus short in length, however, good enough for textile making. Such silk fibre extraction is humane compared to conventional silk harvesting.

8.3.1.3 Reeling, degumming, and fabric making

For reeling, the dried cocoons are immersed in slightly alkaline water to soften the binding material, sericin. The free silk ends of several cocoons are found using a brush and attached to a spool of the reeling machine. The silk threads then dried and re-reeled. The product thus obtained is raw silk which is an international trading commodity.

The sericin is removed during the degumming process. The cocoon pieces are boiled in a 0.02 M Na_2CO_3 solution for 20–30 min (Abbott et al., 2016). To obtain scaffold from the degummed silk, silk washed with distilled water is dissolved in a 9.3 M Lithium Bromide (LiBr) solution at 60 °C for 3–4 h (Abbott et al., 2016). The degummed silk is a raw material for a textile industry where a finished silk fabric is made. The basic principle of fabric weaving is the same for wool, silk and all other fibres.

8.3.2 Life cycle assessment of silk

The sustainability assessment studies have been reviewed. Like wool, sustainability studies were performed using ISO 14,040 guidelines by defining purpose, scope,

Wool and silk fibres from sustainable standpoint

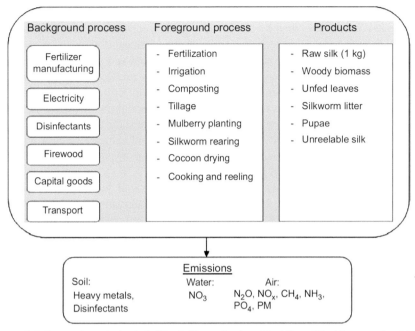

Figure 8.6 System boundaries in LCA of silk production (Karthik and Rathinamoorthy, 2017). with permission from Elsevier.

inventory followed by an impact assessment and data interpretation. Fig. 8.6 shows a typical example of system boundaries for LCA of silk production.

The energy demand is mainly due to electricity and irrigation. The cumulative energy demand of silk is highest among all-natural fibres (Astudillo et al., 2015). The environmental impact on a mass basis is again higher than the other natural fibres (Astudillo et al., 2014). It is because animal fibres need more energy than plant fibres (Astudillo et al., 2015). The reeling requires 3660–5440 KJ wool per kg of fresh cocoon (Mande et al., 2000). The silk degumming process has a maximum environmental impact, especially through water pollution. The pollution can be reduced by recovering sericin through ultrafiltration (Astudillo et al., 2015).

LCA of Indian silk was modelled using SimPro software (Astudillo et al., 2015). The environmental impact was indicated using several indicators such as global warming potential over 100 years (GWP_{100}), renewable and nonrenewable cumulative energy demand (CED), ecotoxicity, agricultural land occupation (ALO), freshwater eutrophication (FE), and blue water footprint (BWF). GWP_{100}, FE, and ecotoxicity are dominated by on-farm emissions. ALO, CED, and BWF are mostly related to mulberry production. The two most energy-consuming steps in silk manufacturing are cocoon production (47%) and cocoon cooking (51%) (Karthik and Rathinamoorthy, 2017). Mulberry cultivation mitigates around 76–81 tonnes of CO_2 per hectare per year. The CO_2-e mitigation is 735 times the weight of the silk fibre (Giacomin et al., 2017).

LCA study of Brazilian silk (Barcelos et al., 2020) also reported that mulberry production has more impact than cocoon production. Soil correction and maintenance was the key factor contributing to human ecotoxicity, freshwater ecotoxicity and freshwater eutrophication. The production of superphosphate and use of phosphorous as fertilizer and barn disinfection emit Sox and NOx in the air. However, mulberry fields can attenuate CO_2-e 735 times the weight of silk per unit of the cultivated area (Giacomin et al., 2017). This has to be taken into consideration while doing the impact assessment. Moreover, silk production requires significantly less capital investment compared to that of synthetic fibre since chemicals and chemical processing plants are not needed for silk. The only lacking point is the rate of production is slow rather not matching with the sped of human progress.

Fig. 8.7 shows LCA analysis of handloom and power loom woven silk fabric (Bhalla et al., 2020). It can be seen that mulberry cultivation needed the highest amount of energy (65%—76%). It is due to the high feed resources needed for mulberry cultivation and the use of fossil fuel in terms of pesticides and fertilizers. Power-loom weaving is the second most energy-consuming source due to the electricity requirement whereas; hand-woven silk fabric requires the least energy. Hence, hand-loom weaving is more sustainable and energy-efficient than the power-loom woven silk fabric. The third-largest energy consumption source is the garment during its use. A large number of petrochemicals and electricity are needed to dry clean, wash, dry, and iron silk fabrics.

8.3.3 Sustainability in silk processing

Natural colours and biomaterials can be added during the sericulture to avoid the dyeing of fabric at a later stage. A soil-based inoculum of Vesicular arbuscular mycorrhizae (VAM) (endomycorrhizae) with a mixed culture of Glomus mosseae and Glomus fasciculatum reduced use of phosphorus by 50% without any loss of yield

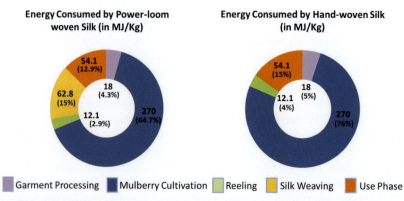

Figure 8.7 Embodied energy per kg fabric: power loom woven silk versus handloom-woven silk (Bhalla et al., 2020).
with permission from Elsevier.

and quality of mulberry leaves (Karthik and Rathinamoorthy, 2017). The dyes can be a part of a diet to obtain coloured silk (Karthik and Rathinamoorthy, 2017); however, the nature of the dye, cytotoxicity studies, and physiological responses need to be critically studied. Hence, such practices are not yet commercial.

Water consumption can be reduced up to 66% by using drip irrigation instead of furrows (Siddalingaswamy et al., 2007) during mulberry cultivation. Biofertilizer applications in a mulberry plantation will reduce the environmental impact. Cocoon drying can be done using solar drying which is an environmentally friendly and renewable energy source. Ultrasound-assisted dyeing and the use of biomordants in the silk fabric dyeing is sustainable, green, and energy-efficient approach (Adeel et al., 2019).

8.3.4 Utilization of silk as a sustainable material

The saris (traditional attire worn by a woman in India) made up of silk are very valuable textile products and are symbols of pride to the privileged classes. The market price of a single handloom silk sari is close to € 65 while that of a power-loom silk sari starts from € 9 (Bhalla et al., 2020). The elegance, drape, and feel of the handwoven silk saree are superior to power-loom-made saree. Beyond conventional textiles, silk is getting attention for developing advanced materials catering to the modern world.

Self-standing silk fibroin thin films have been used as a substrate in photonic and electronic devices (Vepari and Kaplan, 2007), as a biosensor in therapeutic applications such as bone growth (Omenetto and Kaplan, 2008; MacLeod and Rosei, 2013). Silk is a useful sustainable choice in optics, photonics, electronics and biomedical applications. Silk is a transparent, flat and strong material. It is a good material for planar micro-electronic structures (Wang et al., 2011, 2019; Capelli et al., 2011). Developing a biocompatible material has been a challenge due to the complexity of biological responses to various organic and inorganic materials. Silk is a good option as a biocompatible and bioresorbable polymer. It has good mechanical strength. The amino acid degradation products of silk are non-inflammatory. The rate of silk degradation can be modified. It can adapt to protect sensitive electronic functions. Hence, it has potentially useful in bioresorbable electronics (Müller et al., 2011).

Silk can be converted into many useful forms such as fibre, film, fabric, sponge, hydrogel, scaffold, tubes and composites as per the need in the tissue engineering structure (Abbott et al., 2016). For various unique tissue engineering systems like bone tissue, kidney tissue and cortical brain tissue, silk scaffold can be a common base material that provides sustainable development (Abbott et al., 2016).

Like keratin, fibroin of silk is a useful biomaterial. Silk fibroin can be generated using spin-coating, casting, layer-by-layer, and electro-spinning methods (Nguyen et al., 2019). Various structures like particulate, fibre, film and three-dimensional scaffolds have been prepared using silk fibroin. The aid of micro-patterning and 3-D printing can develop silk fibroin-based multi-level structures to mimic biological functions (Qi et al., 2017). In this context, silk has a high potential to remain a sustainable material.

Silk is a strong but biodegradable polymer. Silk can be degraded using enzymes. Silk is susceptible to biological degradation by proteolytic enzymes such as carboxylase, chymotrypsin and actinase (Cao and Wang, 2009). The enzymatic degradation of silk is a two-step process. First is the adsorption of enzymes on the polymer surface and the second is the hydrolysis of the ester bond. The degraded polymer, in form of amino acid, can be easily absorbed in vivo without any immunogenic response. Hence, silk has been the preferred choice in biomedical applications.

8.4 Concluding remarks

Sheep and moths, from which wool and silk are obtained, are part of the natural carbon cycle. They consume organic carbon in plants and convert it into protein fibres. From a sustainability standpoint, both wool and silk have distinct advantages over other fibre textiles. They have very long life spans, fewer washing cycles, the low temperature needed for dyeing and washing, are recyclable, biodegradable and do not contribute to micro-plastic pollution once degraded.

The sheep livestock sector for wool production is under stress due to carbon emission, competition for land and water resources, high fibre cost, low profits, environmental effects, and a very high production rate of synthetic fibres. The LCA studies indicated that wool contributes to GHG emissions. However, no LCA study till now has considered the contribution of carbon mitigation due to sheep. Sheep improves the soil fertility of the agricultural land and enriches the soil with nitrogen through urine and manure. Moreover, wool being a protein fibre can have an affinity towards aldehydes, thus, it can break down hazardous volatile organic compounds in air.

Silk, on the other hand, can reduce CO_2 concentration through the mulberry trees like any other tree soak CO_2. The life cycle of the silkworm is self-sustainable. The energy-intensive drawing and stretching, like in synthetic filaments, is not needed in the case of silk. The relative energy savings and GHG reduction should be considered or converted into equivalent economics.

The key challenge of LCA studies of wool and silk is determining the system boundary. Wool and silk are agricultural produce. They are processed in organized and disorganized industrial sectors. And their utilization is not limited to conventional textiles but also a large number of applications. The waste at each stage is not only recyclable but also yields some revenue. Hence, the entire chain is complex and needs to be carefully examined to avoid misleading interpretation. In most of the LCA studies, GHG emission during fibre production is derived. However, GHG emission over the entire life cycle of fibre (from production to disposal) remained unattended with the probable logic of endless applications of the produced fibre.

Low-emission grazing systems and sustainable management of the rangelands could significantly reduce the effect of climate change impacts (Harle et al., 2007). Effective land-use planning is necessary to meet future demands from the livestock sector. Holistic grazing management is a planned decision-making framework that ensures the decisions are environmentally, economically and socially sustainable. The holistic management practice is discovered by Allan Savory and Jody Butterfield. They successfully implemented the practices in southern Africa and America. It is mainly focused on the short-time intensive livestock rotational grazing within smaller plots of land. Sheep can till and aerate the soil, urinate and defecate in the small area within a short period. This provides nutrient cycling to animals and enrichment of the soil at the same time. The holistic management allows improving profits and livelihood, increasing carbon sequestration, reversing desertification, and improving biodiversity. Flora and fauna can thrive together with this approach without disregarding economic profits to the livestock sector (Butterfield et al., 2006).

For wool and silk sustainability, efforts should be made to reduce GHG emissions at farm and industry levels. And the efforts should be linked with productivity and profits to farm and industry. The economic drive may offset the emissions in the future. An increase in productivity and profitability in wool and silk is the primary factor that can make these fibres sustainable. The circular economy is the method by which by-products are produced from the end product. Wool and silk garments can be converted into protein fibres and can be used to develop second produce. This approach supports the reduce–recycle–reuse model. Wool and silk are perfect examples of circular economy which are robust even today.

Wool and silk are biodegradable, renewable, biocompatible and recyclable fibres. Some of their properties are unique and superior to all synthetic fibres. With the aid of biotechnology and nanotechnology, such properties can be harvested to accelerate newer and unconventional applications of these fibres. It can be a paradigm shift in wool and silk sustainability.

The social life cycle is assessed for the garment industry is assessed in terms of factors such as forced labour, child labour, local employment, and importing country environment (Lenzo et al., 2017). Large brands like NIKE, ADIDAS is using social LCA to address ecological sustainability. Brands are conveying the message that they are socially responsible brands. However, most of the textile apparel and textile industries in the developing world are of small- and medium-scale industries, environmental consciousness is not yet reached. Social LCA of wool and silk can reveal useful information which will help to identify the sustainability potential of these fibres.

The certification system such as organic certification and sustainability certification can be a drive to gain economic advantage and to improve the sustainability of wool and silk. For instance, Grassland Regeneration and Sustainability Standard (GRASS) certification of wool is a certification system in Argentina (Argyropoulos, 2014) that encourages sustainable grazing throughout the Patagonian Grasslands and creates a value chain that aims at maintaining and regenerating these fragile ecosystems through the consistent demand of wool in a sustainable manner. For carbon-neutral products, The Merino Company in Australia has developed a scheme to source ZeroCO$_2$ wool. Eco-labelling is also one of the options. The certification should have mentioned the summarized information about the CF of a product during manufacturing, use, and disposal.

To summarize, the economics and industrial developments of synthetic polymers put wool and silk under stress. The GHG emission and low production rate are the major concerns of wool and silk sustainability. However, the potential of carbon mitigation by wool and silk has been neglected in the LCA studies. Wool and silk are automatic choices for biomedical applications. Green processing and newer applications can thrive on wool and silk sustainability.

References

Abbott, R.D., Kimmerling, E.P., Cairns, D.M., Kaplan, D.L., 2016. Silk as a biomaterial to support long-term three-dimensional tissue cultures. ACS Applied Materials and Interfaces 8, 21861–21868.

Adeel, S., Rehman, F., Iqbal, M.U., Habib, N., Kiran, S., Zuber, M., Zia, K.M., Hameed, A., 2019. Ultrasonic assisted sustainable dyeing of mordanted silk fabric using arjun (Terminalia arjuna) bark extracts. Environmental Progress and Sustainable Energy 38, S331–S339.

Argyropoulos, N., 2014. An Economic Evaluation of Agricultural Management Systems in the Patagonian Grasslands: An Observation of Wool and the Link between Profitability and Conservation. Duke Univ.

Astudillo, M.F., Thalwitz, G., Vollrath, F., 2014. Life cycle assessment of Indian silk. Journal of Cleaner Production 81, 158–167.

Astudillo, M.F., Thalwitz, G., Vollrath, F., 2015. Life cycle assessment of silk production–a case study from India. In: Handbook of Life Cycle Assessment (LCA) of Textiles and Clothing. Elsevier, pp. 255–274.

Auber, L., 1952. VII.—the anatomy of follicles producing wool-Fibres, with special reference to keratinization. Earth and Environmental Science Transactions of the Royal Society of Edinburgh 62, 191–254.

Barber, A., Pellow, G., LCA, 2006. New Zealand merino wool total energy use. In: 5th Australian Life Cycle Assessment Society (ALCAS) Conference, Melbourne, pp. 22–24.

Barcelos, S.M.B.D., Salvador, R., Guedes, M. da G., de Francisco, A.C., 2020. Opportunities for improving the environmental profile of silk cocoon production under Brazilian conditions. Sustainability 12, 3214.

Barwick, S.A., Swan, A.A., Hermesch, S., Graser, H.U., 2011. Experience in breeding objectives for beef cattle, sheep and pigs, new developments and future needs. In: Proceedings of the Association for the Advancement of Animal Breeding and Genetics, pp. 23–30.

Bech, N.M., Birkved, M., Charnley, F., Laumann Kjaer, L., Pigosso, D.C.A., Hauschild, M.Z., McAloone, T.C., Moreno, M., 2019. Evaluating the environmental performance of a product/service-system business model for Merino Wool Next-to-Skin Garments: the case of Armadillo Merino. Sustainability 11, 5854.

Berry, P., 2015. Sheep meat and wool: outlook to 2019-20, agric. Commod 5, 112.

Bhalla, K., Kumar, T., Rangaswamy, J., Siva, R., Mishra, V., 2020. Life cycle assessment of traditional handloom silk as against power-loom silks: a comparison of socio-economic and environmental impacts. In: Green Buildings and Sustainable Engineering. Springer, pp. 283–294.

Brock, P.M., Graham, P., Madden, P., Alcock, D.J., 2013. Greenhouse gas emissions profile for 1 kg of wool produced in the yass region, New South Wales: a life cycle assessment approach. Animal Production Science 53, 495–508.

Brundtland, G.H., Khalid, M., Agnelli, S., Al-Athel, S., Chidzero, B., 1987. Our Common Future, New York, p. 8.

Butterfield, J., Bingham, S., Savory, A., 2006. Holistic Management Handbook: Healthy Land, Healthy Profits. Island press.

Cao, Y., Wang, B., 2009. Biodegradation of silk biomaterials. International Journal of Molecular Sciences 10, 1514−1524.

Capelli, R., Amsden, J.J., Generali, G., Toffanin, S., Benfenati, V., Muccini, M., Kaplan, D.L., Omenetto, F.G., Zamboni, R., 2011. Integration of silk protein in organic and light-emitting transistors. Organic Electronics 12, 1146−1151.

Costanza, R., Patten, B.C., 1995. Defining and predicting sustainability. Ecological Economics 15, 193−196.

Cottle, D.J., Cowie, A.L., 2016. Allocation of greenhouse gas production between wool and meat in the life cycle assessment of Australian sheep production. International Journal of Life Cycle Assessment 21, 820−830.

Dénes, O., Florea, I., Manea, D.L., 2019. Utilization of sheep wool as a building material. Procedia Manufacturing 32, 236−241.

Giacomin, A.M., Garcia, J.B., Zonatti, W.F., Silva-Santos, M.C., Laktim, M.C., Baruque-Ramos, J., 2017. Silk industry and carbon footprint mitigation. In: IOP Conference Series: Material Science and Engineering. IOP Publishing, p. 192008.

Gogoi, M., Kadam, V., Jose, S., Shakyawar, D.B., Kalita, B., 2020. Multifunctional finishing of woolens with lemongrass oil. Journal of Natural Fibers 1−13.

Goodland, R., 1995. The concept of environmental sustainability. Annual Review of Ecology and Systematics 26, 1−24.

Gowane, G.R., Gadekar, Y.P., Prakash, V., Kadam, V., Chopra, A., Prince, L.L.L., 2017. Climate Change Impact on Sheep Production: Growth, Milk, Wool, and Meat. https://doi.org/10.1007/978-981-10-4714-5_2.

Harle, K.J., Howden, S.M., Hunt, L.P., Dunlop, M., 2007. The potential impact of climate change on the Australian wool industry by 2030. Agricultural Systems 93, 61−89.

Hassan, M.M., Carr, C.M., 2019. A review of the sustainable methods in imparting shrink resistance to wool fabrics. Journal of Advanced Research 18, 39−60.

IWTO, 2020. World Fibre Production and Consumption. https://iwto.org/wp-content/uploads/2021/03/202103. https://iwto.org/wp-content/uploads/2021/03/20210315_IWTO_MI_DigitalSample.pdfIWTO.

Jones, A.K., Jones, D.L., Cross, P., 2014. The carbon footprint of lamb: sources of variation and opportunities for mitigation. Agricultural Systems 123, 97−107.

Kadam, V.V., Goud, V., Shakyawar, D.B., 2013. Ultrasound scouring of wool and its effects on fibre quality. Indian Journal of Fibre and Textile Research 38.

Kadam, V., Rani, S., Jose, S., Shakyawar, D.B., Shanmugam, N., 2021. Biomaterial based shrink resist treatment of wool fabric: a sustainable technology. Sustainable Materials and Technology e00298.

Karthik, T., Rathinamoorthy, R., 2017. Sustainable silk production. In: Sustainable Fibres and Textiles. Elsevier, pp. 135−170.

Kaur, A., Chakraborty, J.N., 2015. Controlled eco-friendly shrink-resist finishing of wool using bromelain. Journal of Cleaner Production 108, 503−513.

Kaur, A., Chakraborty, J.N., Dubey, K.K., 2016. Enzymatic functionalization of wool for felting shrink-resistance. Journal of Natural Fibers 13, 437−450.

Kumar, A., Pareek, P.K., Kadam, V.V., Shakyawar, D.B., 2016. Anti-moth efficacy of neem (Azadirachta indica A.Juss.) on woollen fabric. Indian Journal of Traditional Knowledge 15.

Kumar, A., Pareek, P.K., Shakyawar, D.B., V Kadam, V., 2017. Colourless natural antimoth agents for woollens. Indian Journal of Small Ruminants 23, 73−76.

Lenzo, P., Traverso, M., Salomone, R., Ioppolo, G., 2017. Social life cycle assessment in the textile sector: an Italian case study. Sustainability 9, 2092.

MacLeod, J., Rosei, F., 2013. Sustainable sensors from silk. Nature Materials 12, 98–100.

Maia Pederneiras, C., Veiga, R., de Brito, J., 2019. Rendering mortars reinforced with natural sheep's wool fibers. Materials 12, 3648.

Mande, S., Pai, B.R., Kishore, V.V.N., 2000. Study of stoves used in the silk-reeling industry. Biomass and Bioenergy 19, 51–61.

Marino, R., Atzori, A.S., D'Andrea, M., Iovane, G., Trabalza-Marinucci, M., Rinaldi, L., 2016. Climate change: production performance, health issues, greenhouse gas emissions and mitigation strategies in sheep and goat farming. Small Ruminant Research 135, 50–59.

Müller, C., Hamedi, M., Karlsson, R., Jansson, R., Marcilla, R., Hedhammar, M., Inganäs, O., 2011. Woven electrochemical transistors on silk fibers. Advanced Materials 23, 898–901.

Nguyen, T.P., Nguyen, Q.V., Nguyen, V.-H., Le, T.-H., Huynh, V.Q.N., Vo, D.-V.N., Trinh, Q.T., Kim, S.Y., Van Le, Q., 2019. Silk fibroin-based biomaterials for biomedical applications: a review. Polymers 11, 1933.

Omenetto, F.G., Kaplan, D.L., 2008. A new route for silk. Nature Photonics 2, 641–643.

Parlato, M., Porto, S., 2020. Organized framework of main possible applications of sheep wool fibers in building components. Sustainability 12, 761.

Peri, P.L., Rosas, Y.M., Ladd, B., Díaz-Delgado, R., Martinez Pastur, G., 2020. Carbon footprint of lamb and wool production at farm gate and the regional scale in southern patagonia. Sustainability 12, 3077.

Prasad, V.L.M., Ramakrishna, N., Usha, R., Raj, M.P.G., 2010. Analysis of utilization pattern of seri-byproducts among sericulturists in Kolar district. Agriculture Update 5, 274–276.

Qi, Y., Wang, H., Wei, K., Yang, Y., Zheng, R.-Y., Kim, I.S., Zhang, K.-Q., 2017. A review of structure construction of silk fibroin biomaterials from single structures to multi-level structures. International Journal of Molecular Sciences 18, 237.

Rajabinejad, H., Bucişcanu, I.-I., Maier, S.-S., 2016. Practical ways of extracting keratin from keratinous wastes and by-products: a review. Environmental Engineering and Management Journal 15.

Rani, S., Kadam, V., Rose, N.M., Jose, S., Yadav, S., Shakyawar, D.B., 2020. Wheat starch, gum Arabic and chitosan biopolymer treatment of wool fabric for improved shrink resistance finishing. International Journal of Biological Macromolecules 163, 1044–1052. https://doi.org/10.1016/j.ijbiomac.2020.07.061.

Rani, S., Kadam, V., Rose, N.M., Jose, S., Shakyawar, D.B., Yadav, S., 2021. Effect of enzyme treatment on wool fabric properties and dimensional stability. Indian Journal of Fibre and Textile Research 46, 83–90.

Reddy, C.C., Khilji, I.A., Gupta, A., Bhuyar, P., Mahmood, S., AL-Japairai, K.A.S., Chua, G.K., 2021. Valorization of keratin waste biomass and its potential applications. Journal of Water Process Engineering 40, 101707.

Russell, I.M., 2009. Sustainable wool production and processing. Sustainable Textiles 63–87.

Sannapapamma, K.J., Naik, S.D., 2015. Ahimsa silk union fabrics—A novel enterprise for handloom sector. Indian Journal of Traditional Knowledge 14, 488–492.

Shakyawar, D.B., Shanmugam, N., Kumar, A., V Kadam, V., Jose, S., 2018. Utilization of Indian Wool in Decentralized Sector an Overview.

Shanmugam, N., Shakyawar, D.B., Kumar, A., V Kadam, V., Jose, S., 2020. Water absorption and dynamic load bearing properties of coarse wool braided rope mat. Indian Journal of Small Ruminants 26, 225–229.

Siddalingaswamy, N., Bongale, U.D., Dandin, S.B., Narayanagowda, S.N., Shivaprakash, R.M., 2007. A study on the efficiency of micro-irrigation systems on growth, yield and quality of mulberry. Indian Journal of Sericulture 46, 76−79.

Steinfeld, H., Gerber, P.J., Wassenaar, T., Castel, V., Rosales, M., De haan, C., 2006. Livestock's Long Shad. Livestock, Environ. Dev. Initiat. (LEAD). FAO Rome, Italy. http://www.fao.org/docrep/010/a0701e/a0701e00.HTM (accessed June 6, 2021).

Stenton, M., Houghton, J.A., Kapsali, V., Blackburn, R.S., 2021. The potential for regenerated protein fibres within a circular economy: lessons from the past can inform sustainable innovation in the textiles industry. Sustainability 13, 2328.

Thornton, P.K., van de Steeg, J., Notenbaert, A., Herrero, M., 2009. The impacts of climate change on livestock and livestock systems in developing countries: a review of what we know and what we need to know. Agricultural Systems 101, 113−127.

Tinti, F., Barbaresi, A., Benni, S., Torreggiani, D., Bruno, R., Tassinari, P., 2015. Experimental analysis of thermal interaction between wine cellar and underground. Energy and Buildings 104, 275−286.

Kadam, V.V., Meena, L.R., Singh, S., Shakyawar, D.B., Naqvi, S.M.K., 2014. Utilization of coarse wool in agriculture for soil moisture conservation. Indian Journal of Small Ruminants 20, 83−86.

Vepari, C., Kaplan, D.L., 2007. Silk as a biomaterial. Progress in Polymer Science 32, 991−1007.

Volf, M., Diviš, J., Havlík, F., 2015. Thermal, moisture and biological behaviour of natural insulating materials. Energy Procedia 78, 1599−1604.

Wang, C., Hsieh, C., Hwang, J., 2011. Flexible organic thin-film transistors with silk fibroin as the gate dielectric. Advanced Materials 23, 1630−1634.

Wang, C., Xia, K., Zhang, Y., Kaplan, D.L., 2019. Silk-based advanced materials for soft electronics. Accounts of Chemical Research 52, 2916−2927.

Xia, Q., Guo, Y., Zhang, Z., Li, D., Xuan, Z., Li, Z., Dai, F., Li, Y., Cheng, D., Li, R., 2009. Complete resequencing of 40 genomes reveals domestication events and genes in silkworm (Bombyx). Science (80-.) 326, 433−436.

Yu, D., Cai, J.Y., Church, J.S., Wang, L., 2015. Click chemistry modification of natural keratin fibers for sustained shrink-resist performance. International Journal of Biological Macromolecules 78, 32−38.

Zach, J., Hroudova, J., Brozovsky, J., 2014. Study of hydrothermal behavior of thermal insulating materials based on natural fibers. International Journal of Civil, Environmental, Structural, Construction and Architectural Engineering 8, 995−998.

Sustainable protein fibres

9

Asim Kumar Roy Choudhury[1,2]
[1]KPS Institute of Polytechnic, Belmuri, West Bengal, India; [2]Govt. College of Engineering and Textile Technology, Hooghly, West Bengal, India

9.1 Introduction

In ecology, sustainability (from sustain and ability) is the property of biological systems to remain diverse and productive indefinitely. Long-lived and healthy wetlands and forests are examples of sustainable biological systems. In more general terms, sustainability is the endurance of systems and processes. The organizing principle for sustainability is sustainable development, which includes the four interconnected domains: ecology, economics, politics and culture. Sustainability science is the study of sustainable development and environmental science.

Sustainability, and/or sustainable development, refers to those conditions which allow creating a balanced production system and guarantee that the needs of today's generations are met, without compromising the opportunities of future generations. Three key factors of sustainable development were identified: environmental protection, economic growth and social development (WSSD, 2002).

The concept of sustainability/sustainable development, which was born as a paradigm focused merely on ecological aspects, transformed over the years into a dynamic concept that factors in several environmental, economic and social variables.

Today's fibre production strategy is redirected from crude oil to renewable raw materials, eco-friendly and sustainable fibres, that could be biodegraded or recycled. Important raw materials for future textile fibres production could be cheap and worldwide available agricultural by-products, like lignocellulose (from rice straw), wheat gluten, casein protein from milk after butterfat is removed, zein protein from corn after starch manufacture, and soybean protein after beans are pressed and oil is removed (Rijavec and Zupin, 2011).

The scale of human activity has become so large that it is altering ecosystems faster than the possible sustainability model. This fact threatens the integrity of ecosystems and their capacity to provide quality living for future generations (Alves et al., 2009). Demand from consumers for eco-friendly products is growing stronger. Thus, the improvement of environmental performance of products is the main focus for many companies. While organic fibres are meeting part of this need, they are unlikely to be produced in sufficient quantities to meet all the demand for eco-friendly fibre (Poole et al., 2009).

New biodegradable fibres should compete well with organic and other eco-friendly fibres based on their environmental credentials. Once produced, they are able to be processed on conventional textile machinery. They can be used with blending various

---

Sustainable Fibres for Fashion and Textile Manufacturing. https://doi.org/10.1016/B978-0-12-824052-6.00010-X
Copyright © 2023 Elsevier Ltd. All rights reserved.

fibres types and have a large application area. Potential product markets where they may have a competitive advantage are eco-friendly apparel as well as technical and industrial applications.

Globally natural fibres contribute about 48% to the fibre basket with 38% from cotton, 8% from bast and allied fibres and 2% from wool and silk fibres (animal fibres). Different types of animal fibres are available in various corners of this Globe, but wool & silk are the most important. In spite of the very low quantity of production, animal fibres play a significant contribution to in textile production, especially for high-quality winter and luxury garments.

Animal fibres are largely made of protein. The protein fibres can be also be further classified as:

- Animal protein fibres, for example, wool, silk, various hair fibres such as cashmere goat, angora rabbit wool, mohair, alpaca hair, casein from waste cow milk, etc.
- Vegetable regenerated protein fibres, for example, soybean, ground nut, etc.

Wool, silk, mohair, cashmere, etc. are animal protein fibres. Although animal protein fibres such as wool and silk have good physical properties and have been used extensively in the textile industry, they are relatively expensive to use and process. In silk, a large quantity of mulberry leaves is required for the production of a very small quantity of silk resulting in an increased cost of production. In addition, apart from the economic aspects, animal fibres are physically limited in several aspects. Firstly, both wool and silk fibre vary in diameter and their performance profile is limited. Secondly, morphologically, the presence of scales on wool surface results in felting shrinkage and difficulties in dyeing. In contrast, being manmade, regenerated vegetable protein fibres, such as soybean fibre, casein fibre, electro-spun fibre do not have a theoretical limit in fineness to which fibres may be drawn.

In general, protein fibres have moderate strength, resiliency and elasticity. Their moisture absorbency and transport characteristic are excellent. They do not produce static charge. While they are resistant to acids, they are easily attacked by bases and oxidizing agents. They tend to yellow colour on exposure to sunlight due to oxidative attack.

9.2 Animal protein fibres

Animal fibres are the natural fibres that can be sourced from animals. Animal fibres that are extracted from different animals usually have different properties. Furthermore, the types of fibres may also vary from species to species.

After plant fibres, animal-originated fibres are the next in their usage for composite reinforcements. Generally, the animal fibres are made up of protein. Examples include wool, human hair feathers and silk. Animal fibres are more expensive than plant fibres because they are not as readily available as plant fibres, which consequently limit their utilization in many applications. However, they have an edge over the plant fibres because they are not seasonal as the latter (Njoku et al., 2019).

Composites reinforced with animal fibres have found wide applications in biomedical procedures such as implants, bone grafting, sutures and medical devices (Sanjay et al., 2016). Wool or hair fibres obtained from mammals or animals include camel hair, goat hair (cashmere and mohair), sheep wool, angora rabbit hair, alpaca hair (related to camel) and yak hair (Ryder, 1992). The hair fibres are named according to their origin, and their properties differ. A fairly fine soft fibre is obtained from alpaca while a highly prized fine short hair is obtained from cashmere. The undercoat of the camel gives a strong soft fibre, but the outer coat produces a strong coarse fibre. A long lustrous, springy fibre is got from the angora goat (mohair); fine, short and long spiky hairs from angora rabbit; and soft fine hair from yak (Ryder, 1992).

Njoku et al. (2019) reviewed the property and application of natural fibres as reinforcement in the development of composite systems for structural, semi-structural and technological materials.

All animal fibres consist exclusively of proteins and with the exception of silk constitute the fur or hair of animals. They are formed by the natural animal source through condensation of α-amino acids joined together by peptide linkages to build repeating units of polyamide with many substituents on a α-carbon atom. The sequence and the type of the amino acids bonding together the individual protein chains contribute to the overall properties and characteristics of the resultant fibres. There are two major classes of natural animal protein fibres exist. They are as follows:

- Keratin (hair or fur) fibres such as wool from sheep, Cashmere from goat, angora wool from rabbit, alpaca hair from alpaca animal.
- Secreted (insect) fibres such as silk. Domestic silk is known as mulberry silk, while wild silk, originally produced in forest and now cultivated to some extent known as *Tasar* or *Tussah* (*Antheraea mylitta*), Eri (*Philosamia ricini*) and Muga (*Antheraea assamensis*).

Table 9.1 describes microscopic views of wool and silk fibres (Roy Choudhury, 2006). The major difference between the above two is that the keratin fibre proteins are highly cross-linked by disulphide bonds, whereas the secreted silk fibroin fibres tend to have no cross-links and a more limited array of less complex amino acids.

Wool and all other mammalian hairs are animal protein fibres. They are fibrous, α-keratinous, nano-composite materials and as such part of a larger group of biological, functional materials, referred to as keratins.

Table 9.1 Microscopic views of wool, silk (raw) and degummed (treated with soap and soda ash).

Fibre	Longitudinal	Cross-sectional
Wool	Crimped solid rod with fish-like scales on the surface.	Circular or elliptical, variable in diameter.
Raw silk	Rough filaments cemented in pairs, separated in places.	Triangular with rounded corners, in pairs.
Degummed silk	Structure-less, like a hollow glass rod.	Separated fibres, triangular, rounded corners.

Wool and hair have many general properties in common, making them desirable for use in a variety of garments:

- They are generally durable, flame retardant and water repellent.
- They offer good insulation due to their moisture-wicking properties and ability to trap air.
- The bilateral core of keratin causes the fibres to twist and bend, giving wool its natural crimp and resilience.
- They take dye extremely well and both sheep's wool and hair from alpacas offer a wide range of natural colours that require no additional dyeing.
- They generally have a small environmental impact as compared to using synthetic fibres.
- Wool and hair are easily renewable and recyclable.

And while it requires both food, energy, water and medicine to keep animals reared, the environmental impact naturally varies from factory to factory, with some wool and hair processing being organic and free from pesticides and chemicals.

The disadvantages of animal protein fibres are

- One drawback, however, is that wool and hair become weaker when exposed to water, with wool losing about a quarter of its strength when wet.
- Another disadvantage of wool and hair (and other natural protein fibres) is their tendency to become moth food, a problem not shared by synthetics.

That being said, the beneficial properties of wool and hair arguably make it unrivalled by any man-made fibre known today.

9.2.1 Sustainable wool fibres

Wool is a 100% natural fibre. Natural fibres are fibres that come from animals or plants and are totally biodegradable. Wool is made of a natural protein called keratin, similar to that found in human hair. Synthetic fibres on the other hand are man-made and often times petroleum-based.

Wool is renewable. Sheep grow wool continuously making wool a renewable fibre. Sheep produce a new fleece every year, which can be shorn off again the following year.

Wool is naturally an organic fibre, and it has inherent characteristics that make it renewable and biodegradable. That applies to all woollen products. However, within the diverse world of wool, there is also a distinct category that identifies what is officially certified as 'organic wool', which relates to the manner in which the material has been farmed. At present by 'Organic fibre' we mean the materials which have not come across any synthetic and toxic chemicals during their production and processing.

In addition to wool fibre, sheep farming commonly produces meat and sometimes other products such as milk and skins. Where multiple products share the same process, as for wool and meat from sheep, an allocation method is required to partition the environmental impacts between them wherever the system cannot be expanded to account for all products.

9.2.1.1 Organic wool

Organic wool is produced without involving many widely used chemicals, including commonly used veterinary medicines such as preventative treatments against lice and flies and internal parasites. The inability to provide sheep with modern veterinary chemicals and medicines to prevent ill health is a major reason why organic wool is not widely produced and supplies are limited.

Organic textile fibres may include cotton, wool, hemp, flax (linen) and other natural fibres grown according to national organic standards without the use of toxic and persistent pesticides, synthetic fertilizers or genetic engineering. The accredited third-party certification organizations verify that organic producers use only permissible methods and materials in organic production (Organic Trade Association, www.ota.com/).

To be classed as organic, wool must have been sheared from sheep given organic feed and raised without the use of hormones or pesticides. Organic livestock management is different from non-organic management in at least two major ways:

- Sheep cannot be dipped in parasiticides (insecticides) to control external parasites such as ticks and lice.
- Organic livestock producers are required to ensure that they do not exceed the natural carrying capacity of the land on which their animals graze.

This poses problems in the prevention of blowfly (meat-seeking fly strike) on sheep, when the usual sheep dipping is not allowed. In many countries, sheep are dipped in organophosphate or synthetic pyrethroid types of pesticide (Roy Choudhury, 2013).

Wool has a long lifespan. Laboratory tests have shown that wool fibres resist tearing and can bend back on themselves more than 20,000 times without breaking. To make a comparison, cotton breaks after 3200 bends. Due to the long lifespan, wool products are used or worn longer than other textile fibre products.

Conventional wool is far from being eco-friendly as we would expect. Certified (mostly by GOTS) organic wool guaranties that pesticides and paraciticides are not used in pasture land or on the ship themselves and that good cultural and management practices of livestock are used.

Unlike other animals, most sheep are unable to shed. Shearing sheep is usually carried out in the spring so sheep don't become overheated during the summer. If sheep are not shorn, the excess wool impedes the ability of sheep to regulate their body temperatures which could cause sheep to become overheated. Most sheep breeds need to be sheared at least once a year, although some breeds have wool that grows faster and need more frequent shearing.

Wool has water repellent properties and is resistant to dirt, meaning that wool textile products tend to be washed less frequently and at lower temperatures, resulting in a lower impact on the environment.

When wool is being exposed to moisture for long periods, for example, in soil or compost, the fibre will readily decompose. The wool products are completely biodegradable. In conclusion, wool can be seen as a sustainable and planet-friendly fibre.

9.2.2 Cashmere wool

Cashmere wool, usually simply known as cashmere, is a fibre obtained from cashmere goats, pashmina goats and some other breeds of goat. It has been used to make yarn, textiles and clothing for hundreds of years. Common usage defines the fibre as wool, but it is finer, stronger, lighter, softer and approximately three times more insulating than sheep wool.

The reason that the fabric is **so** pricey is that Kashmir goats only produce around 113 g of cashmere fibre each annually. It takes around two goats every year to produce enough fibre to make a single jumper. Clothes made from cashmere are up to eight times warmer than sheep's **wool** and are extremely soft to touch.

Cashmere is perhaps the most widely recognized of all the luxury fibres. Cashmere wool is the best sustainable and renewable fibre with virtues to protect the user from the surrounding rudiments. Fibres from these garments will not peel and will retain their form for many years, even for generations. More than 3000 tonnes of cashmere are made every year with the majority of them from Mongolia, followed by Australia, New Zealand, Iran and Afghanistan (Franck, 2001).

Cashmere is prized for its exceptionally fine texture and is strong, ultra-light and soft. It provides a superior insulative function without bulkiness, and the fibres are highly adaptable and easily spun. Like wool, cashmere has a high moisture content that allows the insulating properties to change with the relative humidity in the air. It is, however, weight for weight, warmer than wool. The finest Cashmere is obtained from the neck area of the goat.

The ever-growing lust for the beautiful and soft cashmere fabrics encouraged the breeding of cashmere goats. During the past century, the production of cashmere wool has increased up to the extent that it has become unsustainable and is posing a threat to the environment. What was once beautiful, unspoilt grasslands are now becoming deserts, ravished by the goats breed for their cashmere clip. This is now creating a devastating effect on the ecological balance of the planet. The impact is more visible in Mongolia. The country produces 90% of the cashmere fabrics sold worldwide (Thangavel et al., 2015).

Mohair is obtained from the Angora goat (sometimes confused with the angora yarn that comes from the angora rabbit). It has a larger undercoat compared to the cashmere goat, but the guard hairs from the topcoat are, unlike with cashmere, often mixed with the hairs from the undercoat. This gives mohair its distinct, frizzy look with the slightly stiff short hairs visible in the final product.

9.2.3 Angora wool

The wool of Angora (Ankara) rabbit is known as Angora wool or Angora originating in Asia Minor and currently bred in Europe, China and Americas. Angora wool is produced only by Angora rabbits, and it is the lightest natural fibre exhibiting original qualities of fineness, lustre, and feel, for the production of high value-added luxury items.

Angora is known for its softness, thin fibres and what knitters refer to as a halo (fluffiness). It is also known for its silky texture. It is much warmer and lighter than wool due to the hollow core of the angora fibre.

Angora is often considered one of the 'noble' fibres. Angora (Ankara) rabbit is the only animal breed that produces the finest and the longest white silky wool amongst other wool-producing animal breeds such as sheep, goat, lama, alpaca, and camel each at outputs of 5000—30,000 tonnes annually (Franck, 2001).

Angora hair is unusually long owing to the prolongation of the active phase of the hair follicle cycle: the hair grows for approximately 14 weeks, whereas that of the rabbit with ordinary (short) hair grows at the same rate but for only 5 weeks. This is due to the presence of a recessive gene in Angora rabbits. The interval between hair collections is a decisive factor in hair length. Though the period of shearing varies due to the length of fibre that is targeted, the Angora (Ankara) rabbits are generally shorn every 3 months, 4 times in a year; which approximates to a total amount of 1 kg. As the Angora wool is shorn from the rabbits, there is no hardship to the rabbits and no rabbits die in the production of Angora wool. The length of Angora hair accounts for its textile value, because it permits cohesion in the thread (Oijala, 2014).

Wool is obtained through different shearing or collection methods such as electric or manual shears, scissors, plucking or depilation. Most commonly used hair collection method is shearing. It is preferred due to its advantages; it is less stressful for the rabbit, less time, and labour-consuming, provides protection against cold, and provides possibility of obtaining more wool through shorter shearing intervals. It takes 10—20 min to shear an Angora (Ankara) rabbit.

- Clipping method (with scissors) increases the amount of sheer wool (less than 10 mm length) because of the post-shearing corrections. It is very important that the skin should not be harmed during clipping. Especially, the nipples are very sensitive to injuries.
- While plucking thick ended, immature hair is collected. This process takes 30—40 min. In China, wool is hand plucked to obtain the maximum amount of wool.
- Depilation has long been the technique of choice in France, synchronizing the reactivation of hair follicles with a well-structured coat with good guide hairs.

A point to be considered very carefully is that Angora (Ankara) rabbit production is labour-intensive and also requires great expertise. The slightest mistake may result in the loss of productive adults: the animals have to be over a year old to return a profit. Hair collection is always a delicate operation and careless sorting irredeemably downgrades the product. Above all, not all climates are suitable: excessive heat and intense light are very bad elements for especially albinos. In cold countries, or in countries with cold winters, the solution is to use buildings that shelter the animals against the rigours of the winter, and to regulate the temperature of the interiors. Recently, denuded animals require special care, however. The feed requirements of Angora (Ankara) rabbits are also very important: a poor deficient diet will always mean qualitatively and quantitatively poor hair production.

The quantity and quality of wool are primarily very much dependent on genes and inheritance of the breed; though factors including feeding, hygiene, age, sex, weight, season, climate and pregnancy also affect the production and quality of wool.

Therefore, selection of pure breed Angora rabbits with high wool production capacity is essential for the sustainability of high-quality wool standard (Hoskins, 2014).

Angora fibres are short, and therefore usually mixed with other soft fibres, such as cashmere and lamb's wool. After processing and spinning the yarn may be used (as with other animal fibres), for blends and also to create novelty effects in woven fabrics, but generally angora is more popular for knitwear yarns (Franck, 2001). Angora is generally viewed as a luxury fibre, and most angora wool products are very expensive, which is reflective of the laborious harvesting process and the small number of cottage industry-style producers. China currently dominates world angora production and is responsible for over 80% of the 3000 tonnes global yield (Thangavel et al., 2015).

9.2.4 Alpaca fibre (hair)

The alpaca (*Vicugna pacos*) is a species of South American camelid mammal. It is similar to, and often confused with, the llama. However, alpacas are often noticeably smaller than llamas. The two animals are closely related and can successfully crossbreed. Both species are believed to have been domesticated from their wild relatives, the vicuña and guanaco. There are two breeds of alpaca: the Suri alpaca and the Huacaya alpaca. The Suri Alpaca, in particular, is a very rare breed, yielding some of the most exclusive alpaca. An average Alpaca will yield approximately 3.5 kilos in a year.

Alpacas are kept in herds that graze on the level heights of the Andes of Southern Peru, Western Bolivia, Ecuador and Northern Chile at an altitude of 3500–5000 m (11,000–16,000 feet) above sea level.

Alpaca hair comes from the Alpaca that produces some of the finest hair available. Ranging from 15 to 40 microns, alpaca fibres can be very fine and soft, but are generally quite itchy at or over 30 microns, and thus less likely to be used for clothing. Generally a little bit stiffer than merino or cashmere, alpaca fibres are sometimes blended with wools like Merino to improve its draping qualities.

Huacayas have crimpy, bundled fibre and a fluffy look when in full fleece. Suri fleece falls in long, lustrous locks or ringlets which move freely as the animal walks. The microscope shows that the height of the scale on suri fibre is lower than that of huacaya. The higher scale on a huacaya fibre also has a steeper edge angle than that of a suri scale. This gives suri fibre a slicker, softer hand than huacaya.

Both kinds of alpaca can be processed in both worsted and woollen methods and both can be woven, knitted, crocheted and felted. Crimpy huacaya makes fabulous, lofty yarn for knitted and crocheted applications. The rarer suri is ideal for sensuous, drapable, incredibly luxurious woven fabrics. Both can be felted, though the nature of the scale structure generally makes felting suri a longer process than felting huacaya.

9.2.4.1 Warmth

Alpaca is known to be extremely warm. One reason is that alpaca has some medullated fibres: in other words, there are tiny hollow pockets in the centres of many individual alpaca fibres. These areas hold the warmth.

Sustainable protein fibres

One study showed that if worn in a 0°F environment, alpaca gives a 50°F comfort range, sheep's wool gives a 30°F and synthetics provide a 20°F comfort range in the same environment (Ziek, 2021).

9.2.4.2 Strength

Alpaca fibre is extremely strong. The average tensile strength of alpaca is 50 N/ktex; 30 N/ktex is considered adequate to run on modern mill machinery. Other sources say alpaca is the strongest mammal fibre. One historian wrote that the Incans braided alpaca/llama fibre with reeds or cotton to make bridges spanning canyons in the Peruvian Andes. Tensile strength makes for strong yarn and a garment that will wear well.

9.2.4.3 Abrasion resistance

The abrasion resistance of alpaca is about the same as that of other mammal fibres. As reported to The Alpaca Fibre Symposium in June 2010 alpaca had a score of 15,000 cycles of abrasion. This meets upholstery standards.

9.2.4.4 Resistance to compression

The alpaca fibre has low resistance to compression and is thus not well suited to uses requiring resistance to compression or high bulk unless blended with fibre that has excellent compression resistance.

9.2.4.5 Water resistance

Alpaca is very water-resistant. Several tests showed alpaca to be virtually water repellent. It was found all but impossible to saturate alpaca fibre to do the test (sheep's wool absorbs up to 35% of its weight in water).

Alpaca wicks moisture away from the body in knitted and woven (but not felted) fabric. This wicking characteristic is important in socks for diabetics. Many fans of alpaca socks reported that the socks are very warm and that their feet don't feel sweaty while wearing them. Thus, while wearing alpaca socks, one's feet remain comfortable in cold, damp conditions.

In the mid-nineteenth century, alpaca umbrellas were considered as best umbrellas in rainy England. They were considered much more desirable than the cheaper silk variety! Other common uses for alpaca were ladies' dresses, gentlemen's suits, military and police and academic robes.

The moisture regain of alpaca and cotton is 8%, silk 9% and wool is 16%. The fact that alpaca has half the moisture retention as wool may be why alpaca is judged to be more comfortable and breathable by wearers.

9.2.4.6 Simple preparatory process

Alpaca does not contain lanolin, making it possible to process without the kind of scouring that sheep's wool requires. It is also a reason alpaca can have an

87%−95% clean fibre yield whereas sheep tend to have a 43%−76% clean fibre yield. Like hair from goats and rabbits, it needs less preparatory steps compared to wool.

Low allergenic

Alpaca tends to be low allergenic. Many people who cannot wear wool, report that they can wear alpaca with no allergic reaction. In some people, this may be because there is no lanolin or lanolin residue in an alpaca garment.

Odour resistance

Alpaca is reported to resist odours better than other fibres, even in socks.

9.2.4.7 Prickle factor

Alpaca tends to have a somewhat lower prickle factor than sheep's wool of the same micron. This is because the individual scales on the shafts of alpaca fibres are smoother and lower than those on the fibres of other mammals. (Scale height on huacaya alpacas is 0.4 microns; that of sheep is 0.8. Suri alpacas' scale height is much lower than the 0.4 of huacayas.) This means that the handle of alpaca feels about 3−5 microns less than that of wool of the same micron. This gives a luxurious feel when you stroke alpaca.

The structure of the scales on alpaca fibre is also a reason that garments made from alpaca have good pill resistance. A test of the Alpaca Blanket project showed alpaca scored three on a pill resistance test (no pill = 5; excessive pill = 1). Three is considered good pill resistance in woven goods.

9.2.4.8 Shine

The scales on alpaca fibre are longer than those on sheep's wool. Light reflects off the scales. This results in the shine or glow alpaca fibre is famous for; the longer the scale (suri is longer than huacaya), the more it reflects light.

9.2.4.9 Flame resistance

Alpaca is a class I Fibre regarding flame resistance—more flame resistant than plant or synthetic fibres. It is marginally flame retardant which means it will self-extinguish. It does not melt onto the skin like synthetics do. These characteristics suggest it has potential for use in blankets, insulation, mattress stuffing and industrial uses for fire fighters or in the military.

9.2.4.10 Wrinkle and shrink resistance

Some textile experts say alpaca is more resilient and wrinkle resistant than cashmere. They also suggest that it has a lower tendency to shrink than wool and cashmere.

9.2.4.11 Durability

People report buying an alpaca sweater 30 or 40 years ago and still wearing it or say that it has been passed down through the generations and is still worn! They say their sweaters still look new after years of wear. Alpaca does retain its fibre characteristics, including softness, brightness and lustre, for decades. (camelid textiles found in 2500-year-old Peruvian ruins are often in surprisingly good condition!)

Incan culture and Peruvian history tell us that an Incan man's wealth was counted in textiles made from alpaca fibre. Alpaca fibres were used to make hats, gloves, sweaters, scarves, pillows, blankets, rugs, bags, puppets, pin cushions, wall hangings and a myriad of other products from this fabulous fibre!

Alpaca comes in 22 gorgeous natural colours ranging from white to true black and including delicate beiges, vicuna-like fawns, luscious rich browns and a full range of greys. No other animal produces so many coloured fibres. Alpaca is the only fibre animal that grows true black fleece. Thus, no dye is required to produce alpaca yarn in this large range of earth colours, making it particularly eco-friendly. Yet, when other colours are desired, alpaca accepts dye beautifully (Ziek, 2021).

A study on alpaca fibres by Radzik-Rant et al. (2018) showed that average fibre diameter, coefficient of variation of fibre diameter as well as staple length of alpaca fibres were obtained from different firms located on different continents. The medullation of these wools seems to be independent of the breeding work, climatic conditions and nutritional conditions.

Table 9.2 shows average diameter of some of the finest hair fibres (Williams, 2007).

Table 9.3 shows morphological features of a few hair fibres as reported by Tridico (2010).

9.2.5 Sustainable silk fibre

Silk is a very fine regular, translucent mono-filament fibre of high lustre, softness and strength and is a highly valued fibre. The filament may be up to 600 m long, but averages about 300 m in length. The diameter may vary from about 12 to 30 microns. The beauty and softness of silk's lustre is due to the triangular cross section of the silk filament. As the silk filament is usually twisted about itself, the angle of surface reflection changes continuously. Consequently, the beam of reflected light is broken, resulting in a softer, more subdued lustre (Gohl and Vilensky, 1980).

Table 9.2 Average diameter of some of the finest hair fibres.

Hair fibre	Fibre diameter (micrometres)
Merino sheep	12–20
Angora goat (mohair)	25–45
Angora wool (rabbit)	13
Cashmere goat	15–19
Alpaca (suri)	10–15

Table 9.3 Longitudinal scale patterns medulla and cross section of various hair fibres.

Name of the fibre	Cuticular scale pattern	Medulla	Cross-section
Wool (fine)	Simple coronal	none	Circular to elliptical
Wool (coarse)	Regular and irregular mosaic, smooth	Wide lattice, unbroken or fragmented narrow	Round to elliptical
Angora goat (mohair)	Irregular waved mosaic and simple wave pattern	Fragmental or unbroken lattice	Circular to oval
Angora wool (rabbit)	Single or double chevron	Uni-serial or multi-serial ladder	Oval to rectangular, dumb-bell ovoid
Cashmere goat	Coronal, distant scale margins smooth	No medulla or simple medulla (medium diameter)	Circular to oval
Alpaca (suri)	Irregular waved mosaic, smooth near margins	Varies	Almost circular, varies,

The bringing-up of domesticated silkworms and the life of wild silkworms are, by nature, sustainable. Silk fabric when produced by weavers on handlooms has a near-zero energy footprint and satisfies most of the guidelines for sustainable fabric production.

Silk fibre, both in principle and when compared with other natural and/or artificial fibres, fully corresponds to the increasingly strict requirements set out for sustainable raw materials.

Silk is an eco-friendly and fully degradable material, with a fully circular production cycle:

- Silkworms only eat natural food: leaves of the mulberry tree or leaves of specific types of plants;
- Mulberry tree plantations allow for maintaining the original natural ecosystem and strengthening it;
- Mulberry tree plantations increase the level of biodiversity in the ecosystem;
- Mulberry tree plantations allow a respectful use of territory especially if compared with other textile crops, such as cotton, or non-textiles, such as cereals, which require enormous amounts of water;
- Mulberry tree plantations are never sprayed with fertilizers and pesticides as it is scientifically proven that the use of these chemical agents causes an exponential fall in the yield of the cocoons bred for the production of silk filament, and might potentially cause the death of silkworms.

Silkworm breeding is a perfect example of a circular economy. As a matter of fact, silk production was given as an example of such a system at the first international UN conference on agro-ecology (FAO, 2018).

- Silk fibre processing has a very low environmental impact in terms of atmospheric emissions;
- The fibre remains natural throughout the entire period of its processing;
- During its life, silk does not release polluting substances and does not contribute to the gigantic problem of plastic micro-fibre pollution typical of synthetic fibres which for now has no solution;
- Silk fibre, as it is a combination of two natural proteins, leaves no residues in the environment at the end of the product's life;
- Silk fibre is 100% recyclable and reusable;
- People or entities involved in silk production are rewarded in different ways, at the same time environmental variables are taken into consideration and the territory is treated with respect;
- For thousands of years, silk breeding has been a source of income and, in many cases, a driving force for many modern economies, including Italy until the first decades of the last century;
- In silk production, silk fibre is used fully as the fibre is quite costly and due to reuse wastage is reduced to a minimum or possibly eliminated;
- Silkworm breeding mostly takes place in rural agricultural environments and represents a significant source of supplementary income for many families who are primarily dedicated to other activities, in the first place agricultural;
- Aggregated revenue levels, even though they come from different producers, are on average considerably higher per unit of time than it would be possible to obtain from other crops and/ or crops or activities. The existing system of silk production supports a considerable amount of productive, commercial and distribution entities of the textile system. The use of the silk protein components contributes to other sectors, too, such as research, cosmetics, medical and biomedical sectors, guaranteeing high added values. The silk production system is an efficient and effective circular economy that can allow the use of relevant economies of scale. Activities related to silk production accompany the entire history of humankind. In some economies, they became an integral part of culture and society.

The first production of silk started long time ago, and became an important part of the agricultural system in many cultures. Affordable ways of production, respectful use of resources and their full recovery are all part of the tradition. Smart use of resources and environmental protection both typical of sericulture and linked to reasonable economic returns; positively stimulate the dynamics of personal and social involvement. The United Nations selected silk breeding as an economic activity to support policies aimed at replacing harmful crops, like opium in Burma, encouraging environmental and social improvement through appropriate economic returns. Respective pilot projects were approved. Silk is part of Italian history which undeniably influenced the culture, traditions and customs in many regions of the Apennine peninsula. It is indispensable for our society. Silk is part of world history. China played a fundamental role and even today is a source of discoveries about the role of this fibre in the history. Silk plays a vital role in modern society both from the point of view of traditional textile use and in terms of being an important cultural reference. For example, the name 'Silk Road' is used to identify the contemporary Chinese infrastructure program in Asia and Europe.

9.2.5.1 Organic silk

In a similar process to farming, organic silk production means that no pesticides, insecticides or harsh chemicals have been used in the production of the silk fibre and in producing the finished cloth.

For organic silk, the process is the same as with conventional silk with the only difference that the silkworms are fed with mulberry tree leaves from organic agriculture. The trees are not treated with fungicides, insecticides or genetic sprays; this has a positive impact on the size of the cocoon and the quality of the silk thread, it also significantly reduces water pollution that results from conventional silk production practices.

9.2.5.2 Organic spun silk

The species used to make organic spun silk is the larvae of 'Antheraea mylitta' (the caterpillar of the wild Tasar silk moth). While conventional and organic silk manufacturing methods involve boiling the cocoons while the silkworm is still inside, spun silk allows the completion of the metamorphosis of the silkworm to the butterfly, so no animal suffers for fashion. This particular method can slow down production, but it is fundamental. It is done according to the principle of non-violence/cruelty-free and manufactured under the most stringent social and natural conditions. Once the butterflies left their cocoons, the cocoons are processed without the use of harmful chemicals, and the fibres are spun using solar-powered machinery. Organic spun silk is a sustainable and cruelty-free alternative to silk.

Organic spun silk is cruelty-free meaning no animal has to suffer or die for fashion. Instead of the domesticated silkworm, the silkworm breeding takes place in the wild under natural conditions. Organic agriculture means the feed is created without the use of any chemicals or treatments using insecticides, pesticides or synthetic fertilisers. The small-scale silk farmers are paid fairly for the cost of their cocoons and are not controlled by brokers or numerous agents.

Regenerative farming practices help improve soil health and store more carbon in the ground.

Workers and surrounding communities are not exposed to carcinogens through toxic pesticide use. The ethical practice supports the sustainable development and economic prospects of communities, so future generations can thrive.

The organic silk production is the more environmentally friendly, nonviolent and sustainable practice of silk cultivation. Zero chemicals or treatments are required for raw silk, which is readily biodegradable. The silkworms are allowed to live out their full lives and die naturally. Organic silk is a highly sustainable crop with cocoons being produced when the silkworms are about 35 days old. The silkworm continues its natural cycle to morph into a moth. Then it lays eggs and dies naturally about 5 days later. Natural silk colours are produced but some organic silk is dyed with natural dyes. Raw silk has the versatility to be blended with just about any other fibre. Blends with cotton and other fibres will produce a silken sheen and added softness (Kaplan, 2014).

Organic silk farming has reaching effects by also promoting the sustainability of mulberry trees, which are the silkworm's food source. One mulberry tree will feed roughly 100 silkworms. One acre of renewable trees sustains silkworm life to produce 30−35 pounds of raw silk. These trees, in turn provide a valuable renewable resource for the local production of baskets, furniture and even folk remedies (Franck, 2001).

9.2.6 Chicken feather fibre

Sixty-seven million tonnes of natural and synthetic fibres are currently in use. It is important to find alternative sources to replace at least partially due to the decreasing availability of resources required to produce those.

Attempts have been made to use agricultural by-products containing proteins such as zein in corn and soya proteins as a source to produce regenerated protein fibres. However, none of the attempts on producing high-quality protein fibres from agricultural by-products has been commercially successful (Yang, 1996).

Worldwide, the poultry meat processing industry generates large quantities of chicken feather by-products that amount to 40×10^9 kg annually. The feathers are considered wastes although small amounts are often processed into valuable products such as feather meal and fertilisers.

Chicken feathers contain about 91% protein, 1% lipids and 8% water. Feathers are made of keratin, which contains amino acids such as cysteine, lysine, proline and methionine. It's the same material that comprises fingernails, claws, beaks, spurs and hair. It contains almost no histidine, lysine or methionine (Griffith, 2002).

The feathers may be hydrolysed, dried and ground to a powder to be used as a feed supplement for a variety of livestock, primarily pigs (Tesfaye et al., 2017). This is a fairly expensive process, however, and results in a protein product of low quality for which the demand is low. Other disposal means such as burning or burying are also occasionally utilized, but these methods are considered environmentally unsound and are therefore largely prohibited. Improper disposal of these biological wastes contributes to environmental damage and transmission of diseases. Economic pressures, environmental pressures, increasing interest in using renewable and sustainable raw materials, and the need to decrease reliance on non-renewable petroleum resources inspired the industry to find better ways of dealing with waste feathers. A closer look at the structure and composition of feathers shows that the whole part of a chicken feather (rachis and barb) can be used as a source of a pure structural protein called keratin which can be exploited for conversion into a number of high-value bio-products. Additionally, several technologies can be used to convert other biological components of feathers into high value-added products. Thus, conversion of the waste into valuable products can make feathers an attractive raw material for the production of bio-products (Tesfaye et al., 2017).

Fig. 9.1 shows the structure of a chicken feather (Bitchin' Chickens, 2020). Each feather has a hard central shaft. The bottom of the mature shaft, a quill, is hollow where it attaches under the skin into the follicle. Around each follicle are groups of tiny muscles that allow the feather to be raised and lowered, allowing the bird to fluff itself up. The portion above the skin, where the smaller barbs extend from is called the rachis.

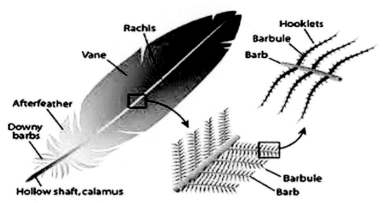

Figure 9.1 Structure of chicken feather.

On both sides of the shaft are rows of barbs, and on each barb are rows of barbules. The barbules have tiny hooks along the edge that lock them together like a zipper to make a smooth feather. At the base of the feathers, there are often barbs that are not hooked together, called downy barbs. One of the purposes of preening is to smooth and connect the feather barbs together. Silkies lack barbules, giving them their fluffy appearance.

Developing feathers have a vein in the shaft, which can bleed profusely if the feather is cut or torn. When these pinfeathers start growing they are tightly rolled and look like pins sticking out of the chicken's skin. They're covered with a thin, white coating that falls off or is groomed off by preening. When the cover comes off, the feather expands to its full length and the vein dries up.

Chicken feather fibres (CFF) have unique structures and properties quite different from those of existing natural or synthetic fibres. CFF cannot be processed as the protein fibres like wool and silk due to their complex structure of the feathers. The secondary structures of feathers, that is the barbs (tips) have the structure and properties that make them suitable for use as natural protein fibres. The feather barbs make them unique fibres due to their low density, excellent compressibility and resiliency, ability to dampen sound, warmth retention and distinctive morphological structure. The density of chicken feathers is about 0.8 g/cm^3 compared to about 1.5 g/cm^3 for cellulose fibres and about 1.3 g/cm^3 for wool (Jones et al., 1998). None of the natural or synthetic fibres commercially available today have a density as low as that of CFF. Such unique properties make barbs preferable for many applications such as textiles and composites used for automotive applications. In addition to the unique structure and properties, barbs are cheap, abundantly available and a renewable source of protein fibres.

Poultry feathers contain about 90% protein and are a cheap and renewable source of protein fibres (Barone and Schmidt, 2005). The secondary structures of the feathers, the barbs are in fibrous form and could be a potential source of protein fibres. More than four billion pounds of chicken feathers are produced in the world every year (Winandy et al., 2003). About 50% of the weight of the feathers is barbs, and the other

50% is rachis (Barone and Schmidt, 2005). Even assuming that 20% of the barbs have lengths greater than 1 inch required for textile applications, about 400 million pounds of barbs will be available as natural protein fibres every year. This means availability of 8% of the protein fibres consumed in the world every year. Since the two natural protein fibres wool and silk are relatively expensive fibres, using the low-cost barbs as protein fibres will make many protein fibre products to be economical and also add high value to the feathers (Reddy and Yang, 2007).

Current applications of chicken feathers are mainly in composites and non-woven fabrics (Evazynajad et al., 2002). Recently, several attempts on using the barbs as 'feather fibres' for composites and non-wovens have been reported (Evazynajad et al., 2002). These feather fibres have been recently characterized for their microstructural properties (Hernandez-Martinez et al., 2005). However, commercially available feather fibres are the barbs in a pulverized form with lengths of about 0.3−1.3 cm (Barone and Schmidt, 2005). Feather fibres do not have the lengths required to be processed on textile machines and are therefore not suitable for making spun yarns and woven fabrics in 100% form or as blends with other natural and synthetic fibres. Being able to produce yarns and fabrics from barbs is important because of the potential for higher value addition and the large textile market.

Although several researchers have reported the structure and properties of feathers from various birds, most of the work has been done on the feather in its entirety and mostly on the feather rachis (Cameron et al., 2003). Limited work has been done on elucidating the structure and properties of feather barbs, especially the chicken feather barbs.

Turkey feather barbs have been characterized for their properties and used as textile fibres and processed to produce blended yarns and non-woven fabrics (Evazynajad et al., 2002).

The study by Reddy and Yang (2007) showed that the structure and properties of chicken feather barbs are useful as natural protein fibres. The unique structure of the barbs, their low density, large availability and low cost makes barbs preferred fibres for several applications such as composites and textiles. The presence of honeycomb structures makes barbs have low density and also provides air and heat-insulating capabilities unlike any other natural fibre. Proteins in chicken feather barbs are of the α-keratin type with about 25% crystalline protein but the α-keratin in barbs probably has a different structure and arrangement than the proteins in the rachis. Chicken feather barbs have the strength of 1.4 g per denier (180 Mpa) and a modulus of 36 g per denier (4.7 GPa), similar to that of wool. This study indicates that the structural interaction of the chicken feather barbs with other fibres could provide unique properties to products made using the barbs with other fibres. Further research is necessary to understand the behaviour and contribution of chicken feather barbs to the processability and properties of various products.

The colour of feathers comes from a combination of pigments and the way that keratin is arranged in layers within the feathers.

Melanins: brown-black pigments add colour to the feather, make them denser and more resistant to wear and breakdown by sunlight.

Carotenoids: yellow, orange or red pigments are synthesized in plants, absorbed by the bird's digestive system and then are taken up by the cells of the follicle as the feather is developing.

Porphyrins: red and green pigments are produced by cells in the feather follicle.

Iridescent greens and blues usually come from the way light reflects off the layers of keratin. Roosters generally have more iridescent colours than hens. When feathers have lost their shine and appear dull, it is often a symptom of a health issue (Bitchin' Chickens, 2020).

9.2.7 Casein fibre

A new generation of innovative fibre and a kind of manmade fibre produced from milk casein through bioengineering method with biological healthcare function, natural and long-lasting antibacterial effect, which got valid certification for Oeko-Tex Standard 100 in April 2004. It is very comfortable with excellent water transportation and air permeability. It is also healthier, light, soft and resistant to fungus, insects and ageing. Constitutionally casein has a striking similarity to wool. It is a phospho-protein built up from a number of amino acids. Its main difference from wool as regards these constituents is in its low sulphur content (0.17) as against 2.7−5.4% in case of wool.

Casein is responsible for the white opaque appearance of milk in which it is combined with calcium and phosphor as clusters of casein molecules called micelles. In comparison with vegetable proteins, casein stands prominent from the point of view of abundance, good colour and the possibility of isolating it without molecular brake-down. Fibres of both fabrics were very similar to wool and could be dyed by the same processes, and like wool they were easily damaged by alkalis. Both had a resilient woolly feel but were not as strong and firm nor as elastic as wool. Moths would not attack and shrinkage was not as much as wool, but fibres mildewed easily when damp.

During the First World War, the Germans discovered the potential of milk casein for making cloth while looking for newer sources of textile fibre. It is regenerated as in the case of regenerated cellulose or viscose. The solutions of casein were forced through the jets into hardening baths forming solid filaments in which the long casein molecules had been given sufficient orientation to hold together in typical fibre form. These early fibres were of little value. They were brittle, hard and lacked the resilience and durability.

During early 1930s, an Italian chemist Antonio Feretti, experimented with casein fibres to try and overcome their drawbacks. He was successful in making casein fibres which were pliable and had many properties associated with wool. Feretti sold his patents to large Italian's rayon firm, Snia Viscosa, who manufactured casein fibres under the trade name of Lanital (lana = wool + ital = Italy) on large scale. Casein fibres have since being produced under various names in a number of countries such as:

- Lanital in Belgium and France.
- Fibrolane in Britain.
- Merinova in Italy.

Sustainable protein fibres 199

- Wipolan in Poland.
- Aralac in America.

Basically wool when used for human clothing has two outstanding defects: it shrinks on washing and is very subject to attack by larvae of moths and beats. Casein fibres are free from these problems. Some moth will attack casein fibre but not like wool.

9.2.7.1 Merits-demerits of casein fibres

Merits of casein fibre are

- It is environmentally friendly.
- It can be called a 'green product' because it does not contain any formaldehyde.
- This fibre has a pH of 7.8, which is close to human skin. So products made with this fibre are more suitable for people.
- Milk protein is hygienic and flexible.
- It is highly smooth, sheen and delicate.
- It is moisture absorbent, permeable, conductive and heat resistant.
- It is easily dyeable in fast colours and requires no special care because of its natural protein base.
- It can be blended with other fibres like cotton, viscose, wool, etc.
- It is renewable, biodegradable and eco-friendly fabric.

Demerits of casein fibre are

- It gets wrinkled easily after washing and needs to be ironed every time after wash.
- It should not be machine washed, as it is not too hard.
- It has a low durability though casein fibres can be laundered with care as wool but they lose strength when wet and must be handled gently.
- It is expensive.
- They cannot be kept damp for any length of time due to quick mildewing (Anonymous, 2018).

9.2.7.2 Manufacture of casein fibre

Wasted and unmarketable sour milk is collected as raw material. Approximately two million tonnes of milk is thrown away every year in Germany alone. This waste milk contains 2.8%−3.2% casein protein and was used as early as in the1930s.

Casein is obtained by the acid treatment of skimmed milk. The casein coagulates as a curd which is washed, frozen and dried, and then ground to fine powder. Thirty-five L of skimmed milk produce about 1 kg of casein. It is processed until it becomes the most basic protein in the milk − a protein called casein.

- The casein is dissolved in water that contains about 2% by weight of alkali to make a viscous solution with 20%−25% protein. The solution is allowed to ripen until it reaches a suitable viscosity and is then filtered and deaerated.
- The filtered casein solution is pumped by a metering pump through a platinum-gold alloy disc or spinneret which has thousands of fine, accurately placed and uniform holes.

- The spinning solution is wet spun by extrusion through spinnerets into a coagulating bath containing sulphuric acid (2 parts), formaldehyde (5 parts), glucose (20 parts) and water (100 parts). The jets of solution coagulate into filaments in a manner similar to the coagulation of viscose filaments.
- The acid neutralizes the alkali used to dissolve the casein. The small continuous fibres are then stretched, treated in various solutions and collected by the spinning machinery.
- But the next process is very critical as the fibre has to be treated chemically to harden it. The process is commonly described as 'hardening', in that it minimized the softening effects of water. Treatment with formaldehyde forms the basis of many hardening techniques.
- In the plant scale bunches of filaments are collected together into a tow as they leave the coagulating bath and are then steeped in formaldehyde solution.
- The filaments are subjected to drawing at this stage. After treatment, the tow is washed and dried, crimped mechanically, and then cut into staple fibre. Otherwise, the tow to top convertor makes tops for blending with wool.
- The tensile strength of the yarn (just like regular thread) is enhanced by stretching the fibre while it is being tanned with aluminium salts and formaldehyde. A further treatment is needed in order to make the fibre resist to boiling bath.

Casein fibres are produced with formaldehyde, the process required huge quantity of water making the process uneconomic, a problem that is yet to be solved.

Fig. 9.4 shows the flow sheet of casein fibre production.

9.2.7.3 Wet processing of casein fibres

Desizing

It is done to break down the size. Enzyme products may be used, preferably at pH 4.0 to 6.0. If water-soluble sizes have been used, desizing is not necessary.

Scouring

It is mainly to remove the impurities present in the fibre. Synthetic detergents should be used, preferably under acidic conditions.

Bleaching

In common with all wet processing, bleaching should be carried out if possible, under weakly acidic conditions with hydrogen peroxide, as casein fibres retain maximum strength and minimum swelling under these conditions. It improves whiteness by removing natural colour and remaining impurities in the fibre. If alkaline processing is used, it must be followed with very careful washing and acidification with acetic acid.

Dyeing

Casein absorbs moisture readily and does not have a highly orientated structure. Dyes can penetrate the fibre without difficulty. In general, casein can be dyed with the dyestuffs used for wool. Acid, basic and direct dyes are used where good washing fastness is not a prime essential.

Drying

After dyeing, loose stock and yarns may be centrifugally hydro-extracted before being dried in conventional dryer. Woven fabrics can be hydro-extracted by open width suction machine or by centrifuging in open width. It is essential to allow an adequate shrinkage from grey to finished dimensions.

Casein blend fabrics can be printed very effectively. Fabrics containing casein may be printed by block, roller, surface roller and modified paper printer methods. Good results necessitate through preparation. If singeing is needed a light treatment with a low-intensity burner will be sufficient. A thorough scour is essential. Casein fibre is generally white and bleaching is not usually necessary. If required, a mild perborate or peroxide bleach should be used under controlled conditions. After treatment, the fabric should be dried on the drums under minimum warp tension followed by stentering to a stable width.

Finishing

Milk protein fibre's products should be after-treated with finish, such as crease-resist finishes and softeners to keep it soft and delicate.

Crease resistant finishing

The crease-resistant finishing agents have more choices and the eco-friendly finishing with good crease resist effect should be selected.

Softening

The milk protein fibre feels hard after crease-resistant finishing at high temperature and tension. In order to make fabric full and soft, softening is needed and softening with a suitable softening agent is an effective method.

Carbonizing

The process is the same for cloth as for loose wool. The vegetable matter is destroyed by soaking the cloth in weak acids and then heating it in an oven. Casein will withstand the carbonizing treatment when carried out with the minimum strength of sulphuric acid necessary for the effective removal of vegetable matter. After treatment, the material should be well rinsed and adjusted to pH 4 with sodium bicarbonate. Carbonizing may be carried out before or after dyeing. If done after dyeing, it eliminates the general tendency of the process to cause unlevel dyeing.

9.2.7.4 Properties of casein fibres

- The mass-specific resistance of milk protein fibre is large.
- Moisture regain of milk protein fibre is about 16%.
- Milk protein fibre is bulky and it is easy to open. The cohesion force is relatively weaker.
- Casein fibres resemble wool in having a soft warm handle.
- The fibres are naturally crimped and yarns have a characteristic warmth and fullness handle.
- Casein fibres provide good thermal insulation. They are resilient like wool.

- The mechanical properties of casein fibres are significantly poorer than those of other bio-polymers, but one order of magnitude higher than foams from microcrystalline chitosan suggesting their use in medical applications, that is wound healing, drug delivery or tissue engineering. Chemical modifications as well as thermal treatment resulted in significantly increased water resistance, showing a way to possible eco-friendly production methods of casein fibres without formaldehyde or cytotoxic citric acid (Bier et al., 2017).
- Organo-phosphorous esters get flame retardancy due to their high phosphorous contents. While burning a polymeric layer of phosphoric acid char is created which block heat transfer. Hellemans (2014) found that casein possesses inherent flame retardancy due to the presence of a large number of phosphate groups. Casein effectively forms a char layer and does not create toxic gases as in the case of phosphate salts like ammonium polyphosphate (APP).

Milk fibre resembles wool in having a soft warm handle. The fibres are naturally crimped, and yarns have a characteristic warmth and fullness of handle. It provides good thermal insulation. They are resilient, like wool.

Casein fibres cannot be distinguished from wool fibres by chemical or burning tests, but only by microscope. because the chemical composition is so similar, casein burns like wool with the odour of burning hair. It has no surface scales like wool, but the surface is smooth and the cross section is round when viewed under a microscope. It is damaged readily by alkalis and mildews.

9.2.7.5 Blending of casein fibres

Casein fibres are blended with many other fibres like cotton, cashmere, silk, etc.

Blend with silk and bamboo fibres

It is cool, free of moisture, sweat exhibitor, comfortable and aerated fibre. It is soft and silky with an attractive sheen. The dazzling grace is reflected in the personality of wearing this fabric (Periasamy and Solanke, 2012).

Blend with wool and cashmere

It is a heat-protective fibre. Milk fibre has a type of three-dimensional arrangement, that is with permeability and humidity-resistant properties, the milk fibre when combined with wool and tepid cashmere, turns out to be an extremely warm material and it is a comfortable and healthy fabric.

Blended fabrics combining milk protein fibre with cashmere increase garment strength and glossiness.

Blend with cotton and cashmere

It is suitable for comfortable undergarments. The milk protein contains ample amino acid and moisture-protecting genes. It is competent enough of resisting micro-organisms. The natural cotton and cashmere fibres also contain similar characteristics and combined with milk fibre, these traits churn together to make healthy and comfortable under clothing (Khanum, 2013).

Care of casein fibres

Casein fibres are very delicate and therefore textiles made from these fibres should be handled carefully to prevent damage.

Washing

Garments containing casein fibres should be washed with care and treated very gently. High temperature and strong acid or alkaline conditions must be avoided. Neutral detergents are preferable for washing. Washing is done by hand or in washing machine (in bag and weak force) under 35°C. Chlorine bleach is harmful (washing with chlorinated washing powder is not recommended).

Drying

Garments should be dried as wool, and care being taken to avoid high temperatures. It should not be wrung out or dried by hanging when the moisture content is over 50% (Hariram, 2011).

Ironing

The full, soft handle of garments containing casein will be maintained if they are slightly damp or almost dry before being ironed or pressed. Wool settings should be used during ironing at medium temperature (or with steam).

Dry cleaning

Casein is not affected by dry-cleaning solvents and garments and can be dry cleaned readily as wool.

9.2.7.6 Uses of casein fibres

The healthy nature of milk fibre is considered a perfect material for manufacturing underwears. Milk casein proteins are considered as a main ingredient of milk protein fibre, which can lubricate the skin. The milk proteins contain the natural humectants factor which can help to maintain the skin moisture to reduce the wrinkles and smoothen the skin − which may help to realize the people of taking milk bath (Chauahn et al., 2018). Casein fibre is used in children's garments, T-shirts, sweaters, women's garments, under clothing's, uniforms, beddings, socks, sportswear's, and eye masks, dog garments, newborn's bath towels, etc.

Table 9.4 shows the properties of soybean protein fibres (SPF) (SoySilk) and milk protein fibres (SilkLatte) (Brinsko, 2010).

9.3 Vegetable protein fibres

Proteins are available in various vegetables. Many of them are fibrous but mixed with several other substances (impurities) and as such they are not directly spinnable to textile yarns and are to be regenerated into pure spinnable fibrous form. In terms of

Table 9.4 Properties of soybean protein fibres (SoySilk) and milk protein fibres (SilkLatte).

Properties	Soyabean protein fibre Yarns from Southwest trading company	Milk protein fibres
Cross-section and longitudinal view	Bean-shaped with pronounced and elongated micro-pores inclusions	Bean-shaped with small Micro-pores inclusions
Birefringence	0.021−0.027	0.016−0.024
Melting point	250−260°C	235−245°C
Chloroform, acetic acid, acetone, DMF	Insoluble	Insoluble
Formic acid	Swell	Swell
Conc. H_2SO_4, conc. HNO_3	Partially soluble	Soluble, gels
Characteristic peaks on FT-IR spectrum	Amide I at 1640 cm^{-1} Amide II at 1530 cm^{-1}	Amide I at 1640 cm^{-1} Amide II at 1530 cm^{-1}

components, there is no difference between animal and plant proteins. They are both made up of amino acids, and they both contain the same 22 amino acids. However, the ratio of these amino acids is different.

Some newly developed regenerated vegetable protein fibres are as follows:

- Soybean fibre
- Groundnut fibres
- Corn fibres

Fibres of regenerated protein were produced commercially in the 1930−1950s and by today's standards they would be considered natural, sustainable, renewable and biodegradable. Casein from milk was used by Courtaulds Ltd. to make Fibrolane and by Snia to make Lanital, groundnut (peanut) protein was used by ICI to make Ardil, Vicara was made by the Virginia−Carolina Chemical Corporation from zein (corn protein) and SPF was developed by the Ford Motor Company (Poole et al., 2009).

9.3.1 Sustainable soybean fibre

Despite the fact that the synthetic fibres (polyester, acrylic, nylon) of various features have become one of the major raw materials of the global textile industry, they have certain serious weaknesses such as follows:

- Synthetic fibres rely on gradually depleting oil resources.
- Some products create pollution during production.
- The comfort properties of chemical fibre-based products are inferior to those of natural fibres.

In order to overcome the fatal weaknesses of chemical fibres, the following three means are being adopted:

- Adopting eco-friendly natural resources from agriculture, animal husbandry and forestry as raw materials which are abundant and cheap.
- Using production processes which are clean and friendly to the environment.
- Searching for skin-friendly and more comfortable products.

The plants' proteins differ from animal proteins in the detailed structure of their molecules. But all proteins are basically similar in chemical design. All protein molecules are in the form of long threads of atoms. Plant as well as animal proteins are therefore able to satisfy the first requirement of a fibre-forming material.

There are three vegetable proteins, which are more important as the source of fibre formation.

- Soya fibre (glycinin) from soybean.
- Groundnut fibre (arachins) from groundnuts.
- Zein fibre from maize.

One of the fibres, which pass all above mention characteristics, is soya bean or soya bean fibre. It is the healthy and comfortable fibre of the 21st century. A Soya Protein Fibre (SPF) was the first textile fibre spun from vegetable protein in 1937. This fibre is often called 'Soy silk' and possess luxurious feel. At present, only industrialized countries have invested heavily in developing these new fibres.

The use of soy textiles is good for our planet, because it is made from fibres that are spun from the leftovers of the soy food industry. Because the plant itself is easily renewable and the fibre biodegrades more quickly than oil-based products like polyester, its use has a minimal environmental impact. This fabric is so biodegradable that you could throw it on your compost pile when it wears out.

Fig. 9.2 shows soybean, its seeds and fibre produced from soybean.

Soybeans contain a large quantity of proteins, approximately 37%—42% compared to peanut (25%), milk (3.2%) and corn (10%). Soya bean proteins comprise of 18 amino acids, predominant being glutamic acid (18.2%), glycine (8.8%), alanine (7.5%), phenylalanine (4.4%), serine (6.4%), aspartic acid (12.8%), glutamic acid (18.2%), proline (5.6%), etc.

Developments in biodegradable fibres from renewable resources in the late 20th century have revived interest in man-made fibre equivalent to wool. The development of a wool-like fibre from soya beans is a story of technological innovation (Blackburn, 2005).

Figure 9.2 Soybean, its seeds and fibre produced from soybean.

Due to increasing prices of petroleum and a growing concern about the environmental damage arising from a slow degradation and poor biodegradability of synthetic fibres, researchers began to search for new possibilities for developing fibres from renewable raw materials, also from soya bean proteins.

Although natural protein fibres such as wool and silk have good physical properties and have been used extensively in the textile industry, they are relatively expensive to use and process. In silk, a large quantity of mulberry leaves is required for the production of a very small quantity of silk resulting in an increased cost of production. In addition, apart from the economic aspects, animal fibres are physically limited in several aspects. Firstly, both wool and silk fibre vary in diameter and their performance profile is limited. Secondly, morphologically, the presence of scales on wool surface results in felting shrinkage and difficulties in dyeing. In contrast, regenerated protein fibres, such as soybean fibre, do not have a theoretical limit in fineness to which fibres may be drawn. In addition, soybean is a competitive production material for fibres in the textile industry since it is abundant and cost-effective.

The fact that proteins are renewable and biodegradable materials has attracted considerable attention from many researchers in the area of textile fibres in the last 2 decades to re-examine the production of fibres from soybean proteins and casein. Soybean proteins have a greater potential for use as textile fibres because of their lower cost than casein proteins derived from milk.

Using animal proteins as raw material for spinning fibres is very expensive. The first attempts to manufacture textile fibres from soybean protein were carried out in Japan by Kajita and Inoue in 1940 and in the USA (Boyer et al., 1945). The wet-spinning process included the extraction of oil to achieve an oil-free meal, extraction of proteins from the meal with alkali, dispersion of the alkaline proteins, fibre formation by passage through a spinneret into an acid coagulating bath and post-spinning treatments.

The first commercial SPF from soybean proteins and polyvinyl alcohol (PVA) were developed in China by G. Li at Huakang R&D Center (Li, 2007). The fibres are the first manufactured fibres, invented by China. The production process for new fibres in laboratory was established in 1993 and commercially promoted in 2000. In 2001, the fibres were standardized and in 2003 they were launched in the market (Rijavec and Zupin, 2011).

SPF is the only botanic protein fibre in the world. Soybean (Glycine max) is a leguminous plant. It is one of the very few plants that provide a high-quality protein with minimum saturated fat. SPF are manufactured fibres, produced from regenerated soybean proteins in combination with synthetic polymer (PVA) as a predominant component (Rijavec and Zupin, 2011). According to textile fibre labelling (FTC, 2010), textiles from SPF can be marked as azlons from soybean. Azlons are manufactured fibres in which the fibre-forming substance is composed of regenerated naturally occurring proteins (FTC, 2011).

Soybeans are very rich in proteins (about 37–42% of dry bean) (Krishnan et al., 2007) in comparison with milk (3.2%), corn (10%) and peanuts (25%). Soybean proteins are used for food and feed and in many industries as adhesives, emulsions, cleansing materials, pharmaceuticals, inks, plastics and also textile fibres. Raw material for spinning textile fibres is obtained from soybean remaining flakes after the extraction of oils and other fatty substances (Yi-you, 2004).

Soybean proteins contain 18 different amino acids. There are about 23% of acidic amino acids (glutamic acid and aspartic amino acid), about 25% of alkaline amino acids (serine, arginine, lysine, tyrosine, threonine, tryptophan) and about 30% of neutral amino acids (leucine, phenylalanine, valine, alanine, isoleucine, proline, glycine). Sulphur-containing amino acids are present also in soy proteins: about 1.0% of cysteine and 0.35% of methionine (Rijavec and Zupin, 2011).

Soy proteins consist of several individual proteins and protein aggregates with a wide range of molecular sizes. However, the most important proteins in soybean are globulins. Globulins can dissolve above or below their isoelectric point (pI) and are insoluble near their pIs. Soy proteins demonstrate maximum solubility at pH 1.5−2.5 and above pH 6.3, whereas minimum solubility is obtained between pH 3.75 and 5.25 (Pearson, 1984). Soybean proteins are mainly composed of two storage proteins glycinin and β-conglycinin with their isoelectric point being between pH four and five are responsible for the insolubility of soybean proteins in that pH range (Nielsen, 1985). β-conglycinin is a heterogeneous group of glycoproteins composed of varying combinations of three subunits, namely, a′,α and β with their molecular weights (MW) being 58,000, 57,000 and 42,000, respectively. The subunits contain hydrophobic regions and link to form compactly folded trimers (Kinsella et al., 1985).

Soybean proteins consist of various groups of polypeptides with a broad range of molecular size: about 90% are salt-soluble globulins (soluble in dilute salt solutions) and the remainder is water-soluble albumins (Zhang and Zeng, 2008). Very important as raw material for producing textile fibres are storage globulins with predominant β-conglycinin (30−50% of the total seed proteins) and glycinin (ca. 30% of the total seed proteins). β-conglycinin is a heterogeneous glycoprotein composed of three subunits (α′, α, β) containing asparagine, glutamine, arginine and leucine amino acids. Subunits are non-covalently associated with trimeric proteins by hydrophobic interactions and hydrogen bonding without any disulphide bonds. Glycinin is a large hexamer, composed of acidic and basic polypeptides linked together by disulphide bonds (Zhang and Zeng, 2008). On the basis of the sedimentation coefficient, a typical ultracentrifuge pattern of soybean proteins has four major fractions: 2, 7, 11 and 15S (Zhang and Zeng, 2008).

The amino acid compositions of soybean, wool and silk fibres indicate that differences do not only occur in the macroscopic view (soybean-globular, wool/silk-fibrous) but also on the molecular scale. Acidic amino acids (glutamic and aspartic acid) are in much higher amounts than in wool, with the predominant amino acid being glutamic acid. However, sulphur-containing amino acids such as cystine are lower than in wool, indicating less cross-linkages through disulphide groups.

Scanning electron microscopy analysis of soybean fibre indicated longitudinal striations on the surface parallel to the axis, varying in length and depth. The cross-sectional shape is kidney bean-like. Recent research on the cross section of soybean correlates well with the previous finding indicating kidney form shape. Studies on cross-sectional shapes of wet-spun fibres have associated the coagulation rate with the cross section. It was suggested that noncircular cross sections occur due to a high coagulation rate in wet spinning (Vynias, 2011).

Soybean protein is a globular protein in its native stage and is not suitable for spinning. Therefore, it has to undergo denaturation and degradation in order to convert the protein solution into a spinnable dope. Denaturation or degradation denaturation of soybean protein can be achieved with any of the following three:

- Alkalis
- Heat
- Enzymes.

Globular proteins are composed of segments of polypeptides connected with hydrogen bonds, electrostatic interactions, disulphide bonds and hydrophobic interactions. Conformational changes in unfolding globular proteins through denaturation process (Zhang and Zeng, 2008) and reducing the inclination of denatured proteins to form aggregates are important for spinnability of a spinning dope with proper relative viscosity. It is also important for later drawing of fibres and crystallization of proteins in fibres. Denaturation is the modification of the secondary, tertiary and quaternary structure of protein. Exposure of soybean proteins to strong alkali/acids, heat, organic solvents, detergents and urea causes the denaturation of native globular proteins, that is converting into unfolded polypeptide chains, which are connected with interchanging of disulphide bonds. Extruded fibres coagulate in a precipitation acid bath and new disulphide bonds are formed.

Oil extraction with solvents used in the mid-20th century was critical for the whole spinning process of soybean fibres, because the chosen temperatures, pH, urea, salts, organic solvents (hexane) and reducing agents influence the degree of denaturation of proteins, degradation of proteins and changing of proteins colour. Protein degradation is detrimental to the production of high-strength protein fibres. Modern method of modifying soybean globular proteins is biochemical using enzymes and auxiliary agent (Swicofil, 2011).

The oil-free protein substance was extracted with dilute alkaline solution and precipitated by adding metallic salts. The protein was then washed in water and added with tartaric acid when the precipitate was wet. Then it was again dissolved in alkaline solution to form a spinning dope. Fibres were spun in an acid bath with an organic coagulating agent (alcohol, formaldehyde, acetone, etc.), where filaments hardened (Kajita and Inoue, 1946). The fibres had natural white to light tan colour. They were crimped, with high resiliency, warmth and soft feel. In comparison to wool they had lower tensile strength, especially in wet state, and lower moisture absorbency (Rijavec and Zupin, 2011).

Low tensile strength of SPFs in wet state limited their commercial application. Fibres were used predominantly in blends with wool, cotton or synthetic fibres in woven and knitted fabrics for apparel and in upholstery, also in cars, despite lower abrasion resistance than wool (Fletcher, 1942). The production of the mid-20th SPFs was ceased at the end of the World War II.

Zhang et al. (1999) used PVA in order to improve the drawability of soybean fibres. PVA is a synthetic polymer with high tensile strength and modulus (Sakurada, 1985) and has been used as reinforcement in keratin fibres (Katoh et al., 2004). Zhang et al. (2003)

investigated the processing of blended PVA/soybean fibre and demonstrated that the incorporation of PVA into the spinning process enhanced mechanical properties.

Soybean is mainly cultivated for its seeds. It is widely believed that the soybean originated in China, 4000–5000 years ago. Soybean has been one of the staple foods of oriental countries for thousands of years. They are rich in a protein which resembles casein. In America, soybean is cultivated in great quantity as a source of edible oils and protein. Many attempts have been made to spin this protein into useful fibres. The Ford Motor Company has pioneered in this field; production by the company began in 1939 and reached more than three tons a week by 1942. The fibre was used for making car upholstery. Production was taken over in 1943 by the Drackett Products Co. of Cincinatti, but stopped after a few years.

At the moment, the United States of America has become the world leader in Soybean production with 46% of the world market. Other countries with high soybean production are Brazil (20%), Argentina (14%) and China (9%).

The advantages of SPF are

- SPF products are degradable and back to the earth by landfill,
- Moisture absorption, ventilation, and warmth cover the superior performances of natural fibres and synthetic ones,
- SPF has the softness and smoothness of cashmere, compared to animal protein,
- Plant protein could be absorbed by human body more easily without side effect,
- The mass-specific resistance of soybean fibre is similar to silk, and lower than other man-made fibres,
- The soybean fibre has good light fastness property and good resistance to ultraviolet radiation, which is better than cotton, viscose and silk.

9.3.1.1 Production of soybean fibre

Soybean has high protein content (about 35%). It provides a cheap and readily available source of protein for fibre production. Forty kg of protein can be extracted from 100 kg of soybean residue. There are five main steps to produce soybean fibre.

9.3.1.2 Extraction of oil

First the beans are cleaned, cracked, decorticated and de-hulled. After a conditioning step at about 70°C, the beans are steeped in hexane to remove the oil. The oil can be extracted by pressing and extraction. The hexane solution is drawn off and the oil is extracted from this solution. The oil-free bean is known as meal. The oil is a valuable by-product and so it is separated out. The resulting oil-free soybean meal is passed through a steam–jacketed pipe for removal of the solvent.

9.3.1.3 Extraction of protein

The separation of protein from the oil-free meal is accomplished by steeping the meal in dilute alkalis like 1% sodium sulphite for 1 hour to one half hour. The protein is dissolved by this treatment and can be filtered out to separate the protein. The protein is precipitated by means of acid. Sulphuric acid is added and the pH of the solution is made to 4.5. This is near the iso-electric point of Soybean protein, that is at its minimum solubility and therefore will be precipitated. Soybean protein is thus separated out as a creamy—white powder.

9.3.1.4   Preparation of spinning solution

The spinning solution is prepared by dissolving the extracted protein in an alkaline solution. Caustic soda is used for dissolution. The solution is filtered and de-aerated by means of vacuum to remove all undissolved particles and air bubbles. After filtration, the solution is allowed for ageing at the requisite temperature. During the ageing period, the solution becomes more viscose and develops the proper consistency for spinning.

9.3.1.5   Fibre formation

Soybean fibre is formed by wet spinning method. The spinning operation consists of forcing the spinning solution through spinneret. The spinneret consists of several holes, and it is emerged in the coagulation bath. The solution emerging from the spinneret holes is referred as liquid jet. The liquid jet precipitated as fine filaments in the coagulating bath containing 2% sulphuric acid and 15% sodium sulphate or sodium chloride for dehydration.

9.3.1.6   After-treatments

Further treatments are required for the development of properties of the fibres like stretching and hardening. Stretching can be done in a separate bath, while the filament is soft and it induces orientation by rearrangement of the molecules and it enhances strength and durability of the filaments. Hardening can be done with formaldehyde to enhance the filaments with more strength and elasticity. The fibre performance can also be stabilized through hydroformylation. It is dried under controlled humidity and temperature, after which it is cut as per staple length.

Fig. 9.3 shows the flow sheet of soybean fibre production.

9.3.1.7   Properties of soybean fibre

The chemical and dyeing properties of pure regenerated SPFs were similar to wool. The soybean fibres are of low strength and are sensitive to moisture to the extent of losing 69% of their tenacity when wet. It has better fineness, low specific gravity, high tensile elongation, and good acidic and alkali resistance. It is similar to natural fibres such as wool and silk. This new fibre is considerably cheaper than real silk (around one-third of the cost of silk) and can partially replace silk. Its moisture absorption performance is equivalent to cotton, and its permeability is greatly better than cotton, ensuring better comfort.

Sustainable protein fibres

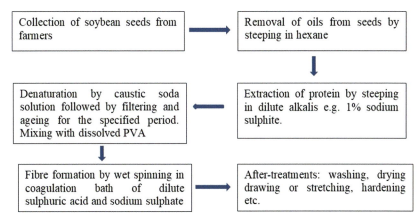

Figure 9.3 Flow sheet of soybean fibre production.

Fabrics from SPF have the following features:

Lustre: SPF has the lustre of silk with excellent drape.

Comfort: It is soft, smooth and light weight.

Absorbency: They have same moisture absorption as cotton and better moisture transmission than cotton.

Easy to dye: The original state of the product has good dyeing, penetration and colour fastness properties.

Strength: They have higher breaking strength than wool, cotton and silk but less than that of polyester.

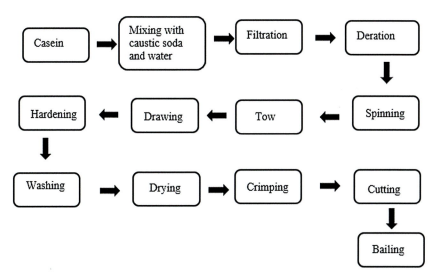

Figure 9.4 Flow sheet of casein fibre production.

Shrinkage: These fibres don't shrink under influence of boiling water.

Easy Care: These fibres are anti-crease, easy wash and fast drying.

Soybean fibre is referred now as a healthy and comfortable fibre in the 21st century.

The breaking strength of single filament of this fibre is over 3.0 cN/dtex — higher than those of wool, cotton and silk, and equivalent to chemical fibres like polyester. The fineness can reach even 0.9 dtex.

Zupin et al. (2008) investigated tensile properties of fabrics with bamboo, PLA and SPF weft yarns and cotton warps. It was determined that among the fabrics woven in plain weave, the fabric with SPF yarn in weft distinguished with the highest tensile strength.

The initial modulus of SPF is quite high, the boiling water shrinkage is low, and so the size and shape stabilities of fabric are good. The anti-crease performance is also outstanding, and it is easily and quickly cleaned and dried. The SPF has good affinity to human skin. The natural colour of soybean fibre is light yellow, closely resemble the colour of silk. With good fastness to light and perspiration, it also has good dyeing brilliance and dyeing fastness in comparison with real silk products. Fabric from pure SPF has got natural colour and pure with abundant fluff on the surface without pilling, excellent hand and drape and softness.

9.3.1.8 Blending of soybean fibre

SPF can be blended with cashmere, wool, silk, cotton and synthetic fibres like polyester. SPF and its blends with various fibres are used in the production of sweater, underwear, towel, bedding, shirt, t-shirt and blanket.

The development of textile process makes the soybean fibre able to be blended with any other fibres at any proportion, without problems in production. Soybean/Cashmere blended fabric has natural softness and puff feeling like cashmere. It is suitable for a cashmere sweater, shawl and coat.

Soybean/silk blended fabric has good lustre, draping, sweat transfer properties. It is suitable for printed silk, knitting underwear, sleepwear, shirts and evening dress.

Soybean/wool blended fabric reduces the shrinkage of wool which is important for manufacturing and processing of sweater. It is suitable for wool sweater, interlock underwear and blanket.

Soybean/cotton blended fabric has soft handle, good moisture absorption and better resistance to bacteria. It is suitable for men and women underwear, t-shirt, infant wear, towel and beddings.

Soybean/synthetic blended fabric has good wrinkle resistance. It is suitable for spring and summer fashion apparel, underwear, shirt and sportswear.

Soybean/lycra blended fabric: by adding a small portion of lycra the fabric becomes more elastic and easier for washing and caring. It is quite active and charming.

Kavusturan et al. (2010) also used comfort fibres like Tencel, bamboo, modal, soybean, 50/50% soybean-Tencel and conventional fibres like viscose and cotton for the production of chenille yarns. In the study, the influence of chenille yarn parameters like

Sustainable protein fibres

pile and core fibre type on fabric abrasion and bending behaviour was investigated. It was indicated that fabrics produced using chenille yarns with soybean and 50/50% soybean-Tencel blend pile fibres showed moderate abrasion resistance compared with the other types.

9.3.1.9 Chemical processing of SPF

SPF has different physical and chemical construction from natural protein fibre, care should be taken in the following processing steps.

Desizing
During pretreatment of SPF oil, lubricants and pigments and other additives, which are added in the production process, are removed. Desizing is done by using enzyme, alkali or oxidizing agents. The enzyme desizing is done typically by using the following recipe:

Amylase enzyme	2–6 gpl.
Glauber salt	2–3 gpl.
Wetting agent	1 gpl.
Temperature	55–60°C.
Time	60 min.

Scouring and bleaching
The natural colour of Soybean is curcuma, that is bright yelow—it is difficult to remove Soybean's pigment. The bleaching of Soybean is done in any of the following three ways:

- Hydrogen peroxide bleaching.
- Reduction bleaching.
- By using both.

Typical recipe for hydrogen peroxide bleaching is as below:

Hydrogen peroxide	20–40 gpl.
Soda ash	3–5 gpl.
Wetting agent	2–3 gpl.
Stabilizer	3–8 gpl.
pH	11
Temperature	92–95°C.
Time	60–80 min.

Typical recipe for reduction bleaching is as below:

Rongolite (Sodium formaldehyde sulphoxylate)	3—8 gpl.
Soda ash	2—4 gpl.
Scouring agent	1—2 gpl.
pH	11
Temperature	92—95°C.
Time	60—80 min.

If required suitable optical brightener is also used.

Dyeing

Soybean can be dyed using reactive, acid, basic and 1:2 metal complex dyes. Dyes are selected according to end uses of fibres and required fastness. Commercially, Soybean fibre is dyed using vinyl sulphone and bifunctional type of reactive dyes. The dyeing method is just like dyeing cotton and viscose with reactive dyes. It can be dyed in the form of loose fibre, tops, yarn, hank and fabric (both knitted and woven).

9.3.1.10 Uses of SPF

The fabric made of SPF shows the lustre of real silk; its drapability is also very good, giving people the sense of elegance; the textile woven with high count yarn has fine and clear grain, suitable for high-grade shell fabric for shirts. The knitting fabric which uses SPF has a soft and smooth handle, and the texture is light and thin, with the sense of blending real silk and cashmere.

9.3.1.11 Yarn

Spinning methods have already been developed for 100% soybean fibre, its blends with natural (cotton, linen, wool, cashmere and silk) and manmade (modified polyester, viscose, Tencel, polynosic, etc.) fibres, and used inplants dealing with cotton, silk and wool. Production of 100% soybean fibre yarns in the range of $21-80^S$ and blended yarns (28/72, 30/70, 45/55, 60/40, 70/30, 85/15, etc.) is possible (Lewis and Sello, 1985).

9.3.1.12 Knit fabric

Soybean protein contained in the fibre remakes a superior, soft hand endowed with both moisture absorbency and permeability, which makes best application in knits and innerwear. Finishes with an anti-bacterial agent, healthcare functionalities are also given. It has great potential in its use in high-grade knits and innerwear.

9.3.1.13 Woven fabric

Weaves made of soybean fibre blended with other natural or chemical fibres have so far been used in shirting and home textiles. A series of such products, too, has already been developed. Their special feature is the lustre and soft hand found in silk. Their economic effects are extremely high. SPF are soft and smooth as well as absorbent it is ideal for products that are worn close to the skin such as underwear, sleepwear, sportswear and children's and infant's clothes, bed sheets, towels and blankets (Jiang et al., 2004).

9.3.1.14 Baby wear

Eco-friendly soybean baby clothing offers many benefits to baby. The breathability, warmth and comfort are outstanding (Huppert, 1944).

9.3.1.15 Biomedical

Soy protein is emerging as a novel material for biomedical applications due to its abundance in nature, ease of isolation and processing, and inherent properties for mediating cell adhesion and growth. Soy protein, a globular, plant-based protein containing isoflavones, is emerging as a promising biomaterial due to flexibility in processing, demonstrated by studies in the food and recently, in biomedical industries. Soy protein is composed of two major proteins, glycinin and b-conglycinin, with hydrophobic components in the molecular structure. Isolated soy protein and soy protein subunits can form hydrogels based on electrostatic interactions, disulfide bonding and hydrogen bonding between subunits within the system. Common methods to induce gelling include thermal treatment, salt-driven cold setting, pressure application, and chemical and enzymatic cross-linking. In addition to specific tissue engineering applications, the use of soy hydrogels may be an ideal carrier for drug delivery applications since soy protein isolate (SPI) has been known to form stable networks in the presence of other ionic compounds and varying pH solutions (Chien et al., 2014).

9.3.2 Groundnut fibres

Due to increasing demand and consequent increasing in costs of natural fibres, nowadays regenerated fibres are getting importance in the textile industries like Viscose, Tencel, Modal, Casein, Soybean fibre, etc. In the coming years, these fibres will occupy the leading positions in textile manufacturing. They can easily be blended with both natural and synthetic fibres.

In 1935, Prof Astbury and Prof Chibnall proposed that fibres can be made by dissolving vegetable proteins in urea and extruding the solution through spinnerets into coagulating baths. The regenerated protein fibres made in the mid-20th century were developed as a substitute for wool. For the production of protein fibre, the main emphasis is given to the commercial availability and their usefulness for the textile purpose. Theoretically, any protein containing substance may serve as a starting material, and the protein may be extracted from it.

One of the most likely sources of vegetable protein for fibre production was groundnuts, which grow as a staple product in many of the hot, humid regions of the world.

Groundnuts (peanuts, monkey nuts) are used in large quantities as a source of the arachis oil (refined peanut oi) required for making margarine. The meal remaining after removal of the oil contains a high proportion of protein. This protein was regarded as a potentially suitable source of vegetable protein. The nuts contain around 25% protein and can be good and can be good and cheap resources for protein fibre.

Experiments were carried out and a process was developed for making the groundnut protein fibre which became known as 'Ardil'. The fibre was first made at Ardeer in Scotland. Commercial production started in 1951.

Groundnuts are cultivated in India, China, West Africa, and southern states of the USA. After harvesting, the nuts are shelled or decorticated. The red skins are removed from the shelled nuts, together with foreign matter such as small stones and nails.

9.3.2.1 Production of groundnut fibres

There are five main steps to produce groundnut fibre (Mahapatra, 2018) which are discussed below.

Extraction of oil
The nuts, which contain about 50% of oil, are crushed and pressed. About 80% of the available oil is squeezed out, leaving the oily groundnut meal which is reduced in breaker rolls and passed through flaking rolls. The thin flakes pass via a series of buckets on an endless chain into an extraction plant. As they pass through the plant, the buckets of the meal are subjected to a thorough washing with a solvent (Hexane) which removes the remainder of the Arachis oil.

Extraction of protein
The extracted meal is heated under low pressure in steam-jacketed pans to remove residual solvent. It is then cooled, screened, weighed and bagged. This special technique for removing oil from groundnut meal was devised to provide protein suitable for fibre production. The groundnut protein is extracted from the meal by dissolving it in caustic soda solution. The residue after extraction is a valuable cattle food. Then acidification of the protein solution precipitates the protein, which is the raw material from which fibre is spun.

Preparation of spinning solution
The spinning solution is prepared by dissolving the extracted protein in aqueous urea, ammonia, caustic soda and solutions of detergents. Caustic soda is used for dissolution. It is allowed to mature under controlled conditions for 24 h. During the maturation, the viscosity of the solution increases and attains the spinning characteristics. The solids content of the protein is between 12% and 30%.

Fibre formation

Groundnut fibre is formed by the wet spinning method. The solution of groundnut protein is filtered and pumped to spinnerets, through which it is extruded at a constant rate into an acid-coagulating bath. The spinneret holes are typical of 0.07–0.10 mm diameter.

The coagulating liquor consists of a solution containing sulphuric acid, sodium sulfate, and auxiliary substances. It is maintained at a temperature between 12 and 40°C.

After treatments

As the filament is being spun, it is stretched to increase the alignment of the protein molecule. It coagulates to a filament that is weak and flabby when wet and brittle when dry. At this stage, the filament dissolves easily in dilute saline solution and in dilute acid and alkali. After leaving the coagulating bath, it is treated with formaldehyde to harden and insolubilize it, and it is then dried and cut into staple fibre.

9.3.2.2 Properties of groundnut fibres

Groundnut fibre is having a circular cross section and smooth, slightly striated surface. It is having tensile strength of 8–12 kg/mm^2 (11,000–14,000 lb/in^2). It does not soften or melt on heating. It chars at 250°C. It is degraded by sodium hypochlorite and sodium chlorite bleaches. Good resistance to organic solvents maybe drycleaned without difficulty.

Groundnut protein fibres are generally similar to wool in that they are protein in structure. They do not have the rough scaly surface of wool fibres and do not undergo felting in the way that wool does.

Groundnut protein molecules carry many side chains, and they cannot pack so closely together as the molecules of silk. Groundnut protein yields a relatively weak fibre, which is much more sensitive to moisture than wool.

9.3.2.3 Blending of groundnut fibre

One of the outstanding characteristics of groundnut protein fibre is soft, wool-like handle. The price of groundnut protein fibre is half the cost of wool fibre. And it is used largely as a diluent fibre which provides wool-like characteristics at low cost. It is used in various worsted units along with wool and polyester. And it is also used with cotton and viscose in various proportions.

9.3.2.4 Chemical processing of groundnut fibre

Groundnut protein fibre has different physical and chemical construction from natural protein fibre, care is taken in the following steps.

Scouring

During scouring, the alkali concentration should be less as compared to other textile fibres and the temperature should be less than 98°C. Wool type scouring conditions are suitable for 100% groundnut protein fibres and kier type boiling should be avoided while processing groundnut blended fabrics.

Bleaching

It should be borne in mind that sodium hypochlorite and sodium chlorite cannot be used for bleaching groundnut protein fibre because they cause degradation. Hydrogen peroxide is the preferred bleaching agent. The dosage has to be decided depending on the quality of groundnut fibre.

Dyeing

Groundnut protein fibre is dyed with the same dyes used for dyeing of wool fibre like acid dyes, metal complex dyes, chrome dyes and selected reactive dyes. Dyes are selected according to end uses of fibres and required dyeing fastness properties. The dyeing method is similar to the dyeing of wool fibres. In general, the affinity for dyes is higher than that of wool. It can be dyed in the form of loose fibre, tops, yarn hank and fabric (both knitted and woven).

9.3.2.5 Uses of groundnut protein fibre (GPF)

In the 1950s, ardil fibre (groundnut protein fibre) was intensively marketed for industrial and domestic uses. Dresses, suits and pajamas were promoted with the slogan 'Happy families wear clothes that contain Ardil'. Ardil fibres were used as garments, carpets or upholstery. Ardil/Wool blends were used for sweaters, blankets, underwear, carpets and felt. Blends with cotton were used for sports shirts, pajamas, dress fabrics and blends with rayon for costume and dress fabrics, tropical clothing, sports shirts and carpets. In course of time, these fibres were subsequently overshadowed by the successful synthetic fibres such as nylon and have been largely forgotten. However, presently more eco-friendly and biodegradable material has been given importance and so the manufacture of these fibres again getting momentum.

9.3.3 Corn protein fibre

Regenerated zein protein fibres made from corn and maize. Zein fibres were commercially sold as Vicara from 1948 to 1957. They were made with the crushed meal of corn and maize after the oil was extracted. The meal is dissolved in an alkali bath then forced through spinnerets to form fine fibrils that were hardened with formaldehyde.

Fabric made from zein fibre is soft, tough and strong and has the warmth of wool. It is resistant to mildew, insects, sunlight, and temperatures to 140°C. Vicara was usually blended with cotton, wool, or rayon. It was used in suits, sweaters, blankets and pile fabrics.

Corn protein, zein, is used to make tough, glossy, grease and abrasion resistance fibre. It is particularly attractive due to its silky lustre, smooth feel, improved washability and improved dyeability. However, it is susceptible to shrinkage and possesses poor wear resistance as compared to synthetic fibres. Zein-based fabrics have been used for fashion application (Lebedyte, 2020).

Zein is a class of prolamine protein found in maize (corn). It is usually manufactured as a powder from corn gluten meal. Zein is one of the best understood plant proteins. Pure zein is clear, odourless, tasteless, hard, water-insoluble, and edible, and it has a variety of industrial and food uses.

The dominant historical use of zein was in the textile fibres market where it was produced under the name 'Vicara' (Lawton, 2002). With the development of synthetic alternatives, the use of zein in this market eventually disappeared. By using electrospinning, zein fibres have again been produced in the lab, where additional research will be performed to re-enter the fibre market (Sellin et al., 2007). It can be used as a water and grease coating for paperboards and allows recyclability (Parris, 2002).

Physical and chemical properties of corn protein are

- Soluble in hot alkaline solutions
- Insoluble in water, dilute acids and most organic solvents
- Fibres are smooth with a circular cross section

The invention by Xiaogen and Haishan (2006) provides a zein fibre, which is provided with good spinning and predominant taking performance, and the producing method. Purifying zein, which is extracted from the stock that is the residue fluid after extracting corn starch, is melt in slag water. It is mixed with various activating agents such as glyoxal. Under the effect of a certain temperature, density, pressure and given various activating agent, original fluid with certain density is made up, Zein fibre is slubbed by the technology of wet method and deaeration spinning is completed. Zein fibre is provided with comfortable hand feeling, good gloss, thin single titer which can be slubbed to 0.7 dtex, light of specific weight and so on. It can be knitted as high-grade cloth material which is various ultra-thin or thickening.

The properties of various sustainable fibres are compared in Table 9.5.

9.4 Green composites

Composite is the synergetic amalgamation of two or more different materials which results in a product with superior properties to its parental components one to five. For the sustainable development purpose, green composites (fabricated from the biodegradable polymers and natural fibres) are the hot topic amongst the researchers (Koronis et al., 2013).

Green composites are named so because of their sustainable and biodegradable properties, so they can be simply decomposed without causing any harm to the environment (Gurunathan et al., 2015).

Several biodegradable polymers that may be either synthetic or natural are commercially available and used as a matrix for the green composite. Proteins are biopolymers which have potential applications in the areas of packaging, automobile, biomedical and agriculture industry, because they have the capability to form films with good

Table 9.5 Properties of various sustainable protein fibres.

Properties	Cotton	Silk	Wool	Casein	Soybean	Groundnut	Corn fibre
Tenacity, g/den	2–5.5	1–1.5	1.5–2.0	0.8–3.0	0.25–0.8	0.7–0.9	1.2 (dry) 0.65 (wet)
Elongation,%	6–10	25–45	25–40	60 = 70	50	40–60	25–35 (wet) 30–45 (dry)
Density, gm/cm^3	1.50–1.54	1.34–1.38	1.33	1.3	1.29	1.31	1.25
Moisture regain, %	9	11.0	14–16	14	8.6	12–15	10
Acid resistance	Bad	Excellent	Excellent	Good	Excellent	Excellent	Good
Alkali resistance	Excellent	Good	Bad	Poor	Good	Poor	Poor
Resistance to moth/fungus	Resistance to moth but not to fungus	Resistance to fungus but not to moth	Resistance to fungus but not to moth	Resistance to moth, but not to mildew.	Good	Resistance to fungus but not to moth	X
Refractive index	1.53	1.54	1.54	x	x	1.53	
Flammability	Burns	Burns	Burns slowly	Very good	Burns	Burns very slowly	Not burn easily

barrier properties in dry conditions. Since biopolymers along with biofibres are compostable, have low cost and are based on renewable resources, they propose significant advantages for the environment.

The experimental investigation by Verma et al. (2019) on the soy protein-sisal fibre-reinforced green composites has led to subsequent conclusions. Firstly, incorporation of the sisal fibre in soy protein matrix resulted in an improvement in tensile strength. 5 wt.% of sisal fibre is found to be optimum fibre wt.% at which highest ultimate tensile strength (UTS) is obtained.

Semi-rigid composites of polyurethane foams (SRPUF) modified with the addition of keratin flour from poultry feathers and flame retardant additives were manufactured. Ten per cent by mass of keratin fibres was added to the foams as well as halogen-free flame retardant additives such as Fyrol PNX, expandable graphite, metal oxides, in amounts such that their total mass did not exceed 15%. Thermal and mechanical properties were tested. Water absorption, dimensional stability, apparent density and flammability of produced foams were determined. It was found that the use of keratin fibres and flame retardant additives changes the foam synthesis process, changes their structure and properties as well as their combustion process. The addition of the filler made of keratin fibres significantly limits the amount of smoke generated during foam burning. The most favourable reduction of heat and smoke release rate was observed for foams with the addition of 10% keratin fibres and 10% expandable graphite. Systems of reducing combustibility of polyurethane foams using keratin fillers are a new solution on a global scale (Wrzesniewska-Tosik et al., 2020).

Chicken Feather Fibre (CFF) have immense importance and great scope for improvement in the advanced composite materials. Different combinations can be made along with CFF as matrix, particulate or fibre and also the amalgamation of all forms can be added in various percentages to design hybrid composites. Highly improved and hybrid materials can be designed through modification to the existing materials. CFF has good fibrous nature, its morphological nature results in the uniform dispersion. With systematic fibre orientation, manipulated sizing, improved mixing with matrix during casting, cost-effective developed composites with enhanced characterization can be achieved (Bansal et al., 2017).

The storage modulus of CFF/reinforced polylactic acid (PLA) composites is higher than that of pure polymer, whereas the mechanical loss factor (tan δ) is lower. The addition of CFF enhanced thermal stability of the composites as compared to pure PLA (Cheng et al., 2009).

9.5 Conclusion

Concerns for the environment and consumer demand are driving research into environmentally friendly fibres as replacements for part of the 38 million tonnes of synthetic fibre produced annually.

The protein fibres are formed by the natural animal source through condensation of α (alpha)-amino acids to build repeating units of polyamide with many substituents on a (alpha)-carbon atom. The sequence and the type of the amino acids bonding together the on individual protein chains contribute to the overall properties and characteristics of the resultant fibres.

In general, protein fibres have moderate strength, resiliency and elasticity. Their moisture absorbency and transport characteristic are excellent. They do not produce static charge. While they are resistant to acids, they are easily attacked by bases and oxidizing agents. They tend to yellow colour on exposure to sunlight due to oxidative attack.

Two natural protein fibres, namely, wool and silk, have significant role from ancient times to modern civilization. These fibres are sustainable, but some of their life cycles are unsustainable. Organically processed wool and silk are fully sustainable.

Wool is known for warmth and is in high demand for winter garments. On the other hand, silk is known for its strength and high lustre. It plays a big role in fashion textiles. Both the fibres are beyond common man's reach as the costs are increasing exponentially due to increasing demand and limiting supply. Silk is the only natural filament (continuous) fibre, the unique nature of which can be substituted by any other proteins. Regenerated cellulose fibre, viscose was manufactured as a substitute of silk, but it is far inferior to silk.

Wool is in high demand. Hence, substitution is in desperate need. Various animal hair fibres like cashmere, angora, alpaca, mohair, etc. are in use since ancient time, but their supply is limited and cannot be increased to a large extent. Chicken feathers are wasted globally, and hence, the efforts are being made to use these for producing textile fibres. A huge quantity of milk is spoiled globally every day. This has become a raw material for a very useful textile fibre called casein fibres. Vegetable proteins have very similar chemical structures to those of animal proteins. Thanks to the development of biotechnology which opens out avenues for newer textile fibres from vegetables like soybean, groundnut, corns, etc. But much more research works in these fields are necessary for their large-scale commercial success.

References

Alves, C., Ferrao, P.M.C., Freitas, M., Silva, A.J., Luz, S.M., Alves, D.E., 2009. Sustainable design procedure: the role of composite materials to combine mechanical and environmental features for agricultural machines. In: Materials and Design, 30, p. 4060.

Anonymous, 2018. Properties, production and use of milk fiber. Textile Today. www.textiletoday.com.bd/.

Bansal, G., Singh, V.K., Gope, P.C., Gupta, T., 2017. Application and properties of chicken feather fibre (CFF) livestock waste in composite material development. Journal of Graphic Era University 5 (1), 16−24. ISSN 0975-1416 (print).

Barone, J.R., Schmidt, W.F., 2005. Polyethylene reinforced with keratin fibres obtained from chicken feathers. Composites Science and Technology 65, 173−181.

Bier, M.C., Cohn, S., Stierand, A., et al., 2017. Investigation of eco-friendly casein fibre production methods. IOP Conference Series: Material Science and Engineering 254, 192004.

Bitchin' Chickens, January 31, 2020. Feather Anatomy. https://bitchinchickens.com/2020/02/13/chicken-feathers-101/feather/.

Blackburn, R.S., 2005. Biodegradable and Sustainable Fibres. Woodhead Publishing Limited, Cambridge-England.

Boyer, R.A., Atkinson, W.T., Robinette, C.F., 1945. Artificial fibres and manufacture thereof. United States Patent 2 (377), 854.

Brinsko, K.M., 2010. Optical characterization of some modern "eco-friendly" fibres. Journal of Forensic Sciences 55 (4)), 915−923. ISSN 1556-4029, April 15, 2011. Available from: interscience.wiley.com.

Cameron, G.J., Wess, T.J., Bonser, R.H.C., 2003. Young's modulus varies with differential orientation of keratin in feathers. Journal of Structural Biology 143, 118−123.

Chauahn, N., Arya, N., Sodhi, S., 2018. Fiber from milk byproducts - A new dimension. International Journal of Current Microbiology and Applied Sciences (IJCMAS) 7 (4), 1257−1264.

Cheng, S., Lau, K.-T., et al., 2009. Mechanical and thermal properties of chicken feather fibre/PLA green composites. Composites Part B: Engineering 40 (7), 650−654.

Chien, K.B., Chung, E.J., Shah, R.N., 2014. Investigation of soy protein hydrogels for biomedical applications: materials characterization, drug release, and biocompatibility. Journal of Biomaterials Applications 28 (7), 1085−1096.

Evazynajad, M., Kar, A., Veluswamy, S., McBride, H., George, B.R., 2002. Production and characterization of yarns and fabric utilizing Turkey feather fibres. Materials Research Society Symposium Proceedings 702. U1.2 1−12.

FAO, April 3, 2018. 2nd International Agroecology Symposium, T.Silk (Thermo Seta). Rome. Available from: https://t.silk.bio/en/news-en/agroecology-nature-biodiversity/. (Accessed 26 December 2020).

Federal Trade Commission (FTC), 2010. Textile Fibre Labeling. April 10, 2011, Available from: http://www.nordstromsupplier.com/NPG/PDFs/Product%20Integrity/Labeling%20 Requirements/Textile%20Fibre%20Labeling.pdf.

Federal Trade Commission (FTC), March 05, 2011. Rules and Regulations under the Textile Fibre Products Identification Act, 16 CFR Part 303.§303.7(g) Generic Names and Definitions for Manufactured Fibres, USA Federal Trade Commission. Available from: http://www.ftc.gov/os/statutes/textile/rr-textl.htm.

Fletcher, H.A., 1942. Synthetic fibres and textiles. Kansas Bulletin 300, 8−10.

Franck, R.R., 2001. Silk, Mohair, Cashmere and Other Luxury Fibres. Woodhead Publishing, UK.

Gohl, E.P.G., Vilensky, L.D., 1980. Textile Science. Longman, Melbourne, Australia.

Griffith, B.A., 2002. Feather Processing Method and Products, United States Patent Application 20020079074.

Gurunathan, T., Mohanty, S., Nayak, S.K., 2015. A review of the recent developments in biocomposites based on natural fibres and their application perspectives. Compos Part A 77, 1−25.

Hariram, A., 2011. A study on casein fiber. https://www.slideshare.net/.

Hellemans, A., March 11, 2014. Milk Proteins protect fabrics from fire, C& en. Available from: https://cen.acs.org/articles/92/web/2014/03/Milk-Proteins-Protect-Fabrics-Fire.html. (Accessed 10 June 2021).

Hernandez-Martinez, A.L., Santos-Velasco, C., Icaza de, M., Castano, V.M., 2005. Microstructural characterisation of keratin fibres from chicken feathers. International Journal of Environment and Pollution 23 (2), 162−178.

Hoskins, T.A., 2014. Cruelty-Free Angora Fur Trade May Be Incompatible with Fast Fashion. Available from: http://www.theguardian.com/sustainable-business/sustainable-fashion-blog/cruelty-freeangora- rabbit-fur-fashion-peta-footage. (Accessed 11 December 2021).

Huppert, O., 1944. Modified Soybean Protein Fibre. US Patent 2364035.

Jiang, Y., Wang, Y., Wang, F., Wang, S., 2004. The ultra structure of soybean protein fibre. Textile Asia 35 (7), 23.

Jone, L.N., Riven, D.E., Tucker, D.J., 1998. Handbook of Fibre Chemistry. Marcel Dekker, Inc., New York.

Kajita, T., Inoue, R., 1946. Process for Manufacturing Artificial Fibre from Protein Contained in Soybean. US Patent 2394309.

Kaplan, D.L., 2014. Kirk-Othmer Encyclopedia of Chemical Technology. Wiley, New York.

Katoh, K., Shibayama, M., Tanabe, T., Yamauchi, K., 2004. Preparation and properties of keratin-poly(vinyl alcohol) blend fibre. Journal of Applied Polymer Science 91 (2), 756−762.

Kavusturan, Y., Ceven, E.K., Ozdemir, O., 2010. Effect of chenille yarns produced with selected comfort fibres on the abrasion and bending properties of knitted fabrics. Fibres & Textiles in Eastern Europe 18 (1), 48 (78).

Khanum, H., Shivaprakash, A.V., 2013. Product development of blended milk casein knitted garment. Academic Journal 41 (2), 50.

Kinsella, J.E., Damodaran, S., German, B., 1985. Physicochemical and functional properties of oilseed proteins with emphasis on soy proteins. In: Altschul, A.M. (Ed.), New Protein Foods, Vol. 5: Seed Storage Properties. Academic Press, New York.

Koronis, G., Silva, A., Fontul, M., 2013. Green composites: a review of adequate materials for automotive applications. Compos Part B 44, 120−127.

Krishnan, H.B., Natarajan, S.S., Mahmoud, A.A., Nelson, R.L., 2007. Identification of glycinin and beta-conglycinin subunits that contribute to increased protein content of high-protein soybean lines. Journal of Agricultural and Food Chemistry 55, 1839−1845. ISSN 1520-5118.

Lebedyte, L., 2020. The Luxury of Manmade Protein Fibres. Available from: www.rsustainable.com/blog-1/man-made-protein-fibres.

Li, G., 2007. Phytoprotein Synthetic Fibre and Method of Manufacture Thereof. US Patent 7271217.

Lawton, J.W., November 1, 2002. Zein: A History of Processing and Use. American Association of Cereal Chemists.

Lewin, M., Sello, S.B., 1985. Encyclopedia of Polymer Science and Engineering. John Wiley and Sons, New York.

Mahapatra, N.N., 2018. Groundnut Protein Fibre, its Properties, and End Use. Textile Today. Posted on March 5. Available from: https://www.textiletoday.com.bd/groundnut-protein-fibre-properties-end-use/.

Nielsen, N.C., 1985. Structure of soy proteins. In: Altschul, A.M. (Ed.), New Protein Foods Vol.5: Seed Storage Proteins. Academic Press, New York.

Njoku, C.F., Daramola, M.O., Alaneme, K.K., 2019. Natural fibres as viable sources for the development of structural, semi-structural, and technological materials -a review. Advanced Materials Letters 10 (10), 682−694.

Oijala, L., 2014. Fibre Watch: Camel Hair for Sustainable Luxury from the Steppes. Available from: http://ecosalon.com/fibre-watch-camel-hair-for-sustainable-luxury-from-the-steppes. (Accessed 21 November 2014).

Periasamy, A.P., Solanke, P.K., 2012. An Overview of Processing and Application of Casin Fibre. Textile Review. https://www.fibre2fashion.com/.

Parris, N., 2002. Recyclable zein-coated kraft paper and linerboard (PDF). Progress in Paper Recycling 11 (3), 24−29. Retrieved November 4, 2019.

Pearson, A.M., 1984. Soy Proteins. In: In: Hudson, B.J.F. (Ed.), Developments in Soy Proteins-2. Elsevier Applied Science Publishers, New York.

Poole, A.J., Church, J.S., Huson, M.G., 2009. Environmentally sustainable fibres from regenerated protein. Biomacromolecules 10, 1.

Radzik-Rant, A., Profelska, O., Rant, W., 2018. Characteristics of alpaca wool from farmed animals located on different continents. Annals of Warsaw University of life Science − SGGW, Animal Science no 57 (2), 151−158. https://doi.org/10.22630/AAS.2018.57.2.15.

Reddy, N., Yang, Y., 2007. Structure and properties of chicken feather barbs as natural protein fibres. Journal of Polymers and the Environment 15, 81−87. https://doi.org/10.1007/s10924-007-0054-7.

Rijavec, T., Zupin, Z., 2011. Chapter 23. In: Krezhova (Ed.), Soybean Protein Fibres (SPF) in Recent Trends for Enhancing the Diversity and Quality of Soybean Products. InTech, ISBN 978-953-307-533-4.

Roy Choudhury, A.K., 2006. Textile Preparation and Dyeing, Science Publishers, USA and Oxford & IBH, India the Second Edition by the Society of Dyers and Colourists Education Charity, India in 2010. www.sdc.org.in.

Roy Choudhury, A.K., 2013. Green chemistry and the textile industry. Textile Progress 45 (1), 3−143 (Textile Institute, UK).

Ryder, M.L., 1992. The interaction between biological and technological change during the development of different fleece types in sheep. Anthropozoologica 16, 131.

Sakurada, I., 1985. Polyvinyl Alcohol Fibres. Marcel Dekker, New York.

Sanjay, M.R., Arpitha, G.R., Naik, L.L., Gopalakrishna, K., Yogesha, B., 2016. Applications of natural fibres and its composites: an overview. Nature Resources 7, 108.

Selling, G., Biswas, A., Patel, A., Walls, D., Dunlap, C., Wei, Y., 2007. Impact of solvent on electrospinning of zein and analysis of resulting fibres. Macromolecular Chemistry and Physics 208 (9)).

Swicofil, A.G., April 27, 2011. Soybean Protein Fibres. Available from: http://www.swicofil.com/soybeanproteinfibre.html.

Tesfaye, T., Sithole, B., Ramjugernath, D., 2017. Valorisation of chicken feathers: a review on recycling and recovery route—current status and future prospects. Clean Technologies and Environmental Policy 19, 2363−2378. https://doi.org/10.1007/s10098-017-1443-9.

Tesfaye, T., Sithole, B., Ramjugernath, D., Chunilall, V., 2017. Valorisation of chicken feathers: Characterisation of physical properties and morphological structure. Journal of Cleaner Production 149, 349−365.

Thangavel, K., Rathinamoorthy, R., Ganesan, P., 2015. Chapter: sustainable luxury natural fibres— production, properties, and prospects in handbook of sustainable luxury textiles and fashion. In: Gardetti, M.A., Muthu, S.S. (Eds.), Environmental Footprints and Eco-Design of Products and Processes. Springer Science+Business Media Singapore. https://doi.org/10.1007/978-981-287-633-1_4.

Tridico, S.R., 2010. Chapter 3. In: Houck, Max M. (Ed.), Natural Animal Fibres: Structure, Characteristics and Identification in Identification Textile Fibres. Woodhead, Cambridge, UK.

Verma, A., Singh, C., Singh, V.K., Jain, N., 2019. Fabrication and characterization of chitosan-coated sisal fibre − phytagel modified soy protein-based green composite. Journal of Composite Materials 53 (18), 2481−2504.

Vynias, D., 2011. Soybean fibre: a novel fibre in the textile industry. In: Ng, T.-B. (Ed.), Soybean - Biochemistry, Chemistry and Physiology. Open-access book by. https://www.intechopen.com/.

Williams, B., 2007. Llama Fibre. International Llama Association. Midwest Manufacturing, Inc. Archived from the original on October 12, 2011. (Accessed 24 October 2011).

Winandy, J.E., Muehl, J.H., Micales, J.A., Raina, A., Schmidt, W., September 1–2, 2003. In: Proceedings of EcoComp 2003. University of London, Queen Mary, pp. 1–6.

World Summit on Sustainable Development (WSSD), 2002. Johannesburg Summit, 26 August–4 September. Available from: https://sustainabledevelopment.un.org/milesstones/wssd.

Wrzesniewska-Tosik, K., Ryszkowska, J., Mik, T., et al., 2020. Composites of semi-rigid polyurethane foams with keratin fibres derived from poultry feathers and flame retardant additives. Polymers 12, 2943. https://doi.org/10.3390/polym12122943.

Xiaogen, H., Haishan, Z., 2006. Zein Fibre and its Preparing Process. Chinese Patent, CN100424242C.

Yang, Y., Wang, L., Li, S., 1996. Formaldehyde-free zein fibre—preparation and investigation. Journal of Applied Polymer Science 59, 433–441.

Yi-you, L., 2004. The soybean protein fibre − a healthy & comfortable fibre for the 21st century. Fibres and Textiles in Eastern Europe 12 (2), 8–9. ISSN 1230-3666.

Zhang, Y., Ghasemzadeh, S., Kotliar, A.M., Kumar, S., Presnell, S., Williams, L.D., 1999. Fibres from soybean protein and poly(vinyl alcohol). Journal of Applied Polymer Science 71, 11–19.

Zhang, X., Min, B., Kumar, S., 2003. Solution spinning and characterization of poly(vinyl alcohol)/soybean protein blend fibres. Journal of Applied Polymer Science 90 (3), 716.

Zhang, L., Zeng, M., 2008. Proteins as sources of materials. In: Belgacem, M.N., Gandini, A. (Eds.), Monomers, Polymers and Composites from Renewable Resources. Elsevier, Oxford, Boston, ISBN 9780080453163, pp. 479–493.

Ziek, B., 2021. What's So Special about Alpaca Fibre? Available from: https://wildhairpress.weebly.com/whats-so-special-about-alpaca-fibre.html.

Zupin, Z., Dimitrovski, K., 2008. Tensile properties of fabrics made from new biodegradable materials (PLA, Soybean, Bamboo Fibres). In: 4th International Textile, Clothing & Design Conference − Magic World of Textiles, Dubrovnyk, Croatia.

Further reading

Avinc, O., Yavas, A., May 3, 2017. Soybean: For Textile Application and Printing, Open Access Book: Soybean − the Basis of Yield, Biomass and Productivity. Available from: www.intechopen.com.

Mahapatra, N.N., 2017. Soybean Fibre - Properties, Processes and Uses. Available from: https://www.textiletoday.com.bd/. Posted on December 21.

Part Three

Sustainable synthetic fibres

Regenerated synthetic fibres: bamboo and lyocell

C. Prakash[1] and S. Kubera Sampath Kumar[2]
[1]Department of Handloom and Textile Technology, Indian Institute of Handloom Technology, Fulia Colony, Shantipur, Nadia, West Bengal; [2]Department of Chemical Engineering (Textile Technology), Vignan's Foundation for Science, Technology & Research, Vadlamudi, Guntur, Andhra Pradesh, India

10.1 Bamboo fibre

10.1.1 Introduction

Bamboo fibre is a natural, ecofriendly fibre. It is one kind of grass with hollow and woody stem. There are so many varieties of bamboo plant across the world. Botanically categorized as a grass and not a tree, bamboo just might be the world's most sustainable resource. It is the fastest growing grass and can shoot up a yard or more in a day. Bamboo reaches maturity quickly and is ready for harvesting in about 4 years. Bamboo does not require replanting after harvesting because its vast root network continually sprouts new shoots which almost zoom up, pulling in sunlight and greenhouse gases and converting them to new green growth.

Bamboo grows in a natural way without the need for petroleum-guzzling tractors and poisonous pesticides and fertilizers. Bamboos are a group of woody perennial evergreen plants in the true grass family Poaceae, subfamily Bambusoideae, tribe Bambuseae. Some of its members are giant, forming by far the largest members of the grass family. Although bamboo is a grass, many of the larger bamboos are very tree-like in appearance and they are often called 'bamboo trees'. The reason bamboos are so different from trees is they lack a vascular cambium layer and meristem cells at the top of the culm. The vascular cambium is the perpetually growing layer of a tree's trunk beneath the bark that makes it increase in diameter each year. The meristems make the tree grow taller (Sharma and Goel, 2011).

Few decades ago, bamboo was traditionally used for making variety of household goods such as furniture, sporting goods, handbags, flooring and cutting board ,etc. Nowadays, due to developments in manufacturing processes, it is possible to produce fibre from bamboo stem, which has remarkable properties for its use in yarns and fabrics.

Bamboo viscose is a new regenerated cellulosic fibre that has recently evinced keen interest primarily because the bamboo plant is a fast-growing renewable resource and therefore, the fibre can arguably be considered a 'sustainable' fibre (Sarkar and Appidui, 2009). Bamboo pulp is refined through a process of hydrolysis-alkalization and multi-phase bleaching before converting in to the fibre. Physical and chemical properties of bamboo fibre are nearly close to viscose. It has good durability, stability, moderate tenacity and thus good spinnability. Bamboo products are further characterized by their good hydrophilic nature, excellent permeability, soft feel, excellent dyeing behaviour and colour vibrancy and overall its antimicrobial property.

It possesses unique properties such as anti-UV radiation, anti-bacterial, breathable, cool soft handle, etc. Due to various micro-gapes and micro-holes in cross section, it has better moisture absorption. Bamboo fabric absorbs and evaporates sweat very easily, and hence gives comfortable feel. It can be spun into 100% bamboo yarn and also be blended with natural and manmade fibres like cotton, polyester, silk, etc. Fabric made of bamboo fibre possesses better moisture transmission property, softer, handle, drapability and easy drying. It has many applications such as apparels, sweaters, bath suits, mats, towels, t-shirts and socks, etc. Bamboo fibre has anti-bacterial function which makes it suitable for underwear, t-shirts and socks etc.

Bamboo has also wide scope in the field of hygienic materials such as sanitary napkin, masks, mattress, bandages, surgical cloths, surgical gown, absorbent pads and food packing, etc. Due to anti-ultraviolet radiation characteristics of bamboo fibre, it is suitable for making summer clothing for the protection of human skin against damages of UV radiation (UVR). Wallpapers, curtains and sofa covers are produced from bamboo fibre because it absorbs UV radiation of different wavelength from atmosphere, hence protecting human skin from UV radiation (Ajay and Avinash, 2012).

Bamboo fibre constitutes a recent kind of natural material that has a huge application possibility in the textile field due to some of its unique properties (Liu and Hu, 2008). For example, it has a unique structure that makes them superior to other natural lingo cellulosic fibres (Ray et al., 2004). In recent years, bamboo fibres have attracted great attention as the most abundant renewable biomass materials that can be used in textiles (Xu et al., 2006) and composite reinforcement (Das et al., 2006). They possess many excellent properties when used as textile materials such as high tenacity, excellent thermal conductivity, resistant to bacteria, high water and perspiration adsorption. Natural bamboo fibres have excellent properties and therefore they have potential for use in textiles; however, they have not received the attention they deserve owing to their coarse and stiff quality. A chemical method for extraction and modification of natural bamboo fibres was used in this article. Regenerated bamboo fibre is 100% cellulose, biodegradable and is claimed to be 'green' and environmentally friendly (Erdumlu and Ozipek, 2008).

Owing to micro-gaps in its structure, bamboo fibre has high air permeability and water absorption properties. Bamboo-based fabrics are antibacterial and very soft, with a low amount of pilling and creasing (Karahan et al., 2006). Fabrics made of bamboo fibre have very good physical properties. When compared to cotton fabrics, bamboo fabrics require a lower amount of dye for the same depth of shade. Moreover, the colorant is absorbed better and faster than in cotton fabrics and dyed bamboo fabric appears better than dyed cotton fabric (Wallace, 2005).

10.1.2 Manufacturing process of regenerated bamboo fibre

The manufacturing process for regenerated bamboo fibre using hydrolysis alkalization with the multi-phase bleaching principle is given below (Erdumlu and Ozipek, 2008; Gericke and Van Der Pol, 2010; Taraboanta, 2012):

10.1.2.1 Preparation

Bamboo leaves and the soft, inner pith from a hard bamboo trunk are extracted and crushed.

10.1.2.2 Steeping

The crushed bamboo cellulose is soaked in a solution of 15%−20% sodium hydroxide at a temperature of between 20°C and 25°C for one to 3 h to form alkali cellulose.

10.1.2.3 Pressing

The bamboo alkali cellulose is pressed to remove excess sodium hydroxide solution.

10.1.2.4 Shredding

The alkali cellulose is shredded by a grinder to increase the surface area and make the cellulose easier to process.

10.1.2.5 Ageing

The shredded alkali cellulose is left to dry for 24 h to be in contact with the oxygen of the ambient air. During this process, the alkali cellulose is partially oxidized and degraded to a lower molecular weight due to high alkalinity. This degradation is controlled to produce chain lengths short enough to produce correct viscosities in the spinning solution.

10.1.2.6 Sulfurization

In this stage, carbon di-sulphide is added to the bamboo alkali cellulose to sulphurise the compound, causing it to jell.

10.1.2.7 Xanthation

The remaining carbon disulphide from the sulphurisation is removed by evaporation due to decompression and cellulose sodium xanthogenate is the result.

10.1.2.8 Dissolving

A diluted solution of sodium hydroxide is added to the cellulose sodium xanthogenate, dissolving it to create a viscose solution consisting of about 5% sodium hydroxide and 7%−15% bamboo fibre cellulose.

10.1.2.9 Spinning

After subsequent ripening, filtering and degassing, the viscose bamboo cellulose is forced through spinneret nozzles into a large container of diluted sulfuric acid solution which hardens the viscose bamboo cellulose sodium xanthate and reconverts it into cellulose bamboo fibre threads which are spun into bamboo fibre yarns. The above-mentioned manufacturing process for regenerated bamboo fibre resembles that of viscose rayon fibre; consequently, regenerated bamboo fibre is also called bamboo viscose.

10.2 Research on bamboo fibre

10.2.1 Research on bamboo at fibre level

Bamboo fibres possess many excellent properties when used as textile materials such as high tenacity, excellent thermal conductivity, resistance to bacteria and high water and perspiration adsorption. Regenerated bamboo fibre, called ecological fibre (Okubo et al., 2004), has many good properties such as resistance to ultraviolet radiation, bacterial and odour (Zhang, 2005), and is used in the production of functional textile products. Bamboo fabric absorbs and evaporates sweat very easily; it therefore gives comfortable feel. It can be spun into 100% bamboo yarn and also blended with natural and man-made fibres like cotton, polyester, silk etc.

Weidong (2005) observed that the structure of natural bamboo fibre is similar to that of other natural fibres. Wang and Gao (2006) reported that regenerated bamboo fibre has low strength and high moisture regain compared with natural bamboo fibre. The morphological structure, IR, fibre orientation and breaking strength of bamboo fibre were investigated by Lipp-Symonowicz (2011), who deduced that on account of the various micro-gaps and micro-holes in its cross-section the fibre has better moisture absorption.

Majumdar et al. (2011) observed that the quality characteristics of blended yarns of bamboo fibres with other fibres depend upon the bamboo content in the blend. The blending of different types of fibres is a widely practised means of not only enhancing the performance but also the aesthetic qualities of textile fabric. These advantages also permit an increased variety of products to be made, yielding a stronger marketing advantage. Bamboo fibre is usually blended with cotton as 50/50 combinations.

Lipp-Symonowicz et al. (2011) compared bamboo and viscose fibres and state that the so-called bamboo fibres are in reality man-made viscose fibres made from bamboo cellulose and that bamboo fibres are comparable to viscose fibres in their morphological structure and properties. Structurally, bamboo viscose is cellulose II with a low degree of crystallinity and high water retention and release ability (Xu et al., 2007). As a result, the fibre possesses desirable comfort properties such as good moisture absorption and permeability; aesthetic properties such as soft handle and pleasing tactile sensation (Shen et al., 2004) and can be processed easily because of its excellent dyeing and finishing abilities.

Another feature of bamboo viscose fibre that has garnered considerable attention claims about its protective abilities, namely, that the fibre possesses UV properties and antimicrobial functions that are inherited from the bamboo plant itself (Xu et al., 2007; Li et al., 2004). Okubo et al. (2004) analysed the mechanical properties of bamboo fibre and reported that the strength of bamboo fibre was equivalent to that of glass fibre.

Waite and Platts (2009) attempted to answer the differences in textile properties between chemically manufactured and mechanically manufactured bamboo textiles. The textile properties examined relate to sustainability: wear and tear and moisture wicking. The SEM images show various similarities and differences among the natural and chemical bamboo fibre. Chemical bamboo fibre displayed a tubular and ribbed (celery-like) longitudinal surface; the cross sections were filled with voids. Mechanical bamboo fibre displayed a bamboo-like longitudinal section, tubular with nodes; the cross section displayed some voids, though much fewer than the chemical bamboo fibres. The results for fibre breaking force and breaking tenacity illustrate that mechanically manufactured bamboo fibre is more than two times stronger than chemically-manufactured bamboo. The breaking tenacity of cotton, wool, viscose rayon, and polyester are all below that of the bamboo fibres tested; therefore, bamboo fibres may be more resistant to wear and tear than conventional fibres. It appears that the processing of fibre into yarn creates some level of strength degradation for mechanically manufactured bamboo. The fabric tear results show that woven fabrics are more resistant to tear forces than that of knitted fabric. Chemical bamboo woven fabric is better at water absorption than mechanical bamboo woven fabric, but chemical bamboo knit fabric takes a very long time to absorb. There also is a difference in absorption properties between bamboo species in textile form, and perhaps the same differences can be found in raw bamboo in nature.

Lipp-Symonowicz et al. (2011) investigation demonstrated that the 'bamboo fibres' available on the market are not natural fibres derived from bamboo, but man-made viscose fibres from bamboo cellulose. 'Bamboo fibres' are visibly finer than those of viscose, and the surface characteristic is also slightly different. Cross sections of 'bamboo fibres' are of a similar shape to those of viscose fibre cross sections, the longitudinal view being also comparable. The chemical structure of 'bamboo fibres' is close to that of regenerated cellulose. The strength and elongation at break of 'bamboo fibres' are lower than that of viscose, but the degree of crystallinity is comparable.

Gericke et al. (2010) confirmed that regenerated bamboo fibres are superior to cotton and viscose rayon fibres. It should be noted though that none of the claims made were proven false—regenerated bamboo fibres can be made into fabrics that are very comfortable and have excellent moisture and temperature management properties. The results of physical tests on other cellulosic fibre fabrics, indicated, however, that the knitted cotton and the viscose rayon fabrics are comparable with regard to properties pertaining to comfort. It was clear that in many cases the performance of the viscose rayon in terms of comfort properties was closer to that of the regenerated bamboo than to that of the cotton fabric.

10.2.2 Advantages of bamboo fibre

- **Smooth, soft and luxurious feel**: It has a basic round surface which makes it very smooth and to sit perfectly next to the skin. Bamboo apparel is softer than the softest cotton, and it has a natural sheen like silk or cashmere. Bamboo drapes like silk or satin yet is less expensive and more durable. Bamboo/organic cotton blends are also extremely soft but heavier in weight
- **Allergy reduced**: Bamboo's organic and naturally smooth fibre properties are non-irritating to the skin, making it ideal for people with skin sensitivities or other allergies and dermatitis.
- **Good absorption ability**: Bamboo fibre absorbs and evaporates sweat very quickly. Its ultimate breathability keeps the wearer comfortable and dry for a very longer period.
- **Temperature adaptability**: Fabrics made from bamboo fibre are highly breathable in hot weather and also keep the wearer warmer in cold season. Bamboo is naturally cool to the touch. The cross section of the bamboo fibre is filled with various micro-gaps and micro-holes leading to much better moisture absorption and ventilation. It is also very warm in cold weather, because of the same micro structure as the warm air gets trapped next to the skin.
- **Antibacterial**: Bamboo is naturally antibacterial, antifungal and anti-static. Bamboo has a unique anti-bacteria and bacteriostatic bio-agent named "bamboo kun'," which bonds tightly with bamboo cellulose molecules during the normal process of bamboo fibre growth. This feature gets retained in bamboo fabrics too. It makes bamboo fabrics healthier, germ free and odour free
- **Thermal regulating**: Bamboo fabrics are warm in the winter and cool in the summer. Bamboo clothing's excellent wicking properties also make it ideal for warm summer days.
- **UV protection**: Bamboo naturally provides added protection against the sun's harmful UV rays.
- **Antistatic**: Due to its high moisture absorption property, bamboo fabric results in the enhancement of anti-static property
- **Green and bio-degradable**: As a regenerated cellulose fibre, bamboo fibre was 100% made from bamboo through high-tech process. They are all three-four year old new bamboo, of good character and ideal temper. The whole distilling and producing process in our plant is green process without any pollution. It produces natural and eco-friendly fibre without any chemical additive. As a natural cellulose fibre, it can be 100% bio-degraded in soil by micro-organism and sunshine. The decomposition process does not cause any pollution environment. Bamboo fibre is praised as "the natural, green, and eco-friendly new type textile material of 21st century'.
- **Breathable and cool**: Bamboo fibre gives human skin a chance to breathe free. Because the cross section of the bamboo fibre is filled with various micro-gaps and micro-holes, it has much better moisture absorption and ventilation. With this unparalleled micro-structure, bamboo fibre apparel can absorb and evaporate humans sweat in a split of second.

10.3 Lyocell

10.3.1 Introduction

Lyo is a Greek word which means dissolve. Lyocell fibre brand name, Tencel, is a natural man-made or regenerated cellulosic staple fibre or filament spun by a solvent spinning process. The solvent spinning technique is both simpler and more environmentally sound, since it uses a non-toxic solvent, an amine oxide that is recycled in the manufacturing process with the 99.5% recovery (Achwal, 2000).

Lyocell is a new generic name given to a cellulosic fibre which is produced under an environmentally friendly process by dissolving cellulose in the tertiary amine oxide N-methylmorpholine N-oxide (NMMO) as shown in Fig. 10.1 (Periyasamy, 2012).

Lyocell is the first in a new generation of cellulosic fibres made by a solvent spinning process. A major driving force to its development was the demand for a process that was environmentally responsible and utilized renewable resources as their raw materials (White et al., 2005). The lyocell fibre is the first man-made fibre which possesses the advantages of the natural ones (comfort in wear by the humidity absorption and biodegradability) and of synthetic ones (high wet and dry resistance and low shrinkage) (Iorga et al., 1999). Lyocell fibre is the latest cohort of manmade cellulosic fibre shaped by extremely advanced environment friendly and engineered technical process. Lyocell fibre not only avert environment contamination but enriches the same by turning barren lands to fertile lands, purifying the air, absorbing solar radiations and consuming less sweet water thus leaving abundant sweet water for supplementary usage. With the best possessions of all the natural and synthetic fibre characteristics, lyocell fibre is indisputably the answer to the requirements of today's customers (Sardana and Viswakarma, 2010). The increasing public awareness and sense of social responsibility related to environmental issues have led the textile industry to manufacture products with improved environmental profiles (Chen and Burns, 2006).

Lyocell is produced from wood pulp and spun into a solvent bath in a closed manufacturing process. The chemicals used for the production of lyocell are significantly less hazardous to the environment. Because the solvent is recycled efficiently and the wood is harvested from tree farms specifically developed for this end use, lyocell can be described as an environmentally friendly fibre (Gharehaghaji et al., 2010).

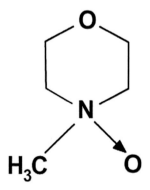

Figure 10.1 Structure of N-methylmorpholine N-oxide (NMMO).

10.3.2 Origin of lyocell

Lyocell was first made in the lab and then in the pilot plant at Coventry during the early 1980s. In 1988 a semi-commercial plant (capacity 30 tonnes/week) was started up at Grimsby, this plant is known as S25 because originally there were 25 spinning ends. The fibre was branded Tencel. In May/June 1992 the first full-scale Tencel factory (SL1) was commissioned in Mobile, Alabama. Lenzing of Austria commenced production of their lyocell fibre in 1997 (White et al., 2001).

Lyocell fibres are now marketed under different brand names depending on the manufacturing companies as shown in Table 10.1.

Today there is only one company, Lenzing, making fibres on a commercial scale. It is currently manufacturing the fibres in Heligenkreuz (Austria), Grimsby (U.K) and Mobile (U.S.A) (Taylor, 1998).

10.3.3 Manufacturing process of lyocell

The major steps for manufacturing the lyocell staple or filament fibres are shown in Fig. 10.2 (White, 2001). The various steps of the process are detailed below: Lyocell

Table 10.1 Manufacturing companies for lyocell.

Brand name	Company
Tencel	Lenzing, Austria
Alceru	Titk, Germany
Newcell	Akzo Nobel, Germany
Orcel	Russian Research Institute

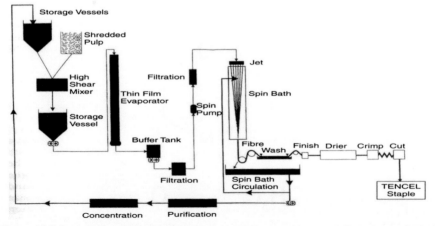

Figure 10.2 Preparation of wood pulp.

fibre offers luxury and practicality while working in tandem with nature. Fabrics made of 100% lyocell or blends have a luxurious, sensual and silky hand and vibrant colours. Unlike the previous generation of cellulosic fibres, the new generation of lyocell has tenacity that withstands rigorous processing. Across the fashion spectrum, it has been embraced by well-known designers and retailers. Lyocell fibre also blends well with other natural and synthetic fibres such as cotton, polyester, lycra or wool adding comfort and performance (Tyagi et al., 2013).

A study on lyocell union fabrics, namely lyocell/silk and lyocell/polyester fabrics by weaving in different fabric constructions and dyeing with reactive dyes, acid dyes and a disperse dye was conducted by Joshi et al. (2010). The resulting dyed fabrics were given a resin finishing treatment and their wash fastness was measured. The dyed and finished fabrics had a smooth, lustrous handle, ideal for lightweight garments.

An effort was made to study the hand properties of silk blends with classic modal, micro-modal, lyocell, cotton, 100% viscose, polyester and cotton of varying blend composition in knitted fabrics by Vijayakumar et al. (2008). Results showed that silk fabrics became stiffer when the percentage of silk increased. Kilc and Okur (2014) observed that the hairiness for 100% cotton yarn is the lowest, 100% tencel is the highest and the blended yarns are in-between. Vijayakumar et al. (2007) blended silk and lyocell fibres effectively using the short staple spinning system. The short staple spinning system reduced the number of machinery passages compared to existing spun silk yarn processing.

Blending of silk with lyocell has many advantages over blending of silk with cotton or any other synthetic fibres. Lyocell is more versatile and uniform than cotton as it is a manmade fibre. It scores over synthetics because it is a biodegradable fibre and thus causes fewer burdens to ecosystem (Gahlot and Pant, 2011).

For manufacturing the high quality of lyocell fibres, pulp in the range 100−1000 units DP (degree of polymerization) are required. For Tencel pulp of 500−550 units DP is used. The trees are cut to 20 ft (6.1 m) lengths and debarked by high-pressure jets of water. Next, the logs are fed into a chipper; a machine that chops them into squares little bigger than postage stamps. Then the chips are loaded into a vat of chemical digesters that soften them into a wet pulp. The wet pulp is washed with water, and may be bleached. Then, it is dried in huge sheets, and finally rolled onto spools; weighing some 500 lb (227 kg) and the sheets have a thickness of poster board paper (Syed, 2010). Fig. 10.2 shows the fibre manufacturing process.

10.3.3.1 Dissolving cellulose

After unrolling the spools of cellulose they are broken into one inch squares. These squares are loaded into a heated, pressurized vessel filled with the 76%−78% N-methylmorpholine N-oxide solution with water, which is reclaimed and recycled in a 'closed loop' spinning process conserving energy and water. Up to 99% of solvent is recovered and reused (Syed, 2010).

10.3.3.2   Filtering

After a short time soaking in the solvent, cellulose dissolves into a clear solution. It is pumped out through a filter, to ensure that all the impurities are removed. Impurities are materials such as pulp feedstock, undissolved pulp fibres or sand and ash (inorganic compound). The solution is passed through two types of filters. One is a media candle filter element which consists of sets of sintered steels and other is candle filter elements associated with spinning machine (White, 2001).

10.3.3.3   Spinning

The filter solution is pumped through spinnerets, pierced with small holes. When the cellulose is forced through it, long strands of fibres come out. These fibres are then immersed in another solution or spin bath containing diluted amine oxide. The filaments are drawn from the air gap with a ratio between 4 and 20, in order to get consistent fibre properties. Below a draw ration of four the tenacity of lyocell fibre is low and above 20 spinning properties are decreased (Fink et al., 2001).

10.3.3.4   Washing

The sets of fibre strands are washed with hot de-mineralized water in a series of washing baths. The level of washing liquor is maintained by counter current feeding. After washing the filaments may be treated with bleaching agent, and finishing agent which may be soap or silicone. The filament could be treated with other chemicals to obtain the specific fibre properties.

10.3.3.5   Drying and finishing

The lyocell fibres next passes to drying perforated drums, where water is evaporated from the filaments. The entire manufacturing process, from unrolling the raw cellulose and to the baling of fibres, takes only about 2 h. After this, lyocell fibres may be processed in a wide assortment of ways. It may be spun with other fibres, such as cotton or wool. The yarn can be woven or knit such as any other fabric, and given a variety of finishes, from soft and sued-like to silky. Additives such as surfactants change the fibre structure and properties (White, 2001).

10.3.3.6   Lyocell cross-sectional view

The cross-sectional view of lyocell fibre under the microscope is shown in Fig. 10.3. The fibre has a circular to oval cross-section. Scanning electron microscopy shows a dense cellulosic network structure of lyocell fibre with small finely distributed voids at the exits ranging in dimension 5—100 nm. The cross section is uniform throughout except a small boundary layer which shows high density (Fink, 2001).

The round, smooth and uniform cross section of lyocell fibre allows a very close packing in the yarn structure. The high strength of lyocell yarn is due to the high cohesion between the parallel fibres (Taylor, 1998). The structural properties of lyocell fibre (Sunol et al., 2007) are summarized in Table 10.2.

Regenerated synthetic fibres: bamboo and lyocell

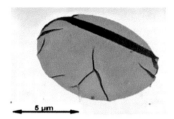

Figure 10.3 Cross-sectional image of a lyocell fibre.

Table 10.2 Structural characteristics of lyocell fibre.

Properties	Behaviour
Cross section	Round/oval
Morphology of cross section	Homogenous, dense
Crystallinity	High
Crystal width	Small
Crystal length	Larger
Crystalline orientation	High
Amorphous orientation	High

10.3.3.7 Lyocell longitudinal view

The rod like structure gives crisp, firm cotton like handle (Taylor, 1998) (Fig. 10.4). Lyocell fibre has little striated longitudinal surface (Wang et al., 2003).

10.3.3.8 Pore structure

It is revealed from the morphological analysis that lyocell fibre has a thin, nano-porous skin of about 150 nm thickness with low porosity in centre of fibre. The area between the skin and core layer of fibre shows intermediate porosity (Manian et al., 2008).

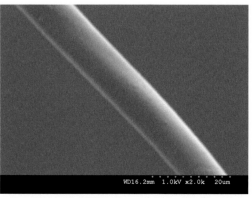

Figure 10.4 Longitudinal view of lyocell fibre.

Lyocell is semi-permeable with a highly porous skin. The size of pores becomes more compact towards the centre of the fibre. The compact or tiny pores in the centre act as barrier for diffusion into the fibre (Rous et al., 2006).

10.3.3.9 Fibrillation

The standard lyocell or Tencel fibres have higher fibrillation tendency than the Tencel LF and A100 fibres. Under wet condition lyocell fibres show an interesting property 'fibrillation'. Similarly, abrasive action on lyocell also develops micro-fibrils (or tiny fibres) on the surface. Fibrillation is the peeling back or splintering of the fibre ends making a view of tiny hairs on the surface of the yarn or on the fabric (Pearson and Taylor, 1997).

10.3.3.10 Types of lyocell

Basically, lyocell fibres are of two types as listed below. However with respect to applications, they can be further subdivided into many other forms.

10.3.3.11 Standard lyocell

Standard lyocell and Tencel A100 differ mainly in their fibrillation properties. Standard Tencel has more capability to fibrillate as compared to the A100. It is used for variety of purposes such as denim, menswear, women wears and home textiles. It gives soft, sued-like peach touch to smooth, clean and silky effect by controlling or manipulating fibrillation (Farrington and Oldham, 1999).

10.3.3.12 Cross-linked lyocell

Cross-linked lyocell; Tencel A100 and Tencel LF, introduced in 1998, as non-fibrillating lyocell fibres. Possessing the same softness, drape and bulk characteristics as standard lyocell. Lyocell LF is used in knitwear and woven fabric for shirting, sportswear and home furnishing textiles. Tencel A100 type is sensitive to alkalis but type LF is sensitive to acids (Bredereck and Hermanutz, 2005).

10.3.3.13 Properties of lyocell

Lyocell feels like silk, and drapes luxuriously. Compared to cotton, lyocell wrinkles less, is softer, more absorbent, and much more resistant to ripping. In material physical properties, lyocell is more like cotton than rayon. Like other cellulosic fibres, it is breathable, absorbent, and very comfortable to wear. In fact, lyocell is more absorbent than cotton or silk, but slightly less absorbent than wool, linen or rayon (Periyasamy et al., 2011).

10.3.3.14 Durability properties

Lyocell fibre shows some key advantageous characteristics over other cellulosic fibres; for instance, a high dry and wet tenacity and high wet modulus, but one disadvantage of this fibre is generating of fibrillation during the wet state, which causes the formation of longer and more oriented crystalline regions and smaller but more oriented amorphous regions in the fibre structure (Periyasamy, 2012). These properties allow customers great scope for making strong yarns in blend with virtually all the other commercially available staple fibres. They also lead to excellent efficiencies in converting these yarns to woven and knitted fabrics (Borbely, 2008). The properties of various fibres are given in Table 10.3 (Schuster et al., 2004).

Okubayashi et al. (2006) studied the mechanism of pill formation in lyocell fabric including fuzz formation and fibrillation in wet state. The fuzz was mainly generated by mechanical abrasion in dry condition while the fibrillation was induced by mechanical abrasion in wet condition. The pilling was formed only on the fabric treated with wash and dry treatments.

10.3.3.15 Moisture related properties

The moisture regain of lyocell is 13% and it transports moisture effectively. Garments made with lyocell therefore have a high degree of comfort and is pleasant to touch (Hada et al., 2011). Firgo et al. (2006) explored the natural hydrophilic properties of lyocell—a man-made, solvent spun cellulosic fibre to enhance the moisture handling and other properties of sportswear fabrics while at the same time improving the aesthetics.

10.3.3.16 Thermal comfort properties

Lyocell has special functional and wellness properties, cool and smooth surface, wear comfort properties due to excellent moisture transport, and temperature control. Lyocell fibres have high moisture uptake because of high porosity and ability to swell. Additional effects of high water uptake in lyocell are less growth of bacteria and thus less odour formation. Lyocell also has no electrostatic charge because there is no moisture bound inside the fibre. Therefore, there is no build-up of electrostatic charge, which may lead to unpleasant electrical shocks from textiles (Rous et al., 2006).

10.3.3.17 Comfort properties

Lyocell is skin friendly as it high in purity. Lyocell imparts excellent colour depth and lustre to fabrics which remain true, even after repeated washing. Lyocell fabric gives unique drape and fluidity (Hada et al., 2011). As with all manufactured fibres, the

Table 10.3 Comparison of properties of lyocell with different cellulosic fibres.

Property	Lyocell	Viscose	Cotton	Polyester
Dry tenacity (cN/Tex)	38—42	22—26	20—24	55—60
Wet tenacity (cN/Tex)	34—38	10—15	26—30	54—58
Dry elongation (%)	14—16	20—25	7—9	44—45
Wet elongation (%)	16—18	25—30	12—14	44—45

luster, length, and diameter of lyocell can be varied depending on the end use. Lyocell can be used by itself or blended with any natural or manufactured fibre. It can be processed in a variety of fabrications and finishes to produce a range of surface effects. With its ability to fibrillate under certain conditions, lyocell offers unusual combinations of strength, opacity and absorbency (Kadolph, 2013).

10.3.3.18   Chemical properties

The chemical properties of lyocell fibre are similar to other cellulosic or regenerated cellulosic fibres. Its surface acquires a net negative charge in water. It has resistance to alkali and to weak acids also. Lyocell on burning turns to grey ashes. It has high heat stability as compared to other regenerated cellulosic fibre such as viscose. Treatment with caustic soda at particular strength (more than 8% NaOH) affects its swelling behaviour and other physical properties (Syed, 2010).

Lyocell fibres show a notable decrease in tensile strength when treated with strong alkali solutions higher than 7.5%. In addition, IPS lyocell fibres with more softwood content display less reduction in tensile properties after strong alkali treatments (Chae et al., 2003).

10.3.3.19   Uses of lyocell

Lyocell has good resiliency: it does not wrinkle as badly as rayon, cotton, or linen, and some wrinkles will fallout if the garment is hung in a warm moist area, such as a bathroom after a hot shower. A light pressing will renew the appearance, if needed. Also, slight shrinkage is typical in Lyocell garments. Lyocell is more expensive to produce than cotton or rayon, but is included in many everyday items. Staple fibre is used in apparel items such as denim, chino, underwear, other casual wear clothing and towels. Filament fibres are used in items that have a silkier appearance such as women's clothing and men's dress shirts. Lyocell can be blended with a variety of other fibres such as silk, cotton, rayon, polyester, linen, nylon, and wool (Periyasamy et al., 2011).

10.3.3.20   Environmental impact of lyocell

In the last few years, the interest in man-made cellulose fibres has grown as a consequence of increased environmental awareness and the depletion of fossil fuels. However, an environmental assessment of modern man-made cellulose fibres has not been conducted so far. Shen and Patel (2010) assessed the environmental impact of man-made cellulose fibres by taking five staple fibre products, that is, Lenzing Viscose Asia, Lenzing Viscose Austria, Lenzing Modal, Tencel Austria, and Tencel Austria 2012. The indicators for resources include Non-Renewable Energy Use (NREU), Renewable Energy Use (REU), Cumulative Energy Demand (CED), water use, and land use. The results showed that Lenzing Viscose Austria and Lenzing Modal offer environmental benefits in all categories (except for land use and water use) compared to Lenzing Viscose Asia. Tencel Austria 2012, Lenzing Viscose Austria, Lenzing Modal, and Tencel Austria are the most favourable choices from an environmental point of view among all the fibres studied.

10.4 Conclusions

From sustainable perspective the new regenerated fibres, bamboo and lyocell are gaining significant importance in industry as the sustainable fibre. The merits of bamboo, especially the mechanically extracted fibre for use in fashion and textile are high growth of mambo plant, high functionality of the fibre and low environmental impact. Although, the chemical extraction method for bamboo fibre is similar to viscose, which polluted the planet, the mechanical extraction is free from it. With the advantages of bamboo such as comfort, high moisture regain, and biodegradability, the bamboo fibre is in the interest of a number of fashion brands. In future the fashion wardrobes will witness more garments as the marketing of the fibre grows. It has been reported that the bamboo fibres have unique properties such as excellent feel, naturally antibacterial, comfort, UV-shielding, and moisture-management characteristics, for which bamboo is being preferred.

In terms of commercially success, bamboo is not yet achieved their full potential and cleaner production processes are evolving. Till date, there are only a small number of manufacturing plants in China that manufacture natural bamboo fibre. Ecologically pioneer textile manufacturing companies like Litrax and Lenzing have already introduced greener manufacturing processes into bamboo textiles. Future research and development should focus on the scaling production of the eco-friendly, natural bamboo fabric. With abundant sources of raw materials, relatively low cost, and unique performance of bamboo fibre it is only a matter of time to develop green and pure bamboo textiles. Further, bamboo textile industry has the potential to provide livelihood for millions of people worldwide.

In the last few years, the interest in man-made cellulose fibres has grown as a consequence of increased environmental awareness and the depletion of fossil fuels. Lyocell has special functional and wellness properties, cool and smooth surface, wear comfort properties excellent moisture transport, and temperature control properties. A major driving force is the low environmental impact and utilised renewable resources as their raw materials (White et al., 2005). The lyocell fibre is the first man-made fibre which possesses the advantages of the natural ones (comfort in wear by the humidity absorption and biodegradability) and of synthetic ones (high wet and dry resistance and low shrinkage. Hence, from sustainability standpoint, both bamboo and lyocell has several advantages which will make it a future sustainable fibre in the fashion and textile world.

References

Achwal, W.B., 2000. 'Lyocell Fiber', Colourage 47 (2), 40−42.

Ajay, R., Avinash, K., 2012. Physical and UV protection properties of knitted bamboo fabrics. Textile Review 7 (10), 24−26.

Borbely, E., 2008. Lyocell, the new generation of regenerated cellulose. Acta Polytechnica Hungarica 5 (3), 11−18.

Bredereck, K., Hermanutz, F., 2005. Man-made cellulosics. Review of Progress in Coloration and Related Topics 35 (1), 59–75.

Chae, D.W., Choi, K.R., Kim, B.C., Oh, Y.S., 2003. Effect of cellulose pulp type on the mercerizing behavior and physical properties of lyocell fibers. Textile Research Journal 73 (6), 541–545.

Chen, H.L., Burns, L.D., 2006. Environmental analysis of textile products. Clothing and Textiles Research Journal 24 (3), 248–261.

Das, M., Pal, A., Chakraborty, D., 2006. Effects of mercerization of bamboo strips on mechanical properties of unidirectional bamboo–novolac composites. Journal of Applied Polymer Science 100 (1), 238–244.

Erdumlu, N., Ozipek, B., 2008. Investigation of regenerated bamboo fibre and yarn characteristics. Fibres and Textiles in Eastern Europe 4 (69), 43–47.

Farrington, D.W., Oldham, J., 1999. Tencel A100. Journal of the Society of Dyers and Colourists 115 (3), 83–85.

Fink, H.P., Weigel, P., Purz, H.J., Ganster, J., 2001. Structure formation of regenerated cellulose materials from NMMO-solutions. Progress in Polymer Science 26 (9), 1473–1524.

Firgo, H., Suchomel, F., Burrow, T., 2006. Tencel high performance sportswear. Lenzinger Berichte 85, 44–50.

Gahlot, M., Pant, S., 2011. Properties of oak tasar/viscose blended yarns. Indian Journal of Fiber and Textile Research 36 (2), 187–189.

Gericke, A., Van Der Pol, J., 2010. A comparative study of regenerated bamboo, cotton and viscose rayon fabrics. part 1: selected comfort properties. Journal of Family Ecology and Consumer Sciences 38 (1), 63–73.

Gharehaghaji, A.A., Zadhoosh, A., Khodabakhsh, M.N., 2010. Study on the mechanical damage of lyocell fibers in ring and rotor yarn spinning. Proceedings of seventh international conference on TEXSCI 1–6.

Hada, S., Mahara, D., Vaghela, D., 2011. Spin to Weave the World with Lyocell.

Iorga, I., Popescu, A., Drambei, P., Mihai, C., 1999. Certain aspects regarding the finishing of lyocell-containing woven fabrics in the Romanian textile industry. In: Proceedings of Joint Conference on Fibers to Finished Fabrics, pp. 115–122.

Joshi, H.D., Joshi, D.H., Patel, M.G., 2010. Dyeing and finishing of lyocell union fabrics: an industrial study. Coloration Technology 126 (4), 194–200.

Kadolph, S.J., 2013. Textiles. Prentice-Hall, Sydney.

Karahan, A., Öktem, T., Seventekin, N., 2006. Natural bamboo fibres. Tekstil Ve Konfeksiyon 16 (4), 236–240.

Kilic, M., Okur, A., 2014. Comparison of the results of different hairiness testers for cotton-Tencel blended ring, compact and vortex yarns. Indian Journal of Fibre and Textile Research 39, 49–54.

Li, R., Liu, G., Liu, C., Wang, R., 2004. 'Research on Bamboo Pulp Fiber's Textile Performances'. Proceedings of the textile institute 83rd world conference, pp. 14–20.

Lipp-Symonowicz, B., Sztajnowski, S., Wojciechowska, D., 2011. New commercial fibres called 'bamboo fibres'—their structure and properties. Fibres and Textiles in Eastern Europe 1 (84), 18–23.

Liu, Y., Hu, H., 2008. X-ray diffraction study of bamboo fibers treated with NaOH. Fibers and Polymers 9 (6), 735–739.

Majumdar, A., Mukhopadhyay, S., Yadav, R., Kumar Mondal, A., 2011. Properties of ring-spun yarns made from cotton and regenerated bamboo fibres. Indian Journal of Fibre and Textile Research 36 (1), 18–23.

Manian, A.P., Rous, M.A., Lenninger, M., Roeder, T., Schuster, C., Bechtold, T., 2008. The influence of alkali pretreatments in lyocell resin finishing—fiber structure. Carbohydrate Polymers 71, 664—671.

Okubayashi, S., Zhang, W., Bechtold, T., 2006. High durable cellulosic textiles—strategies for high resistance to fibrillation and pilling. Lenzinger Berichte 85, 98—106.

Okubo, K., Fujii, T., Yamamoto, Y., 2004. Development of bamboo-based polymer composites and their mechanical properties. Composites Part A: Applied Science and Manufacturing 35 (3), 377—383.

Pearson, L., Taylor, J.M., 1997. Fabric Treatment, European Patent EP 0 705 358 B1.

Periyasamy, A.P., 2012. Effect of alkali pretreatment and dyeing on fibrillation properties of lyocell fiber. In: Proceedings of the RMUTP International Conference. Textiles & Fashion, pp. 1—11.

Periyasamy, A.P., Dhurai, B., Thangamani, K., 2011. An Overview of Processing and Application of Lyocell. Available from: http://www.fiber2fashion.com/industry-article/33/3237/an-overview-of-processing-and-application-of-lyocell1.asp [7 March 2011].

Ray, A.K., Das, S.K., Mondal, S., Ramachandrarao, P., 2004. Microstructural characterization of bamboo. Journal of Materials Science 39 (3), 1055—1060.

Rous, M.A., Ingolic, E., Schuster, K.C., 2006. Visualisation of the fibrillar and pore morphology of cellulosic fibers applying transmission electron microscopy. Cellulose 13 (4), 411—419.

Sardana, A., Viswakarma, S., 2010. Green Story of Lyocell Fiber.

Sarkar, A.K., Appidi, S., 2009. Single bath process for imparting antimicrobial activity and ultraviolet protective property to bamboo viscose fabric. Cellulose 16 (5), 923—928.

Schuster, K.C., Rohrer, C., Eichinger, D., Schmidtbauer, J., Aldred, P., Firgo, H., 2004. 'Environmentally Friendly Lyocell Fibers', Natural Fibers, Plastics and Composites, pp. 123—146.

Sharma, V., Goel, A., 2011. Bamboo fiber versus cotton fiber: a comparative study. Man Made Textiles in India XXXIX (9), 313—318.

Shen, L., Patel, M.K., 2010. Life cycle assessment of man-made cellulose fibers. Lenzinger Berichte 88, 1—59.

Shen, Q., Liu, D.S., Gao, Y., Chen, Y., 2004. Surface properties of bamboo fiber and a comparison with cotton linter fibers. Colloids and Surfaces B: Biointerfaces 35 (3), 193—195.

Sunol, J.J., Saurina, J., Carrasco, F., Colom, X., Carrillo, F., 2007. Thermal degradation of lyocell, modal and viscose fibers under aggressive conditions. Journal of Thermal Analysis and Calorimetry 87 (1), 41—44.

Syed, U., 2010. The Influence of Woven Fabric Structures on the Continuous Dyeing of Lyocell Fabrics with Reactive Dyes. PhD thesis, Heriot-Watt University.

Taraboanta, I., 2012. Study on bamboo fiber process of creating properties-application. Annals of the University of Oradea Fascicle of Textiles 13 (1), 242—245.

Taylor, J., 1998. Tencel—a unique cellulosic fiber. Journal of the Society of Dyers and Colourists 114 (7—8), 191—193.

Tyagi, G.K., Goyal, A., Chattopadhyay, R., 2013. Physical characteristics of tencel-polyester and tencel-cotton yarns produced on ring, rotor and air-jet spinning machines. Indian Journal of Fiber & Textile Research 38 (3), 230—236.

Vijayakumar, H.L., Muralidhar, J.S., 2008. Design and Development of an Equipment to Measure the Hand Value of Apparel Fabric.

Vijayakumar, H.L., Muralidhar, J.S., Ramesh, S.N., 2007. Development of Silk/lyocell Blended Yarn on Short Staple Spinning System.

Waite, M., Platts, J., 2009. Engineering sustainable textiles: a bamboo textile comparison. Energy, Environment, Ecosystems, Development and Landscape Architecture 3 (5), 362—368.

Wallace, R., 2005. Commercial Availability of Apparel Inputs: Effect of Providing Preferential Treatment to Apparel of Woven Bamboo Cotton Fabric. US Patent, p. 332465007.

Wang, Y., Gao, X., 2006. Comparing on characteristics and structure between natural bamboo fiber and regenerated bamboo fiber. Plant Fibers and Products 28 (3), 97—100.

Wang, Y.S., Koo, W.M., Kim, H.D., 2003. Preparation and properties of new regenerated cellulose fibers. Textile Research Journal 73 (11), 998—1004.

Weidong, Z.S.Y., 2005. Research and development situation of bamboo fiber and its products. Cotton Textile Technology 11 (2), 2—6.

White, P., 2001. Lyocell: The Production Process and Market Development. Woodhead Publishing Limited, Cambridge.

White, P., Hayhurst, M., Taylor, J., Slater, A., 2005. Biodegradable and Sustainable Fibers. Woodhead Publishing, Cambridge.

Xu, X., Wang, Y., Zhang, X., Jing, G., Yu, D., Wang, S., 2006. Effects on surface properties of natural bamboo fibers treated with atmospheric pressure argon plasma. Surface and Interface Analysis 38 (8), 1211—1217.

Xu, Y., Lu, Z., Tang, R., 2007. Structure and thermal properties of bamboo viscose, Tencel and conventional viscose fiber. Journal of Thermal Analysis and Calorimetry 89 (1), 197—201.

Zhang, Y., 2005. The characteristics of bamboo fibers and spinning technology. New Textile Technology 13 (3), 56—58.

Sustainable polyester and caprolactam fibres

Sanat Kumar Sahoo and Ashwini Kumar Dash
Department of Textile Engineering, Odisha University of Technology and Research, Bhubaneswar, Odisha, India

11.1 Introduction

The increase in textile waste creates problems globally and it is the main contributor to greenhouse gas emissions (Nørup et al., 2019). The study shows that approximately more than 1.2 billion tons of carbon gas are released from the textile manufacturing unit per year (MacArthur, 2017). For recycling, research has been done to utilize the waste and minimize the carbon gas emission to the environment. The traditional process for the manufacturing of polyester and caprolactam fibre and the wet processing required for the fabric and garment manufacturing should be changed and the implementation of a new process will reduce the carbon footprint and be sustainable. Some researchers found that by using a recycling method, the PET fibres are recovered from textile waste and few values added products are produced from this as a by-product (Hu et al., 2018; Subramanian et al., 2020). For the environmental profits, these bio recycling methods are predictable to give the best result to conduct an organized valuation of environmental effects related to these developing approaches. For a particular time, limit, it is very difficult to assess the value of improvement in environmental impacts that are connected with the developing technologies (Chopra et al., 2020). The LCA is recommended to study the impacts of the environmental life cycle during the textile supply chain, and particularly throughout the recycling techniques. The caprolactam (nylon 6) and polyester (PET) are widely used synthetic fibres, those are produced conventionally by using a melt-spun process using typical polymers and the filaments. But the modified speciality yarns are also produced by the melt-spinning method by changing some parameters and equipment. These fibres are having some special characteristics as compared to conventional ones differing as antistatic fibres, deep-dyeable nylons, flame resistance fibres, cationic-dyeable polyesters, microporous fibres, hollow fibres, ultra-fine fibres, micro fibres etc., which gives the best result in sustainability. The newly developed fibres add value to the product with numerous advantages.

Researchers found these newly developed fibres but their effective commercialization requires the industrial ability to produce adequate products. From the technological point of view, it requires profitable accessibility of special raw materials at a reasonable cost with superior equipment to produce the speciality products and well-trained or process known technicians (Anderson, 2020). Due to using new technology and highly sophisticated machinery, the production cost will be greater as compared to conventional ones. To tackle this, a suitable market with good numbers of customers is required to appreciate the products and also are ready to pay the higher prices for them. This type of customer exists in developing countries and is also available in more advanced production units with such products for several applications. This chapter describes the newly commercialized polyester and nylon fibres and their sustainability by using different raw materials, manufacturing processes and their industrial uses. Some newly discovered principles, and techniques, which are related to industrial application, mostly during the production of yarns, and in wet processing will also be discussed.

11.2 Polyester fibre

11.2.1 History

In the early 90s era, researchers thought to develop a new type of fibre which can be manufactured by using polymers. At first, one scientist named W. H. Carothers of DuPont, USA used the mixer of ethylene glycol and terephthalic acid in the condensation method to develop polyethylene terephthalate (PET) fibre (Lewin & Pearce, 1998). But unfortunately, this work was not succussed and this project was revived by Whinfield and J. Rex Dickson, British scientists who patented the formation of polyethylene terephthalate (PET) or PETE in 1941 (Kricheldorf, 2014). The commercialization of PET was going faster after world war-II by the name Terylene with the help of Imperial Chemical Industries (ICI) of the UK and in name of Dacron in the US by DuPont (MCINTYRE, 2004). DuPont bought all the legal rights from the ICI to sell the PET as both filament and staple fibre in worldwide. The funding collection was started from the starting of 1950 to the end of 1954 for the set-up of the manufacturing unit of the PET fibre (Brown & Reinhart, 1971). But the full-scale production was started in the year 1953 and expanded at an increasing rate to 3 billion pounds nearly in 1970 (Sharp, 1964). From then, till now its sustainability rate is riding the ladder due to its physical properties, chemical properties and especially its durability in apparel and industrial applications.

11.2.2 Classification

Polyester can be classified into two types (1) Saturated (2) Unsaturated. In saturated type, the backbones of polyesters are saturated. In this, low molecular weight liquids are used as plasticizers for production. But in the unsaturated type, the backbones are comprised of alkyl thermosetting resins characterized by vinyl unsaturation and they are mainly used in reinforced plastics (Dholakiya, 2012).

11.2.3 Types

The type of polyester is dependent upon the end-use and the manufacturing process and techniques used. There are various types of polyesters manufactured worldwide and used in daily life, but three types of polyesters are mainly produced and applied in industry. These are ethylene polyester, fibrefill, PCDT polyester and plant-based polyester. Apart from these there are other categories of polyester fibres invented but these are not successful for industrial types. The recycled polyester known as r-PET is manufactured from existing types by using the recycling process. By this, sustainability can be achieved with polyester.

11.2.3.1 Ethylene polyester

The most common and highly produced form of polyester fibre is ethylene polyester (PET) (Van Uytvanck et al., 2014). It is the composition of ethylene glycol, which is derived from petroleum and with dimethyl terephthalate at high temperatures. For producing stable fibrous compounds, the primary component ethylene helps as a polymer to create bonding with other compounds. The end use of fibre is categorized into four types and is mainly dependent upon the manufacturing process.

11.2.3.2 Fibrefill

One special type of polyester filament is called fibrefill. It looks like a ball of cotton, by using the process of combed and fluffed of polyester like a softball. Polyester and other recycled materials are used for the manufacturing of fibrefill. Fibrefill stuffing is inexpensive to make, and it's insulating and filler properties make it ideal for different projects and purposes like pillows, outerwear and stuffing for stuffed animals (Schuhmann & Hartzell, 1989). The sustainability of this type is high due to its recyclability and further use in the home and automobile sector. These are having continuous filaments with the most possible high in volume.

11.2.3.3 PCDT polyester

The PCDT polyester has the same producing method as the PET polyester, however, the chemical structure is totally different from the PET polyester. This special style of polyester is created by condensing terephthalic acid with 1, 4-cyclohexane-dimethanol to form poly-1, 4-cyclohexylene-dimethylene terephthalate or the PCDT Polyester (Punyodom et al., 2006). It is having elasticity more than PET polyester. This deficiency makes PET polyester more durable. Thus, it is usually most well-liked for heavy applications like upholstery and curtains.

11.2.3.4 Plant-based polyester

The plant-based polyesters are produced from ethylene glycol reacted with dimethyl terephthalate. The primary polymer ethylene in PET and PCDT polyester is petroleum-based but in plant-based polyester, the ethylene is collected from plant sources like sugar cane (Pawar et al., 2018). The plant-based fibre is said to have the same durability and ease of processing as conventional polyesters.

Toray, Japan's biggest textile company by sales, plans to quadruple supplies of environmentally friendly materials by fiscal 2030 compared with fiscal 2013 as demand for such products grows. Virent, USA based textile company is able to create a biologically derived version of terephthalic acid, which constitutes 70% of polyester content. Poisonous parts of sugarcane and corn apparently are processed to create the chemical. Toray and Virent envision the fabric being employed in automotive interiors, sportswear and alternative product.

11.2.4   Manufacturing process

There are various types of polyester manufacturing processes adopted globally. But the sustainability of this fibre depends upon the purity of its raw material and the quality of the manufacturing process. PET products are 100% recyclable and are the most recycled plastic and polyester waste worldwide showing their sustainability. The low diffusion constant makes PET far more appropriate than alternative plastic materials to be used as a recovered recycled material. Post-consumer PET bottles and wastes are collected and processed through a series of chemical treatments to interrupt the PET into its raw materials or intermediates that are additional won't turn out recycled PET (r-PET) flakes intermediates which are further used to produce recycled PET (r-PET) flakes. PET bottles and waste containers that find their way to the landfill pose no risk of harm or leaching. Since the chemical compound is inert, it's proof against attack by micro-organisms, and won't biologically degrade.

11.2.4.1   Conventional process

This process is producing the polymer with high molecular weight and zero colour-forming particles. In a general way, PET fibre is produced in a batch polymerization process (Jaffe et al., 2020). But in some cases, for large-scale production and direct spinning, the continuous polymerization is adopted (Luzio et al., 2014). Continued process minimizes the competitive reactions and eliminates the solid polymer, subsequent blending and cost-saving by re-melting it. In comparison with batch processing, it leads the thermal degradation which helps for making high molecular weight with high strength industrial yarns for the use of tyre cords.

In batch-wise polymerization process, successful PET polymerization is monomer purity and the absence of moisture in the reaction vessel. And it is the condensation product of terephthalic acid and ethylene glycol (Nakayama et al., 2020) (Fig. 11.1).

The first stage of PET polymerization produces bis(hydroxyethyl)terephthalate (BHET) (Wilfong, 1961). In the next step of polymerization is melt condensation to high molecular weight. In this reaction step a molecule of glycol, building polymer molecular weight is split off from an ester interchange reaction between two molecules of BHET. The melt-polymerization temperature is maintained are at or above 285°C and viscosities are in the order of 3000 P, with uniform stirring inside the vessel (Aguilar & Cristina, 2013; Chrissafis et al., 2006) (Fig. 11.2).

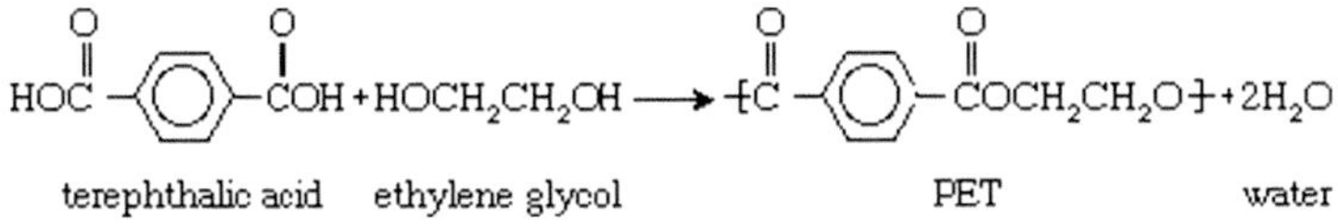

Figure 11.1 Condensation process of terephthalic acid and ethylene glycol.

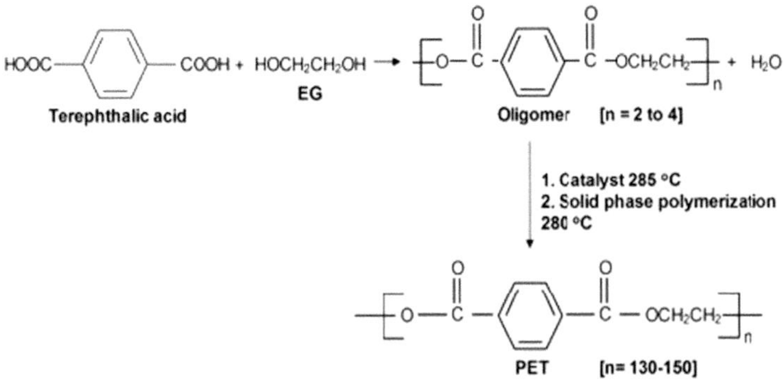

Figure 11.2 The polymerization process of terephthalic acid and ethylene glycol.

The polymer can be pelletized for following melt spinning (batch processing) after achieving the target molecular weight or fed directly into a spinning machine for the conversion of fibre (continuous processing). In the continuous polymerization (CP) or spin-draw process, the spun fibre is fed directly to a draw frame. By using solid-state polymerization, the molecular weight of PET pellets can be supplementarily increased. In this process at about 160°C, the dried PET chips are first crystallized to avoid the as-polymerized chip from sticking together (Ribeiro et al., 2014). Then chips are heated just below the melting point under high vacuum. Also, extreme dryness is required to advance the molecular weight uphill to values of inherent viscosity (Luo & Zhu, 2011).

There is a wide range of reactions used in the industry to produce polyester. Some important reactions are acids and alcohols, alcoholysis and or acidolysis of low-molecular-weight esters or the alcoholysis of acyl chlorides, which can produce PET. The typical polycondensation reactions are shown in the following figure for polyester production (Fig. 11.3).

Whinfield and his team also invented Polybutylene terephthalate (PBT) and Polytrimethylene terephthalate (PTT) in addition to PET and also melt-spun them into fibres and measured their fibre properties (East, 2009). Butane-1,4-diol was used as the glycol in place of ethylene glycol to make PBT. It was found that PBT was easier to melt-spin due to its lower melting point than PET. It was found that some properties like elasticity and resilience of PBT are more than PET. The PBT fibre looks very white and stayed white even after several uses. Because butanediol is more expensive than ethylene glycol, the cost of PBT is more costly than PET and nylon (Fig. 11.4).

The last polyester fibre that the Whinfield team worked on was polytrimethylene terephthalate (PTT). It was made from propane-1,3-diol, variously called PDO or

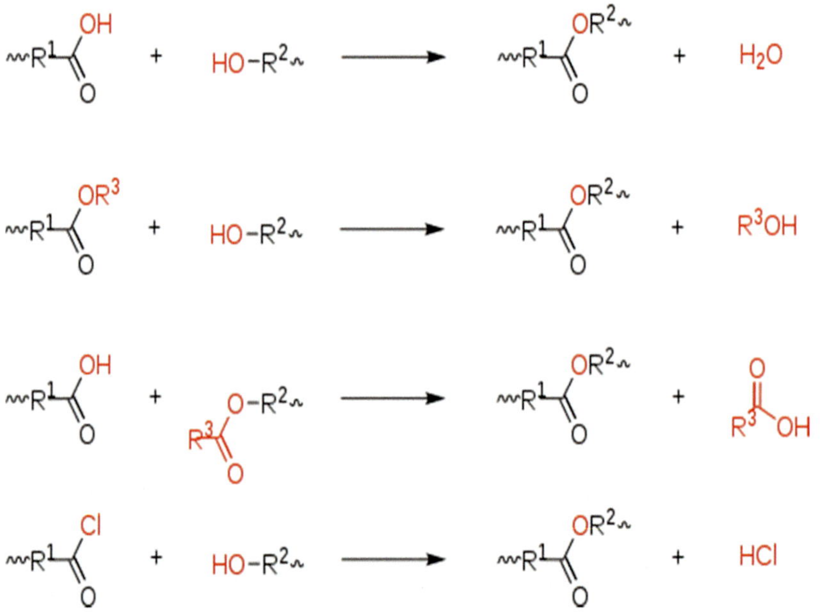

Figure 11.3 The different polymerization processes of PET.

trimethylene glycol (East, 2009). PDO is said to be considerably cheaper than butanediol, even though still more expensive than ethylene glycol restricts the popularity of PTT in the market. In the commercial world, it was very difficult to survive on PBT and PTT as compared to PET, though this ware has some best physical properties.

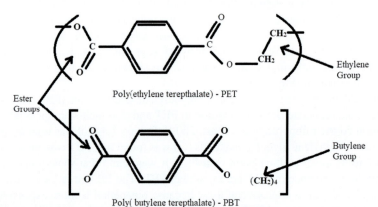

Figure 11.4 The polymerization process of PET and PBT.

11.2.5 Characteristics of polyester

There are some basic properties of polyester described below (Scheirs & Long, 2005):

- It is easy to wash, dry and wear polyester fabrics.
- These are strong and durable and can resist many chemicals.
- It is having good abrasion resistance but produces a pill.
- Also its wrinkle, stretching or shrinking behaviour is good.
- It can withstand very high heat to melt.
- It is hydrophobic in nature but absorbs oil and grease, which is very difficult to clean.
- Polyester does not absorb water, but newly developed polyesters with polypropylene and micro-fibres can absorb water away from the skin.

11.2.6 Properties

11.2.6.1 Physical properties

Moisture regains %, specific gravity and tenacity

The moisture regains ranges between 0.2 and 0.8% of polyester is very low (Bendak & El-Marsafi, 1991). They do not have wicking ability because of it. Its specific gravity depends upon the type of polyester fibre ranging from 1.38 or 1.22 used in industrial purposes (Elbehiry et al., 2021). The fabrics made from polyester fibres are medium in weight. Polyester fibres are having good tenacity due to the hydrogen bonds and their crystalline nature. The range is 5−7 g/den, and varies according to the type of fibre (Elbehiry et al., 2021). Its tenacity can't change even if it is wet.

Heat effect

The melting point of polyester fibre ranges from 250 to 300°C and is a poor conductor of heat with low resistance toward heat (Ducheyne, 2015). It can easily shrink and melt on heating. The shape can be changed by heating them.

11.2.6.2 Chemical properties

Effect of alkalis

At high temperatures, polyester fibres have good resistance to weak alkalis. But at room temperature, it has moderate resistance to strong alkalis. It can be degraded with strong alkalis at elevated temperatures (Nateri & Goodarz, 2013).

Effects of acids

There is no effect of acids, it may be weak or strong at low or high temperatures. But it has some effect if it is in contact with acid for several days. It can degrade the fibres with exposure to boiling hydrochloric acid, and 96% sulphuric acid (Bendak & El-Marsafi, 1991).

Effects of solvents

Generally, polyester fibres are resistant to organic solvents. For cleaning and stain removal some chemicals are used but they do not damage it. Hot m-cresol destroys the polyester fibres. Bleaching agents and oxidizing agents do not damage polyester fibres (Bendak & El-Marsafi, 1991).

Sunlight and micro-organisms

It is highly resistant to acidic pollutants and can withstand the sun's UV rays present in the atmosphere. It can resist bacteria and other micro-organisms (Vilela et al., 2014).

Colour fastness

It is difficult to penetrate dye molecules in the polyester fibre during the dying process. It retains its colour after regular wash (Clark, 2011).

11.2.7 Applications of sustainable PET fibres

Polyester is the world's highest used fibre and also top in the rank in the synthetic category, it can be used as fibre, yarn and fabric form. As per data collected, its percentage of manufacturing on a global basis is an average of 58% of total fibre production (Opperskalski et al., 2019). It can be useful in the non-woven sector for producing blankets, comforters, carpets, cushioning and insulating in pillows, upholstery padding and upholstered furniture by using staple or filament fibres. In apparel and home furnishings products, knitted or woven fabrics made from polyester or by blending with other fibres, thread or yarn are used. The maximum use of sewing thread used in the industrial sector is the polyester or polyester blended thread. In technical textile and industrial areas polyester fibres, yarns and ropes are used in air bag, tire reinforcements, safety belts, tapes, fabrics for conveyor belts, geotextiles, filtration and plastic reinforcements with high-energy absorption (Deopura et al., 2008) (Fig. 11.5).

11.2.8 Dyeing and finishing

PET is normally dyed with disperse dyes because it is not having reactive dye sites, which act primarily through diffusion and are bound in the fibre through secondary bonds. Thermosol is a continuous dyeing process mostly used to apply to disperse dyes to PET fabrics. The fabrics are padded with the dyes, dried and then heated (205° to 240°C for 30−60 s) to diffuse and fix the dyes in the PET fibres (Brown & Reinhart, 1971). The 50: 50 and 65: 33 blends of PET and cellulosic fabrics are commonly dyed by this procedure (Ahmed et al., 2020). There are some modified polyesters introduced by several manufacturers are having anionic dye sites.

These fibres can be dyed with dispersing and cationic dyes; when used in blends with regular polyester, unique multi-colour effects can be obtained. Thermally stable, melt-soluble dyes have also been made to colour the polymer prior to spinning (Avny & Ludwig, 1986).

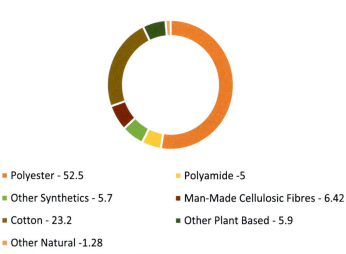

Figure 11.5 Global fibre production in 2019.

The chemical finishing of polyester-cellulosic blends is a very important part of a durable press. It is not a finish for PET, but a system whereby the cotton in the PET blends is given a resin cross-linking treatment to improve its recovery power and wrinkle resistance. The best-known durable press system involves the following five steps: (i) the dyed PET and cotton blend is impregnated with an amino-plast resin derivative, (ii) the fabric is dried, (iii) the garments are cut and sewn, (iv) creases are pressed into the garments and (v) the garments are placed in an oven and the resins cured (Paul, 2014). This is referred to as a delayed or post-cured system. Curing resins before cutting and sewing the fabric, as is the case with durable press sheets, is referred to as pre-curing. The concept of using a blend with a higher percentage of PET to produce a no-resin durable press has been recently introduced.

11.2.9 LCA of sustainable fibres

11.2.9.1 Effluent treatment

Wastewaters from the polyester fibre manufacturing plant generally contain strong and toxic parameters. By oxidation process using H_2O_2, raw wastewater responded well. And it was observed that H_2O_2 at the stoichiometric dosage provides a 30% COD removal within a short reaction time. The removal efficiency could be increased by increasing oxidant doses by up to 70% (Meric et al., 1999). The dilution did not improve or affect the oxidation efficiency. The Chemical oxidation treatment is a costly method; however, it secured a significant COD for the process wastewater investigated. Due to the low wastewater volume involved in the treatment, it proved to be worthy of consideration. To treat the wastewater biologically, it was found to be oxidizable without dilution, but with a medium of H_2O_2 dosage, the removal efficiency is up to 60% COD (Meric et al., 1999). It requires individual solutions, requires several stages of treatment, they are cost-effective, expensive chemicals required and the energy requirement is high for chemical oxidation for high aeration volumes. It was also found that by using by Cyanobacteria and Archaea, fibreglass-reinforced polyester plastic; polyester-polyurethane and poly(methyl methacrylate) can be degraded (Cappitelli & Sorlini, 2008).

11.2.10   Impact on environment

Though polyester is a synthetic petroleum-based fibre, therefore it is very difficult to renew carbon-intensive resource (Karthik & Rathinamoorthy, 2018). For one year's consumption of polyester, nearly 70 million barrels of oil are used around the world. Now it is the most commonly used fibre in making clothes. Researchers found that decomposing polyester takes more than 200 years (Balzani & Armaroli, 2010). Producing fibre requires refining the crude oil into petroleum. This process releases toxins into the environment which can harm living things both in the water and on land (Speight, 2014). The further refinement processes to produce ethylene that is used to make polyester exits more toxins into the environment. The environment and the ecosystems of the area surrounding polyester fibres producers consume the harmful synthetic by-products at the time of transforming ethylene to PET. Also, a number of dyes and treatment processes used by polyester fabric manufacturers release effluent into the water.

Polyester does not naturally degrade in the environment like wool, cotton, or silk. It's impossible to know accurately how long polyester will remain in the Earth's ecosystems before it degrades (Hearle & Morton, 2008). Synthetic fabrics like polyester may take centuries to fully break down due to natural environmental conditions agreed by environmental scientists.

Overall, at every stage in its production polyester harms the environment. The newly invented plant-based polyester fibre would seem to be a step toward reversing this unfortunate state of affairs. But it's unclear whether this product will gain grip on the textile market enough to make an impact on petroleum-based PET.

11.2.11   Recycling

The sustainability of polyester is dependent upon the recycling process of its product and industrial waste. Recycled polyester is made from recycled plastic bottles and the waste collected from the polyester yarn and fabric cuts out the need for petroleum and coal extraction. Recycled polyester literally starts at the dump to collect plastic bottles and wastes from the textile industry that doesn't belong in landfills. From there, the plastic bottles and wastes are shredded into flakes by a machine. Then those flakes are collected and processed through melt spinning extrusion to make them into yarn.

The recycling of polyester can be done either by physical (or) chemical methods. At the time of direct melt processing, an extruder is used for pelletization into chips that produce value-added products. Chemical recycling is useful and by adopting this process it can remove any additive contamination and any type of PET waste. Hydrolysis (Pitat et al., 1959) or methanolysis (Anon, 1959) or glycolysis (Macdowell, 1965) can be used in the chemical recycling method. The hydrolysis process is carried out under high-pressure water or steam. It may be done with the help of aqueous sodium hydroxide solution at 180−250°C under pressures of 1.4−2 MPa or strong inorganic acids such as sulphuric acid at 85−150°C and nitric acid (Al-Sabagh et al., 2016). The methanolysis process is the de-polymerization of PET waste to produce a large quantity of methanol. This process is carried out in the presence of a catalyst at high pressure and temperature for 3−5 h. It releases an excess of glycols such as MEG at elevated temperatures (180−250°C) into either the monomers namely PTA or DEG, MEG or oligomers (Karayannidis & Achilias, 2007). By polycondensation reaction, these monomers or oligomers are reconverted into PET or by polyaddition reaction to unsaturated polyesters. carpets, needle-punched non-woven, tennis ball covers, etc. and for all the required applications where transparency and colour are not of prime importance.

11.3 Caprolactam or nylon fibre

11.3.1 History and demand

The concept of the caprolactam was explained in the late 1800s and by the hydrolysis of the process cyclization of ε-aminocaproic acid, it was produced (Ebata & Morita, 1959). Its demand grows rapidly to fulfill the requirements of the world population. The maximum amount of caprolactam is preferred for the use of filament and fibre. It is the main monomer for the production of nylon 6, and the basic research of this product was started by Wallace Corothers at DuPont in 1928 (Strom, 2017). Paul Schlack, a German researcher developed a polymer at IG Farben known as Nylon 6 or polycaprolactam to repeat the properties of nylon 6,6 without changing the patent on its production in the year 1939 (McIntyre, 2005; Travis et al., 1998). Reports say that the global caprolactam market was equal to 4.603 million tons in 2013. The global market is predicted to reach 6.564 million tons, by 2023 (McCarthy, 2016). The production of caprolactam depends upon the end use of the material. An average of 60−70% is used for the manufacturing of nylon fibre, the rest is utilized by nylon resin with other applications (Fig. 11.6).

11.3.2 Manufacturing process

11.3.2.1 Caprolactam

For the manufacture of nylon 6, Caprolactam is preferred to w-aminocaproic acid because it is easier to make and purify. Due to its commercial implication, many

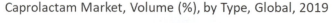
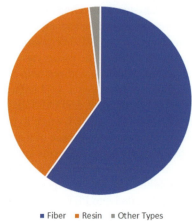

Figure 11.6 Global caprolactam market in 2019.

methods have been developed for the production of caprolactam. It was calculated that synthesizing from cyclohexanone produces 90% of caprolactam (1), then it is first converted to its oxime (2). This oxime treated with acid makes the Beckmann rearrangement to produce caprolactam (3) (Dahlhoff et al., 2001) (Fig. 11.7).

The preparation of cyclohexanone is generally done either from phenol or from cyclohexane. Using phenol, it is comprised of a two-stage process. In the first stage, the cyclohexanone is formed at about 180°C by the reaction of phenol and hydrogen in the presence of a nickel catalyst. After that dehydrogenation stage is followed by the hydrogenation stage. This stage forms cyclohexanone in the gaseous phase at about

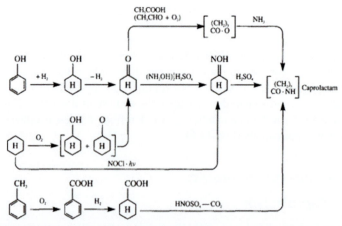

Figure 11.7 Various processes for caprolactam synthesis.

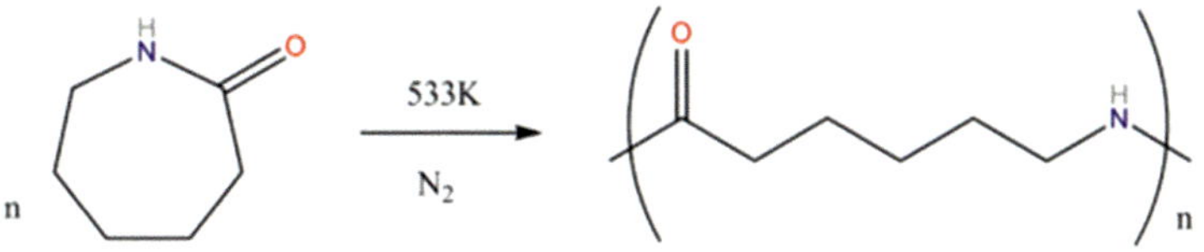

Figure 11.8 Ring-opening polymerization of nylon 6.

400°C by the reaction of cyclohexanol in the presence of a copper catalyst (Brydson, 1999).

For the manufacturing of Nylon 6, almost all caprolactam produced is used. The conversion from caprolactam to nylon 6 involves a ring-opening polymerization (Fig. 11.8).

11.3.2.2 Nylon 6

Typically, nylon 6 is a nylon Z type of polymer. The Z represents the number of carbon atoms present in the monomer. For the production of nylon 6, purity of the caprolactam is required having permanganate number 71; molar ratio %; iron content of 0.5 ppm (Deopura et al., 2008) (Fig. 11.9).

For the manufacture of nylon 6 both batch and continuous processes have been used. In a typical batch process a vessel in which the caprolactam, a molecular weight regulator, e.g., acetic acid and water as a catalyst are charged. And these are also reacted under a nitrogen blanket at 250°C for about 12 h. The product consists of both high molecule and low molecule monomers having a ratio of 9:1. The low molecular weight materials may be removed by leaching and/or vacuum distillation to achieve the best physical properties. Though it takes more time for production and the cost of operation is higher than the other routes. Also, the quality of nylon 6 may vary from batch to batch. The water partition coefficient will change during the polymer produced in melt spinning, causing polymerization of the residual caprolactam. Before the production of nylon 6 in melt spinning, it is necessary to be brought down to a minimum (less than 1%). To avoid problems during melt-spinning, the residual water content must be very low (0.01−0.04%) (Hufenus et al., 2020) in the polymer.

In the continuous process, the reservoirs are used to maintain the reactants at a temperature of about 250°C which continuously feeds reaction columns. The VK tube (from the German *Vereinfacht Kontinuierlich*) is 8−10 m high and its diameter depends upon the required production capacity (Gupta & Kothari, 2012). To move the reactants towards downward flow, inclined baffles are placed in the tube. To facilitate the escape of volatiles bubble caps are occasionally provided in the baffle plates. The feed zone is placed at the top position and acts as a reflex system. Several inlets and

$$n\ [HN-(CH2)5-CO] \rightarrow H[-HN-(CH2)5-CO-]\ n\ OH$$

Figure 11.9 Polymerization process of nylon 6.

outlet pipes are provided for feed and outgoing vapours. An extruder having high capacity is placed at the bottom of the tube. It helps to pump out continuously the molten nylon 6 from the VK tube in the form of a strand. Then these are being cooled to a temperature just above Tg (about 50°C) is guided to a cutter (Jiang, 2016). There is also a feed pipe for nitrogen at the top.

The feed consists of molten CL with the requisite amount of water (catalyst), and viscosity stabilizer (acid or base). As the molten enters the feed zone, the water bubbles try to escape from the molten mixture due to its boiling nature. The material moves downwards and the step addition process is complete within 6 h. After polymerization is complete, the highly viscous melt passes through the extruder, from which it comes out in the form of strands, which are cooled and cut into chips.

In modern technology, the development of a mould for the polymerization casting of nylon 6 in situ. A typical system involves anionic polymerization, in which it uses acetic caprolactam is required about 0.1–1 mol.% as a catalyst and sodium salt of caprolactam has 0.15–0.50 mol.%. Normally the initial reaction temperature ranges between 140 and 180°C but this rises by about 50°C during polymerization (Wendel et al., 2020). These casting techniques can produce mouldings up to one ton in weight. Other than this method, reaction injection moulding techniques, have also been adapted for nylon 6 as a variation of the polymerization casting technique.

11.3.2.3   Nylon 6,6

However, the foremost common caprolactam fibre is termed as nylon 6,6. To produce this fibre we have to mix two molecules produced from petroleum: adipic acid and hexamethylene diamine. Nylon 6,6 is synthesized by a polycondensation reaction between hexamethylene diamine and adipic acid. In a reactor, equivalent amounts of hexamethylene diamine and adipic acid are combined with water which produces nylon salt, an ammonium/carboxylate mixture. Then, either in batch wise or continuous polymerization process nylon salt goes into a reaction vessel (US patent 2130523 & Carothers) (Fig. 11.10).

The amide bonds from the acid and amine are formed by the polymerization reaction after removing the water. After that, the molten nylon 6,6 is formed. At this point, it can either be extruded or granulated. Also, the filaments are directly produced by

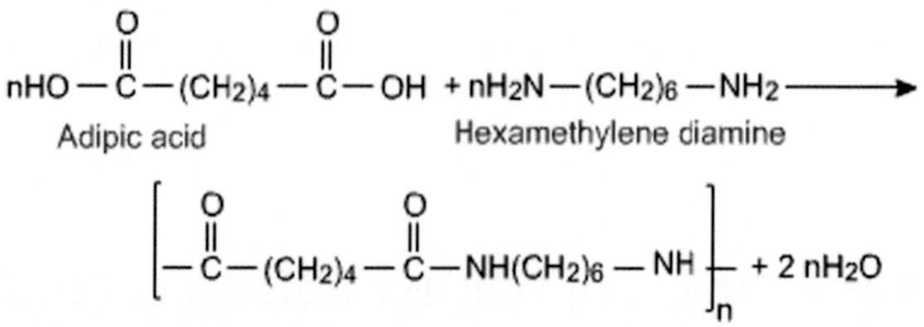

$$nHO-\overset{\overset{\displaystyle O}{\|}}{C}-(CH_2)_4-\overset{\overset{\displaystyle O}{\|}}{C}-OH \ +nH_2N-(CH_2)_6-NH_2 \longrightarrow$$

Adipic acid   Hexamethylene diamine

$$\left[-\overset{\overset{\displaystyle O}{\|}}{C}-(CH_2)_4-\overset{\overset{\displaystyle O}{\|}}{C}-NH(CH_2)_6-NH\right]_n + 2\ nH_2O$$

Figure 11.10 Polymerization process of nylon 6,6.

cooling the fibres extruded from the spinneret. By combining different starting molecules, we can produce different types of nylon, but the process will remain the same.

11.3.2.4 ECONYL®

ECONYL® is a regenerated-nylon yarn. It is manufactured by recycling the nylon with an indefinite number of recycling processes without affecting the quality of the material. It is invented by to research associations, Prada and the textile yarn producer Aquafil, an Italian company. Recycling discarded plastic and nylon waste, which are collected from landfill sites and oceans across the whole planet are used as the raw material for producing ECONYL® (Biloslavo et al., 2020).

This type of fibre is produced in three steps and is discussed below.

1. Collection of materials: It is the gathering of wastes from industry, fishing nets, carpets and discarded nylon are sorted and cleaned to collect the recuperated nylon.
2. Regeneration and purification: By using the de-polymerization procedure, the recuperated nylon is recycled to give the raw material purity.
3. Production stage: The recycled materials are polymerized into polymers and by melt-spinning filaments are produced in the plants Ljubljana, Slovenia and Arco and Italy.

11.3.3 Characteristics of nylon

- It has the best shiny property.
- Moderate elastic recovery but can be extensible by blending with spandex.
- Very strong in nature and used for technical textile.
- It has damage resistant to oil and many chemicals.
- Nylon has good resilient property. It does not absorb water which helps to dry it quickly.
- They are wrinkle and abrasion resistant but pills are generated due more to abrasion.

11.3.4 Properties

11.3.4.1 Physical property

Tenacity, density and moisture regain %

All the aliphatic polyamides including nylon 6 are having linear polymers and thus thermoplastic. The polymers are soluble only in a few liquids of similar high solubility parameters due to the high cohesive energy density and their crystalline state. At room temperature in dry conditions and at low frequencies, the electrical insulation properties are quite good.

The colour of the Nylon 6 is generally white but its colour can be changed by adding dye molecules in a solution bath prior to production to get a different colour. Its tenacity is $6-8.5$ gf/D with a density of 1.14 g/cm^3. Its elongation at break is $15-45\%$ and the MR% value ranges from $3.5-5\%$ (Kojima et al., 1993).

Heat effect

It has melting point due to its high intermolecular attraction to polymers. Its melting point is at 215–228°C and can protect heat up to 150°C on average (US patent 2130523 & Carothers). The glass transition temperature of Nylon 6 varies as per its state of filament and atmospheric condition. It is also found that Tg also depends on a number of (CH_2) units present in the polymer. As (CH_2) unit increases in the polymer Tg value decreases. The Tg value of amorphous nylon 6 varies from around $-10°C$ in wet conditions to around 90°C in dry conditions. The range of Tg values for nylon is from 40 to 55°C and may increase to around 90°C in dry conditions (Parodi et al., 2018).

11.3.4.2 Chemical property (Vagholkar, 2016)

Effects of alkali and acids

Nylon is substantially inactive to alkalis. By being attacked by mineral acids, nylon 6 is disintegrated or dissolved almost. But it is inactive against dilute acetate acid and formic acids even at a boil. It can be dissolved by using concentrated formic acid. Nylon 6 can resist diluting boiling organic acid.

Effects of organic solvent and bleaches

Nylon cannot change its behaviour by most of the solvents. It can be dissolved by phenol meta-cresol and formic acid. But solvents used in washing and dry cleaning do not damage it. There are no changes formed in the nylon due to the attack of oxidizing and reducing bleaches. But it may be harmed by chlorine attack and with the contact of strong oxidizing bleaches.

Effects of light and biological impact

No discolouration. Nylon 6 losses its strength gradually on prolonged extension. Neither micro-organism nor moth, larvae attack nylon.

Effects of electrical and flammability

It has high insulating properties lead to static charges on the fibre. It burns slowly at a high temperature.

11.3.5 Application of sustainable caprolactam fibres

The global consumption of caprolactam fibre is approximately five million tons per year. It is mainly used for the production of Nylon 6 filament, fibre and plastics. Several pharmaceutical drugs are manufactured by using caprolactam including pentylenetetrazol, meptazinol and laurocapram (Dahlhoff et al., 2001).

Nylon 6 fibres, produced from the caprolactam are used in both apparel and industrial sectors. Light-weight woven and knitted garments from fine yarn are produced from nylon 6. Socks are generally made from nylon due to it does not require ironing. Sarees made from fine yarn are widely used and for decorative purposes furs from nylon are also popular for their long life and recovery behaviour. Nylon is used blending with wool to prepare ladies' stockings for its durability.

In technical textiles, nylon is chosen amongst the best carpet fibres due to its best aberration resistance and resilience. Nylon filaments produce safety belts in the automobile sector, hoses, light-weight canvas for luggage and typewriter ribbons. For making hot air balloons, polyurethane-coated nylon fabrics are used. Sterilizable mouldings from nylon plastics have found application in medicine and pharmacy (Deopura et al., 2008). Film for the packaging applications and combs as a beauty product is produced from nylon due to its durability. Some other applications like brush tufting, wigs, sports equipment, surgical sutures, braiding and outdoor upholstery are manufactured with nylon mono-filaments. Some uses include cable sheathing, flexible tubes for petrol and other liquids, rods for subsequent machining, piping for chemical plants and conveyer composite belts. In aerospace and tennis racket applications, carbon-fibre-reinforced nylon 6 mixtures have been commercially used. More recently interest has been shown in the appearance of exceptionally tough nylon plastics.

Multifilament nylon yarns find extensive applications for reinforcing rubbers such as tires, for trucks and aeroplanes are made from nylon tire cords. Due to excellent elastic recovery and high wet strength, fishing nets and sail cloths are mostly made from nylon twine. The underside of the conveyor belts by using nylon/polyester blend to improve wear resistance counter to the machine frame. Due to diameter swelling behaviour nylon is suitable for the umbrella cloth.

11.3.6 Dyeing and finishing

The range of dyes suitable for nylon include direct, acid or metal-complex, basic, reactive, disperse or reactive disperse and soluble vat dyes (Shenai & Parameswaran, 1980). For the production of cheaper quality with dull shade, direct dye can be applied. For Nylon 6, acid dyes are recommended for dyeing because of its ionic bonds. It results bright shades with moderate to poor wash fastness. For deep shade, metal-complex dyes are used. Basic dyes can be used but they are rarely used due to their high cost and produces both poor light and wash-fast shades. Cationic dyes give excellent results but these are costly, for this reason, these are not used in industry. Some selective reactive dyes produce bright shades with poor fastness while dispersed reactive dyes produce bright-fast shades. Disperse dyes are occasionally used on nylon as dyeing lacks wash-fastness (Chakraborty, 2015).

The recommended dye for nylon 6 is acid dyes, which are highly water-soluble and have better light-fastness than basic dyes. For dyeing the 100% nylon fabric, the dye material with wetting agent and acetic acid is mainly required. The recipe for the acid dyeing process is different for different types of fibre and also for the separate quality of dyed materials to be produced. One of such recipes is given here:1% Owf (stock solution—1%), Wetting agent: 1 g/L (stock solution—1%), Acetic acid: 0.8 g/L (stock solution—1%) to maintain the pH 4.5—5.5 at 100°C for 60—70 min is carried out for better results. Then washing is followed by drying for the best washing- and colour-fastness result. The fastness of dyed nylon can be improved by using tannic acid

(2%) and tartar emetic (potassium antimony tartarate, 2%) at 60–90°C for 30 min in post-treatment (Chakraborty, 2015).

11.3.7 LCA of caprolactam fibres

In current years, due to the regular increase of the production quantity of caprolactam plant, its wastewater discharge amount exceeds the limit of the sewage treatment plant. The wastewater from the production unit contains cyclohexanone, benzene, cyclohexane, cyclohexanone fat, caprolactam, organic acid, ammonia nitrogen, etc., which is considered by a large amount of water, poor biodegradability, variable water quality and difficult treatment (Wang, 2020). Traditionally anaerobic-aerobic (A/O)—membrane bioreactor (MBR) process is used for the waste-water treatment. Nowadays the advanced caprolactam wastewater is treated by electrocatalytic oxidation technology in the chemical plant, to control the COD, NH_3 and chroma value of effluent to reach the first-class standard. Caprolactam product is an irritant and is slightly toxic, with an LD50 of 1.1 g/kg (rat, oral) (Acid). In 1991, U.S. has included it as a hazardous air pollutant on the list by the U.S. Clean Air Act of 1990. But as per the request of the manufacturers, it was subsequently removed from the list in 1996 (Caldwell et al., 1998).

Flavobacterium sp. and Pseudomonas sp. stated that Nylon 6 oligomers can be degradable, but not polymers. It is also studied that by oxidation process by using certain white-rot fungal strains can also degrade Nylon 6. Nylon 6 has strong inter-chain interactions of hydrogen bonds which decreases its biodegradability as compared to aliphatic polyesters (Negoro, 2000).

The treatment of waste-water, which contains ε-caprolactam from a Nylon 6 manufacturing plant from the activated sludge was studied. The activated sludge showed better results by synthesizing from ε-caprolactam-utilizing bacteria with respect to BOD removal, sludge volume index (SVI) and transparency of treated water. For this wastewater, the optimum BOD loading was estimated to be in the range of 0.35–0.40 kg BOD/kg MLSS/day. Production of excess activated sludge was 10% of the BOD loaded (Kageyama & Kosuke, 1988).

11.3.8 Recycling

It is very difficult to recycle the solid waste of nylon 6 (fibres, yarns) to produce filament yarn. To reuse the solid waste, the de-polymerization process is adopted to make it into the monomer. DMF which is a high boiling solvent can be used for the cleaning waste and dissolving it. The dissolving process is carried out by using two separate temperature ranges. The waste in the filament can dissolve without reacting with nylon 6 at low temperatures (70–100°C) (Moran & Porath, 1980). Then at an elevated temperature, the cleaned polymer is dissolved in a fresh solvent. In another process, nylon solid waste is treated with normal water to get a product with a low degree of polymerization. An autoclave is used in which the waste is charged and nitrogen is mixed within the air, and steam is applied to the surrounding jacket. It is a slow process in which the polymer is hydrolyzed into the desired degree of polymerization for reuse

(Datye, 1991). Generally, the de-polymerized process of waste produces the monomer caprolactam or aminocaproic acid.

In industrial use, nylon 6 is heated with superheated steam to 200–400°C to convert the solid waste into caprolactam. If required, it can be treated under pressure, in the presence of non-volatile organic or inorganic acids and alkalis (Datye, 1991).

11.3.9 Recovery of caprolactam

It is needed for economic reasons to reuse the caprolactam and low molecular weight oligomers by recovering them. It is studied that only about 4% of caprolactam is lost in the drainage due to low molecular weight at the time of the recycling process (Datye, 1991).

Without any purification treatment, the caprolactam refined from the extractables is pure enough for reuse. During the depolymerization process, numerous acidic and basic impurities are moulded by the ruin of the polyaramid and spin finishes on the fibre waste (Reichert et al., 2020). The rate of side reactions increases, as the temperature of depolymerization increases. It produces black tarry oil, volatile bases and other impurities in large volumes. It is very necessary to meet the standard specifications for the recovery of the caprolactam so that it can be used for polymerization (Datye, 1991).

11.4 Conclusions

For sustainability purposes, all the synthetic fibres including PET and Caprolactam should be biodegradable by modifying their structure or spinning technique. Their raw material for the production requires pure raw material for producing the best quality of fibre. The emission of waste-water and gas to the environment should be recycled or renewable for further use. If some chemicals in the waste-water is emitted into the environment, then it will harm the ecosystem. The BOD and COD values should be maintained as well as the solid waste generated in the industry. If the total waste produced in the society and from the industry can be reusable then the pollution can be controlled which will be beneficiary to the society. For the above requirement, some physical and mechanical properties are to be enhanced by interfering with the polymerization and spinning process.

11.5 Sources for further information

To gather more knowledge for further information and reading in the form of specialized reviews, monographs, symposia series, book chapters, books are entirely dedicated to regarding the sustainable polyester and caprolactam fibre.

1. Deopura, B.L., et al. (Eds.), 2008. Polyesters and Polyamides. Elsevier.
2. Brydson, J.A., 1999. Plastics Materials. Elsevier.

3. Gupta, V.B., Kothari, V.K. (Eds.), 2012. Manufactured Fibre Technology. Springer Science & Business Media.
4. McIntyre, J.E., (Ed.), 2005. Synthetic Fibres: Nylon, Polyester, Acrylic, Polyolefin. Taylor & Francis US.
5. Eichhorn, S., et al., (Eds.), 2009. Handbook of Textile Fibre Structure: Volume 1: Fundamentals and Manufactured Polymer Fibres.

References

B. Acid, "Scientific Committee On Consumer Products SCCP."

Aguilar, L., Cristina, 2013. Bio-based polyesters from cyclic monomers derived from carbohydrates.

Ahmed, N., Nassar, S., El-Shishtawy, R.M., 2020. A novel green continuous dyeing of polyester fabric with excellent color data. Egyptian Journal of Chemistry 63 (1), 1—14.

Al-Sabagh, A.M., et al., 2016. Greener routes for recycling of polyethylene terephthalate. Egyptian Journal of Petroleum 25 (1), 53—64.

Anderson, D.M., 2020. Design for Manufacturability: How to Use Concurrent Engineering to Rapidly Develop Low-Cost, High-Quality Products for Lean Production. CRC press.

Anon, 1959. Recovery of dimethyl terephthalate from crude polyesters. GB Patent 823, 515.

Avny, Y., Ludwig, R., 1986. Chemical modification of polyester fiber surfaces by amination reactions with multifunctional amines. Journal of applied Polymer Science 32 (3), 4009—4025.

Balzani, V., Armaroli, N., 2010. Energy for a Sustainable World: From the Oil Age to a Sun-Powered Future. John Wiley & Sons.

Bendak, A., El-Marsafi, S.M., 1991. Effects of chemical modifications on polyester fibres. Journal of Islamic Academy of Sciences 4 (4), 275—284.

Biloslavo, R., et al., 2020. Business Model Transformation Toward Sustainability: the Impact of Legitimation. Management Decision.

Brown, A.E., Reinhart., K.A., 1971. Polyester fiber: from its invention to its present position. Science 173 (3994), 287—293.

Brydson, J.A., 1999. Plastics Materials. Elsevier.

Caldwell, J.C., et al., 1998. Application of health information to hazardous air pollutants modeled in EPA's cumulative exposure project. Toxicology and Industrial Health 14 (3), 429—454.

Cappitelli, F., Sorlini, C., 2008. Microorganisms attack synthetic polymers in items representing our cultural heritage. Applied and Environmental Microbiology 74 (3), 564—569.

Chakraborty, J.N. (Ed.), 2015. Fundamentals and Practices in Colouration of Textiles. CRC Press.

Chopra, J., et al., 2020. Environmental impact analysis of oleaginous yeast based biodiesel and bio-crude production by life cycle assessment. Journal of Cleaner Production 271, 122349.

Chrissafis, K., Paraskevopoulos, K.M., Bikiaris, D.N., 2006. Effect of molecular weight on thermal degradation mechanism of the biodegradable polyester poly (ethylene succinate). Thermochimica Acta 440 (2), 166—175.

Clark, M. (Ed.), 2011. Handbook of Textile and Industrial Dyeing: Principles, Processes and Types of Dyes. Elsevier.

Dahlhoff, G., Niederer, J.P.M., Hoelderich, W.F., 2001. ε-Caprolactam: new by-product free synthesis routes. Catalysis Reviews 43 (4), 381—441.

Datye, K.V., 1991. Recycling Processes and Products in Nylon 6 Fibre Industry.

Deopura, B.L., et al. (Eds.), 2008. Polyesters and Polyamides. Elsevier.

Dholakiya, B., 2012. Unsaturated polyester resin for specialty applications. Polyester 7, 167−202.

Ducheyne, P., 2015. Comprehensive Biomaterials, 1. Elsevier.

East, A.J., 2009. The structure of polyester fibers. Handbook of Textile Fibre Structure. Woodhead Publishing, pp. 181−231.

Ebata, M., Morita, K., 1959. Hydrolysis of ε-amincaproyl compounds by trypsin. The Journal of Biochemistry 46 (4), 407−416.

Elbehiry, A., et al., 2021. FEM evaluation of reinforced concrete beams by hybrid and banana fiber bars (BFB). Case Studies in Construction Materials 14, e00479.

Gupta, V.B., Kothari, V.K. (Eds.), 2012. Manufactured Fibre Technology. Springer Science & Business Media.

Hearle, J.W.S., Morton, W.E., 2008. Physical Properties of Textile Fibres. Elsevier.

Hu, Y., et al., 2018. Valorisation of textile waste by fungal solid-state fermentation: An example of circular waste-based biorefinery. Resources, Conservation and Recycling 129, 27−35.

Hufenus, R., et al., 2020. Melt-Spun fibers for textile applications. Materials 13 (19), 4298.

Jaffe, M., Easts, A.J., Feng, X., Polyester fibers, 2020. Thermal Analysis of Textiles and Fibers. Woodhead Publishing, pp. 133−149.

Jiang, Y., 2016. Melt Spinning of High-Performance Nylon 6 Multifilament Yarn via Utilizing a Horizontal Isothermal Bath (HIB) in the Thread Line.

Kageyama, M., Kosuke, T., 1988. Activated sludge treatment of wastewater from a nylon 6 manufacturing plant. Water Science and Technology 20 (10), 49−55.

Karayannidis, G.P., Achilias, D.S., 2007. Chemical recycling of poly (ethylene terephthalate). Macromolecular Materials and Engineering 292 (2), 128−146.

Karthik, T., Rathinamoorthy, R., 2018. Sustainable Biopolymers in Textiles: An Overview. Handbook of Ecomaterials, pp. 1−27.

Kojima, Y., et al., 1993. Mechanical properties of nylon 6-clay hybrid. Journal of Materials Research 8 (5), 1185−1189.

Kricheldorf, H., 2014. Important Polycondensates. Polycondensation. Springer, Berlin, Heidelberg, pp. 69−91.

Lewin, M., Pearce, E.M. (Eds.), 1998. Handbook of Fiber Chemistry, Revised and Expanded.

Luo, X., Zhu, J.Y., 2011. Effects of drying-induced fiber hornification on enzymatic saccharification of lignocelluloses. Enzyme and Microbial Technology 48 (1), 92−99.

Luzio, A., et al., 2014. Electrospun polymer fibers for electronic applications. Materials 7 (2), 906−947.

MacArthur, E., 2017. Beyond Plastic Waste, 843-843.

Macdowell, J.T., 1965. Reclaiming linear terephthalate polyesters. US Patent 3 (222), 299.

McCarthy, B.J., 2016. An overview of the technical textiles sector. Handbook of Technical Textiles 1−20.

MCINTYRE, J.E., 2004. "Formerly University of Leeds, UK." Synthetic Fibres: Nylon, Polyester, Acrylic. Polyolefin, p. 1.

McIntyre, J.E. (Ed.), 2005. Synthetic Fibres: Nylon, Polyester, Acrylic, Polyolefin. Taylor & Francis US.

Meric, S., et al., 1999. Treatability of strong wastewaters from polyester manufacturing industry. Water Science and Technology 39 (10−11), 1−7.

Moran, R., Porath, D., 1980. Chlorophyll determination in intact tissues using N, N-dimethylformamide. Plant Physiology 65 (3), 478−479.

Nakayama, Y., et al., 2020. Synthesis, properties and biodegradation of periodic copolyesters composed of hydroxy acids, ethylene glycol, and terephthalic acid. Polymer Degradation and Stability 174, 109095.

Nateri, A.S., Goodarz, M., 2013. Making polyester/cotton blend fabrics silk-like through alkaline oxidation. Research Journal of Textile and Apparel.

Negoro, S., 2000. Biodegradation of nylon oligomers. Applied Microbiology & Biotechnology 54 (4).

Nørup, N., et al., 2019. Evaluation of a European textile sorting centre: material flow analysis and life cycle inventory. Resources, Conservation and Recycling 143, 310−319.

Opperskalski, S., et al., 2019. Preferred Fiber and Materials Market Report 2019. Textile Exchange, Lamesa, TX, USA.

Parodi, E., Peters, G.W.M., Leon, E., Govaert, 2018. Structure−properties relations for polyamide 6, part 1: influence of the thermal history during compression moulding on deformation and failure kinetics. Polymers 10 (7), 710.

Paul, R. (Ed.), 2014. Functional Finishes for Textiles: Improving Comfort, Performance and Protection. Elsevier.

Pawar, A.B., More, S.P., Adivarekar, R.V., 2018. Dyeing of polyester and nylon with semi-synthetic azo dye by chemical modification of natural source areca nut. Natural Products and Bioprospecting 8 (1), 23−29.

Pitat, J., Holcik, V., Bacak, M., 1959. A method of processing waste of polyethylene terephthalate by hydrolysis. GB Patent 822, 834.

Punyodom, W., et al., 2006. Radiation-induced crosslinking of acetylene-impregnated polyesters. II. Effects of preirradiation crystallinity, molecular structure, and postirradiation crosslinking on mechanical properties. Journal of Applied Polymer Science 100 (6), 4476−4490.

Reichert, C.L., et al., 2020. Bio-based packaging: materials, modifications, industrial applications and sustainability. Polymers 12 (7), 1558.

Ribeiro, C.E.G., et al., 2014. Production of synthetic ornamental marble as a marble waste added polyester composite. Materials Science Forum 775. Trans Tech Publications Ltd.

Scheirs, J., Long, T.E. (Eds.), 2005. Modern Polyesters: Chemistry and Technology of Polyesters and Copolyesters. John Wiley & Sons.

Schuhmann, J.G., Hartzell, G.E., 1989. Flaming combustion characteristics of upholstered furniture. Journal of Fire Sciences 7 (6), 386−402.

Sharp, R.R., 1964. The growth of the man-made fiber industry. Financial Analysts Journal 20 (1), 43−48.

Shenai, V.A., Parameswaran, R., 1980. Text. Dyer Printer, p. 21.

Speight, J.G., 2014. The Chemistry and Technology of Petroleum. CRC Press.

Strom, E.T., 2017. Wallace carothers and polymer chemistry: a partnership ended too soon. In: The Posthumous Nobel Prize in Chemistry, 1. Correcting the Errors and Oversights of the Nobel Prize Committee. American Chemical Society, pp. 121−163.

Subramanian, K., et al., 2020. Environmental life cycle assessment of textile bio-recycling—valorizing cotton-polyester textile waste to pet fiber and glucose syrup. Resources, Conservation and Recycling 161, 104989.

Travis, A.S., et al. (Eds.), 1998. Determinants in the Evolution of the European Chemical Industry, 1900−1939: New Technologies, Political Frameworks, Markets and Companies, 16. Springer Science & Business Media.

US patent 2130523, Carothers W.H., "Linear Polyamides and Their Production", Issued 1938-09-20, Assigned to EI Du Pont de Nemours and Co.

Vagholkar, P., 2016. Nylon (chemistry, properties and uses). Chemistry 5 (9).

Van Uytvanck, P.P., et al., 2014. Impact of biomass on industry: using ethylene derived from bioethanol within the polyester value chain. ACS Sustainable Chemistry & Engineering 2 (5), 1098−1105.

Vilela, Carla, et al., 2014. The quest for sustainable polyesters—insights into the future. Polymer Chemistry 5 (9), 3119−3141.

Wang, T., 2020. Study on the treatment of caprolactam wastewater by electrocatalytic oxidation process. In: IOP Conference Series: Earth and Environmental Science, 546. IOP Publishing. No. 5.

Wendel, R., et al., 2020. Anionic polymerization of ε-caprolactam under the influence of water: 2. Kinetic model. Journal of Composites Science 4 (1), 8.

Wilfong, R.E., 1961. Linear polyesters. Journal of Polymer Science 54 (160), 385−410.

Part Four

Fibres derived from waste

Orange fibre

12

Subhankar Maity[1], Pranjul Vajpeyee[1], Pintu Pandit[2] and Kunal Singha[2]
[1]Department of Textile Technology, Uttar Pradesh Textile Technology Institute, Kanpur, Uttar Pradesh, India; [2]Department of Textile Design, National Institute of Fashion Technology, Patna, Bihar, India

12.1 Introduction

Natural fibres like cotton, silk, and wool have been utilized for the maximum by humans to a maximum possible timeline. Other natural fibres of uncommon origin such as jute, flax, hemp, ramie, coir, sisal, nettle, etc. are also extracted from plants and explored for various applications, including clothing. The nineteenth century proposed the replacement of natural fibres with synthetic fibres with excellent performance properties that have ruled for another 100 years now (Aishwariya, 2018). The petrochemical-derived fibres like polyester, polypropylene, etc. are the source of micro-plastic pollution, non-bio-degradable, toxic, and a threat to human civilization. The cost of production of natural fibres is high, and they have a limited production rate. Today, agricultural wastes and other organic wastes are seen as a potential renewable, biodegradable material to be made into textiles (Aishwariya and Amsamani, 2018).

Orange is a highly poly-embryonic, even surfaced and tight-skinned fruit, which has been part of the human diet for ages due to its high nutritional and medicinal values. Consumption of orange fruit generates a large amount of peels as waste, which could be a potential cause of environmental pollution, if not handled properly (Mohapatra et al., 2017).

Orange peel primarily consists of cellulosic fibres and pectin. Further, it has been reported that orange peel extract comprises several bioactive compounds, such as flavonoids and limonoids, which are known to act as anti-cancer and anti-oxidant agents (Mohapatra et al., 2017).

These are also known to enhance vitamin absorption and possess a broad spectrum of medicinal properties, including anti-inflammatory, anti-allergenic, anti-tumour and anti-microbial activities (Mauro et al., 1999; Mohapatra et al., 2017). For example, hesperidin is a predominant flavourless flavonoid present in the peel and membranous parts of orange, which shows such bioactivity (Bilbao et al., 2007; Mohapatra et al., 2017). Rutin is another major flavonoid that can chelate heavy metals like iron and also enhance vitamin absorption (Tundis et al. 2014; Mohapatra et al., 2017). Limonoids are organic compounds commonly found in the peel of citrus fruits and exhibit several beneficial health effects (Mohapatra et al., 2017). This chapter explores the orange peel as a raw material for fibre extraction, the fibre extraction process, structure and chemical composition of the orange peel and the prepared fibre, morphology and

---

Sustainable Fibres for Fashion and Textile Manufacturing. https://doi.org/10.1016/B978-0-12-824052-6.00004-4
Copyright © 2023 Elsevier Ltd. All rights reserved.

properties of the fibre, characterization of the fibre by Fourier transform infrared spectroscopy (FTIR), differential scanning calorimetry (DSC), thermogravimetry analysis (TGA) and scanning electron microscopy (SEM). Moisture management behaviour, anti-bacterial efficacy and solubility behaviour of the fibre are also discussed.

12.2 The orange fruit

Orange production was approximately 51.8 million metric tonnes in 2014 (Kadam et al., 2020). The commonly consumed fruits worldwide are banana, apple, grapes, strawberry and orange. In many countries, orange juice is an essential entity in breakfast to cater to the everyday recommended dose of vitamin C. Oranges are the richest source of vitamins, minerals and energy, which has the ability to give an instant refreshment (Wang et al., 2020). The regular consumption of orange helps in the effective functioning of the heart, kidney and aids infertility. There are other health benefits for skin, teeth, and bone, along with maintaining the normal blood pressure and cholesterol levels in the body (Deng et al., 2020).

The top producers of orange are Brazil, USA, India, China, Mexico, Spain, Egypt and Italy. Brazil produces 1.8 million tonnes of orange per year. Approximately 15−25 million tonnes of peel waste are found on landfill without recycling or composting. The transported waste is thrown in landfills or incinerated. These peels dumped on a site can be a threat to the environment and human health. They may decompose, give away foul odour, microbial infestation, attract flies and risk of spreading diseases during the rainy season. The baseline of the huge tonnes of orange peel waste is a menace. Two ecologists Daniel Janzen and Winnie Hallwachs in 1997 dumped 12,000 tonnes of orange peel over degrading lifeless soil in Amazon (Fig. 12.1). The selected area was a victim of burning trees in the forest for installing oil refineries. Sixteen years later, when the couple returned, they were amazed by the bio-diversity, rich landscapes, biomass that are a result of citrus peel waste (Dockrill, 2017). This is an example of the power of organic waste.

12.3 Orange peel waste as a textile raw material

In 2012, Adriana Santanocito, a native of the Sicilian city of Catania, found a similar organic waste from her city, which was popular as the largest consumer of orange as juice. On further research, she found that globally, 700,000 tonnes of citrus peel waste every year are thrown away in landfills without proper recycling alternatives (Fig. 12.2). She decided to conduct a pilot study to convert the peel into textiles for her university project. The grounded orange peel waste was then processed, and cellulose was separated. This was further sent to a spinning industry in Spain for making the yarn. This is blended with silk and cotton to make satin and poplin material (Aishwariya, 2018). They also make fabrics from 100% orange peel waste fibre which is similar to viscose (textile made from wood-also called rayon), available in light shades

Orange fibre 275

Figure 12.1 Amazon rainforest before-during and after dumping the orange peel waste. The organic waste turned into compost and returned the soil nutrients making it fertile vegetation once again (Aishwariya, 2020).

Figure 12.2 Processing of orange peel waste into textiles (Pollution, Inventors, Process) (Aishwariya, 2020).

that can be dyed, printed and even washed like other conventional materials. The fabric was sold at €30—€40 per metre.

Enrica Arena, the hostel mate of Adriana, who assisted in documenting the research to be drafted in English, joined the research soon after (Fig. 12.3). Fabric made using orange peel waste textiles resembled silk in terms of quality, softness, shiny surface and colour. Nanotechnology and microencapsulation were employed to retain the beneficial properties of citrus fruits onto fabric and the technology-aided in maintaining these properties till 20 washes. The biodegradable material was good to be blended

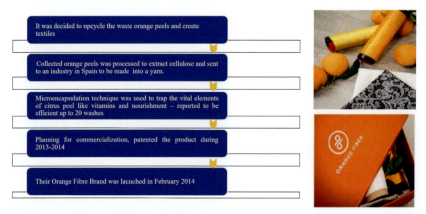

Figure 12.3 Journey of orange peel waste into textile (Orange Fibre brand) (Aishwariya, 2020).

with cotton, silk, elastane and pineapple. The duo decided to present this in various forums to demand, scope and market for the developed textile (Aishwariya, 2018).

12.4 Structure and chemical composition of the orange peel

Orange as a fruit is known for its nutritional and medicinal values, which are mainly associated with the various compositions of the material. Citrus peels contain several valuable compounds such as soluble sugars, insoluble carbohydrate fibres, organic acids, essential oils, flavonoids, and carotenoids. Table 12.1 gives detailed information on the chemical composition (Bicu and Mustata, 2011).

Table 12.1 Chemical composition of citrus peel [14].

Name of constituent	Value in percentage
Cellulose	10.4
Hemicelluloses	8.9
Lignin	1.33
Ash	2.87
Volatile oils	14.92
Simple sugar	15.0
Pectin	28.7
Protein	6.62
Flavonoids (hesperidin, quercetin, etc.)	3.86

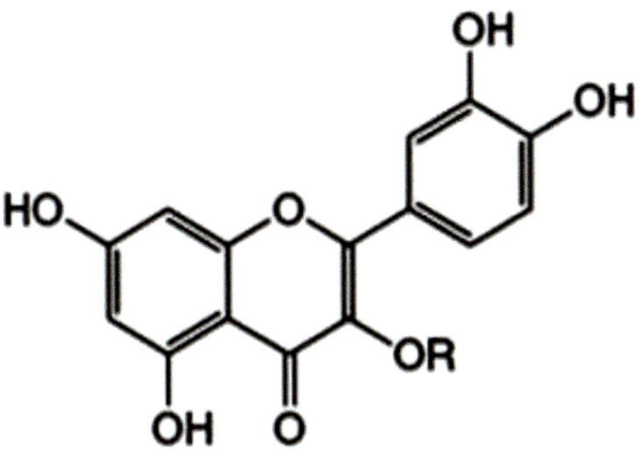

Orange fibre

Figure 12.4 Structure of flavonoid.

The flavonoid (Fig. 12.4) molecules present in the orange peel are mainly responsible for antibacterial and anti-oxidant activities and stabilizing the free radicals involved in oxidative processes of various reactions taking place inside the human body.

12.5 Fibre extraction method

Extract of orange peel is collected by cold pressing of the fresh orange peels, and the extract is used to prepare the fibrous assembly. The extract collected from the peels is spread on a plastic surface and a rubbery surface is pressed onto the liquid for extraction of fibrous assembly. When the two surfaces are separated, the fibrous assembly is formed between the two surfaces, which are then collected. The demonstration of the fibre extraction process is shown in Fig. 12.5.

Approximately 75–80 mg of fibrous assembly can be extracted from the peels of a single fresh orange. The extraction process may be manual as described above. However, the process is mechanized for more scientific way. A modified 4 bar slider-crank mechanism has been developed for the extraction of the fibrous assembly from orange peel extract (Fig. 12.6).

Figure 12.5 Diagram of manual extraction process of fibrous assembly from orange peel extract (Mohapatra et al, 2017).

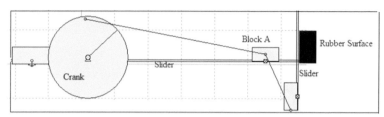

Figure 12.6 Schematic diagram of modified 4-bar slider crank showing different parts (Mohapatra et al., 2017).

In this mechanism, the orange peel extract is spread on the surface of block *A*. Then, block *A* is moved forward by the rotation of the crank and is pressed on the rubbery surface. When the two surfaces are separated, the fibrous assembly is formed between the two surfaces, which are then collected. The width of the fibrous assembly mainly depends on the area on which the peel extract has been spread on the plastic surface and the area on which the plastic surface is pressed on the rubbery surface. The length of the fibrous assembly depends on the distance between the two surfaces when they are separated after pressing and, of course, on the velocity of block A, which is regulated by the rotation of the crank.

12.6 Preparation of film from orange peel extracts

The material used for preparing the film is waste orange peels. The peels are extracted from fresh limes and cold-pressed to extract liquid from them and poured on a plastic

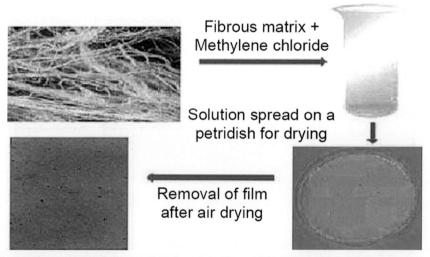

Figure 12.7 Solution casting method for preparation of film from orange peel extract (Mohapatra et al, 2015b).

container. The fibrous matrixes are isolated from the peel extracts by pressing the plastic surface on a rubbery surface through a manual mechanical process. After that 100 mg of fibrous matrix is dissolved in 10 mL of methylene chloride at room temperature with $65 \pm 2\%$ relative humidity in a test tube (Fig. 12.7). The time period required for the dissolution of fibrous matrix in methylene chloride is expeditious. Then, the solution is spread on a Petri dish for air-drying for a minimum of 8 h. Finally, a yellowish non-transparent film of 1 mm thickness is prepared (Fig. 12.2) (Mohapatra et al., 2015a, b).

12.7 Fibre morphology and properties

Optical microscopy and SEM images of the orange peel extract fibres reveal a typical cylindrical appearance of the fibres resembling common man-made fibres as shown in Fig. 12.8. The mean diameter of the individual fibres is measured around 6.19 ± 2.89 μm. There is variation present in the diameter of the fibre which may be due to the variation in the extraction forces applied during fibre processing. The fibres have low tensile strength and modulus as compared to other textile fibres. Initial modulus of the fibrous assembly is about 0.16 cN/Tex and tenacity is of 0.068 cN/Tex. The breaking elongation lies in between 2.2% and 2.5%. The average moisture regain and moisture content values of the fibrous assembly calculated are 12% and 10%, respectively.

12.8 Chemical composition of orange fibre

The fibre is composed of $10 \pm 0.5\%$ of cellulose, $15 \pm 0.2\%$ hemicelluloses, $2 \pm 0.3\%$ of lignin and $1.5 \pm 0.2\%$ of ash content. EDX analysis has been performed

Figure 12.8 (A) Optical microscopy and (B) SEM image of fibres derived from orange peel extract (Mohapatra et al., 2017).

to identify vital elements present in the fibre. It is detected that the major elements are carbon ($\sim 91\%$) and oxygen ($\sim 8\%$). XRD studies reveal that the fibre is having very low crystallinity of about only 8%, and most of the part is amorphous in nature. The fibre has polar chemical groups, such as hydroxyl, carboxyl and carbonyl.

12.9   Burning behaviour of orange fibre

The orange fibre has distinguished burning behaviour from natural fibres. It is observed that the fibres shrinks like synthetic fibre on approaching flame. It burns with a yellow flame like cotton and continues to burn rapidly like rayon with a sweet smell like polyester. After extinguishing it forms a hard round bead formed like acrylic and acetate fibres. The burning behaviour of the fibre is shown in Table 12.2.

12.10   Solubility behaviour of orange fibre

The fibre can be dissolved in various solvents like methylene chloride, a mixture of heptanes and carbon tetrachloride and cyclohexane at room temperature. The fibre also dissolves in phenol and 40% sulphuric acid at boil. The solubility of the fibre experimented with different solvents and results is shown in Table 12.3.

12.11   Moisture absorbency behaviour of orange fibre

Moisture absorbency behaviour of freshly prepared fibre is investigated. It has been observed that the fibre gradually absorbs moisture from the atmosphere and achieves equilibrium after some time, as shown in Fig. 12.9. The moisture regain and moisture content of the fibre are measured to be 11% and 9%, respectively. The good moisture absorption capacity of the fibre is attributed to the presence of various hydrophilic groups (including $-$OH) and highly amorphous structure of the fibre.

Table 12.2 Burning behaviour of orange fibre.

Incidence	Observations
On approaching flame	The fibre shrinks like synthetic fibres
In the flame	Burn quickly with yellow flame like cotton fibre
Out of the flame	Continues to burn rapidly
Residue	Hard round bead is formed
Smell	Sweet chemical smell of ester

Table 12.3 Solubility of orange peel fibre.

Name of the solvent	Solubility
Methylene chloride (in room temperature)	Fibre dissolves immediately
Cyclohexane	Broken into segments and dissolved
Sulphuric acid (40%)	Partially dissolved and colour changes to yellow and fully dissolved at boiling temperature
Phenol	Partially dissolved in room temperature and fully dissolved at boiling temperature
Nitric acid (40%)	Does not dissolve but colour changes to golden yellow
Formic acid (40%)	No change
Hydrochloric acid (40%)	No change
Dimethyl formamide	No change
Acetone	No change
Glacial acetic acid	No change
Ethylene diamine (hydrate)	No change

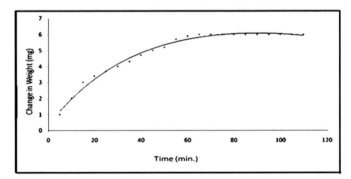

Figure 12.9 Moisture absorption behaviour of orange peel fibre (Mohapatra et al, 2015b).

12.12 FTIR spectroscopy

FTIR spectroscopy analysis of the fibre shows various functional groups and bond vibrations present in the chemical skeleton of the fibre as shown in Fig. 12.10. The most intense band observed at 3322 cm^{-1} can be assigned to the stretching of $-$OH groups of the carbohydrates and lignin. The absorption band at 2913 and 2851 cm^{-1} is observed due to asymmetrical and symmetrical stretching vibrations of C$-$H groups. The band at 1732 cm^{-1} can be assigned to the carbonyl (C=O) stretching. The next intense band at 1030 cm^{-1} denotes the presence of C$-$O$-$R link, while the distinctive band around 1265 cm^{-1} denotes aliphatic chains ($-$CH$_2-$ and $-$CH$_3$) forming the basic chemical structure of this lignocelluloses fibre. Most of the bands have contributions from both carbohydrates (cellulose and hemicellulose) and lignin. The flavonoids

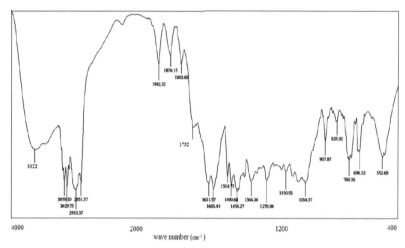

Figure 12.10 FTIR spectra of orange peel fibre (Mohapatra et al, 2015b).

are present in the fibre and which is revealed by the characteristics vibrations of carbonyl group in the frequencies like 1603 and 1631 cm^{-1}. Other typical absorption bands characteristics of pure cellulose are that at wave number 1631, 1031, 907, 3322 cm^{-1}.

12.13 Thermal characterization of orange fibre

The thermal degradation behaviour of the fibre can be investigated by DSC and TGA. In the DSC study, endothermic transition is observed at about 50–60°C range of temperature which is denoting glass transition temperature of the fibre. High amorphous phase of the fibre is responsible for prominent glass transition of the orange fibre. During TGA analysis, the first mass loss can be observed at about 190°C, which is approximately 4% of initial weight. This mass loss can be attributed to the loss of moisture and other volatile components of the fibre. The second step mass loss of approximately 8% can be observed from 190 to 350°C, which can be attributed to the decomposition of hemicelluloses. The third decomposition starts between 350 and 410°C, which is about 28% attributed to the degradation of cellulose and lignin. After 410°C, there is no further loss of mass or degradation is observed till 700°C.

12.14 Anti-microbial efficacy of orange fibre

Orange peel extracts have active components like flavonoids and limonene, which acts as anti-oxidant, anti-obesity as well as anti-carcinogenic agents and also show a

Table 12.4 Anti-microbial efficacy of orange peel's film.

No.	Film thickness [mm]	*E. coli* reduction [%]	*S. aureus* reduction [%]
1.	1	83.75 ± 2.5 SD	94.71 ± 4.35 SD
2.	1.5	90.00 ± 4.0 SD	95.21 ± 3.54 SD
3.	2	93.00 ± 4.3 SD	96.00 ± 2.7 SD
4.	2.5	95.22 ± 3.3 SD	96.25 ± 2.1 SD

tendency to inhibit tumour growth. Only in the last decade, various studies have focused on several bioactive compounds, specifically limonoids and flavonoids which play a major role in preventing chronic diseases.

The fibre is highly efficient against both test bacteria with a reduction rate of 83.75 ± 2.5 for *E. coli* ($n = 5$) and 94.71 ± 4.35 for *S. aureus* ($n = 5$), which indicated the excellent antibacterial property. A possible reason for this antibacterial nature is the presence of different flavonoids and other polyphenolic compounds. The antibacterial effect of the film prepared from orange peels' extract with different bacterial species in terms of reduction of bacterial count is shown in Table 12.4.

All the samples were very effective against both test bacteria with a reduction of over 83% for *E. coli* and 94% for *S. aureus*, thus indicating excellent antibacterial properties. The reason behind the antibacterial property may be due to the presence of different flavonoids and other polyphenolic compounds present in the films (Mohapatra et al., 2015a, 2015b)

12.15 Benefits of textiles made of orange peel extracts

Textiles are rightly called as second skin due to their close positioning to the body; whatever is placed over the skin affects the health of the person by releasing the contents into the bloodstream. Synthetic chemicals of organically derived materials are definitely a choice considering the health benefits for an individual. Fabrics made from orange peel waste have been proved to perform like vitamin C infusers along with moisturizing capabilities. The natural oils of orange peel are retained in the fabric and are believed to give nourishment to the skin. The amazing quality of orange fibre can be its rich source of vitamin C and essential oils that can be like a moisturizing cream on the skin. The reduced carbon footprint by using a material that could be an organic waste dumped on landfills is reverse engineered into a material of value and utility (Fig. 12.11). In order to make the clothing more affordable, a blend with yarns made from pineapple leaf. The blend-ability of orange fibre with cotton, silk, elastane is good; hence the product range is also diversified with a prospective future.

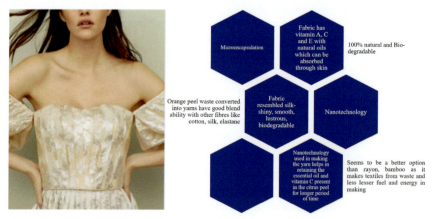

Figure 12.11 Textiles made of orange peel fibres (Aishwariya, 2020).

12.15.1 Future perspective

Textiles from recycling orange peel waste are considered to be a sustainable alternative as it uses a lesser carbon footprint compared to conventional textile production. On the other side, handmade paper making from orange peel wastes is gaining momentum. There are various waste materials are utilized for making papers, including grapes, cherries, orange peel, almond skin, lavender, coffee dust, corn, olives, kiwis, hazelnuts, etc. Thus, this route of textile fibre production and paper making can be one of the eco-friendly solutions. Other than textile and papermaking, the orange peel waste can be used as cattle feed, compost, as a soil amendment, perfumes, cosmetics, essential oil, textile dyeing, printing, print transfer medium, and for fragrance finishing, mosquito repellent, anti-microbial finishing on textiles. It is also used in polystyrene production, limonene extraction, purifying water, treating textile effluent wastewater, as bioadsorbent and as a source of carbon, biochar and biofuel. This chapter demonstrates only the preparation of textiles out of that, which can open up more possibilities in the near future.

References

Aishwariya, S., 2018. Waste management technologies in textile industry. Innovative Energy and Research 07, 175−179.

Aishwariya, S., 2020. Textiles from orange peel waste. Science and Technology Development Journal 23 (2), 508−516. https://doi.org/10.32508/stdj.v23i2.1730.

Aishwariya, S., Amsamani, S., 2018. Exploring the potentialities and future of biomass briquettes technology for sustainable energy. Innovative Energy and Research 7 (221), 2576-1463.

Bicu, I., Mustata, F., 2011. Cellulose extraction from orange peel using sulfite digestion reagents. Bioresource Technology 102 (21), 10013—10019.

Bilbao, M.D.L.M., Andrés-Lacueva, C., Jáuregui, O., Lamuela-Raventos, R.M., 2007. Determination of flavonoids in a Citrus fruit extract by LC—DAD and LC—MS. Food Chemistry 101 (4), 1742—1747.

Deng, L.Z., Mujumdar, A.S., Yang, W.X., Zhang, Q., Zheng, Z.A., Wu, M., Xiao, H.W., 2020. Hot air impingement drying kinetics and quality attributes of orange peel. Journal of Food Processing and Preservation 44 (1), e14294.

Di Mauro, A., Fallico, B., Passerini, A., Rapisarda, P., Maccarone, E., 1999. Recovery of hesperidin from orange peel by concentration of extracts on styrene—divinylbenzene resin. Journal of Agricultural and Food Chemistry 47 (10), 4391—4397.

Dockrill, P., 2017. How 12,000 Tonnes of Dumped Orange Peel Grew into a Landscape Nobody Expected to Find.

Kadam, A.A., Sharma, B., Saratale, G.D., Saratale, R.G., Ghodake, G.S., Mistry, B.M., Sung, J.S., 2020. Super-magnetization of pectin from orange-peel biomass for sulfamethoxazole adsorption. Cellulose 27 (6), 3301—3318.

Mohapatra, H.S., Chatterjee, A., Kumar, P., 2015a. Characterization of fibrous assembly from lime peel extract. International Journal of Pharmacology, Phytochemistry and Ethnomedicine 1, 27—36.

Mohapatra, H.S., Chatterjee, A., Kumar, P., 2015b. Characterization of film for medical textiles application. Tekstilec 58 (4).

Mohapatra, H.S., Dubey, P., Chatterjee, A., Kumar, P., Ghosh, S., 2017. Development of a fibrous assembly from orange peel extract: characterization and antibacterial activity. Cellulose Chemistry and Technology 51 (7—8), 601—608.

Tundis, R., Loizzo, M.R., Menichini, F., 2014. An overview on chemical aspects and potential health benefits of limonoids and their derivatives. Critical Reviews in Food Science and Nutrition 54 (2), 225—250.

Wang, M., Shi, R., Gao, M., Zhang, K., Deng, L., Fu, Q., Gao, D., 2020. Sensitivity fluorescent switching sensor for Cr (VI) and ascorbic acid detection based on orange peels-derived carbon dots modified with EDTA. Food Chemistry 318, 126506.

Coffee fibres from coffee waste   **13**

Ajit Kumar Pattanayak
The Technological Institute of Textile & Sciences, Bhiwani, Haryana, India

13.1 Introduction

The fast-changing fashion trend is the major factor behind the spurring global demand for textile articles. The consistent growth of garments is supported by the constant improvement of the purchasing power of the consumers. The oversupply of textile goods is mostly attributed to the fast-fashion change and economic empowerment of consumers (Ross, 2019). The textile industry also witnessing a growing trend of technical textile products which mostly consist of synthetic fibres. These products have specific functionalities and apply for various applications like filtration, geotextiles, medical textile, etc. The total consumption of textile articles stood at USD 1000.3 billion in 2020 and is expected to grow at a rate of 4.4% from 2021 to 2028 (Textile, 2021).

It is reported that the polyester fibres control around 52% of the textile fibres which are manufactured from the petroleum base and 24% of the market is controlled by cotton fibres. The polyester fibres are manufactured from the petroleum base and hence have a great impact on the environment. The cotton fibres are natural fibres but consume lots of water, fertilizers and pesticides during the cultivation. As per the usage, natural fibres are mostly used for fashion and apparel products. The polyester fibres are suitable for fashion and apparel products, home furnishings and technical textiles due to their excellent mechanical properties. Nylon is the third-largest product used in the textiles industry. It is widely used in apparel and home-furnishing applications owing to its high-resilience, elasticity and moisture-absorbing properties. The other product segment includes polyethylene (PE), polypropylene (PP), aramid and polyamide. Properties such as high resistance against acids and alkalis at high temperatures and minimum moisture retention have increased the demand for PE in the market.

The huge demand for textiles leads to huge greenhouse gas (GHG) emission, and global GHG emission is moving faster than the total gross domestic product of the world (Hoffmann, 2015). The huge consumption of textiles also leads to huge textile waste generation. It is estimated that 87% of the disposed textiles either move to the landfills or incinerated and the rest 13% are recycled. The textile fabrics are mostly recycled by the mechanical shredding process, whereas the current practice of polyester recycling is carried out by thermomechanical process. But the recycled product lacks many ways like weaker mechanical properties. Hence ,more effective recycling process must be invented for textile recycling. The leading fibre manufacturer, Tenjin developed a recycling process to regenerate the polyester fibre from the used polyester

---

Sustainable Fibres for Fashion and Textile Manufacturing. https://doi.org/10.1016/B978-0-12-824052-6.00018-4
Copyright © 2023 Elsevier Ltd. All rights reserved.

clothing as per their commitment to the society. The Teijin's recycling process can able to recycle their polyester from Patagonia into fresh polyester (Humblet, 2006). Recycling of plastics and textile waste got an impetus after Carbios, a French company that specializes in green chemistry announced to recycle polyester from the polyester blended fabrics (Jkaybay, 2020). The Carbios demonstrated the technology of recycling waste PET bottles into either new recycled PET (rPET) bottles or even rPET fabric. This company also developed the technology to covert the polyester textile waste into rPET fabric(Tournier et al., 2020).

Apart from plastic waste, bio-mass generation and its disposal create a new challenge for the environment. In an instant, the burgeoning food waste in modern societies is becoming a major challenge for the government for its disposal. The estimated amount of food waste in the European countries exceeds 173 kg per person in the year 20,212 (EU FUSIONS, 2021). Similar problem is found for the most loved coffee beverage. International Coffee Organization (ICO) reported that 159.66 million bags of coffee were produced and consumed in the year 2018 (Severini et al., 2020). So, the consumption of coffee per rata basis is 60 kg approximately. The coffee is associated with coffee waste such as coffee husk, coffee pulp, coffee silver skin (CS) and spent coffee grounds (SCG). The SPG is a typical waste and generated after brewing of coffee. The polysaccharides, phenolic compounds and dietary fibre are the major components of the SPG. So the disposal of SPG causes huge emission of methane gas if dumped in landfill. Hence, the SPG must be utilized in an effective way to reduce environmental pollution.

The reduce, recycle and reuse are the existing new mantra of sustainability. Going beyond the recycling and the value addition of the waste or reuse of the waste can be a boon for the sustainability effort. For example, Cocona fibre was created by incorporating the carbonizing recycled coconut shells into polyester fibres (Brasquet et al., 1996). Similarly, PET composites are developed by incorporation of Carbonized bamboo (Lou et al., 2007). Many more innovative fibres and yarns are developed and marketed for their sustainable nature and eco-friendliness. One such fibre is 'S.Cafe'. This fibre is developed and patented by the Taiwan-based Singtex Industry. This manufacturer claims that the fibre has excellent fast-drying, odour control and UV protection properties (Subic et al., 2009).

13.2 Coffee botanicas

The coffee bean is derived from the plantation crop. This plantation crop belongs to the Rubiaceae family. The crop is mostly cultivated in the tropical climate round the year. The coffee leaves are shiny, wavy and dark green in colour with conspicuous veins (Murthy and Madhava Naidu, 2012). The major parts of the coffee leaves are shown in Fig. 13.1. The first full crop of coffee beans requires about 5 years and produces full ripe coffee beans for 15 years. These beans can be processed by two basic methods such as wet and dry processing methods.

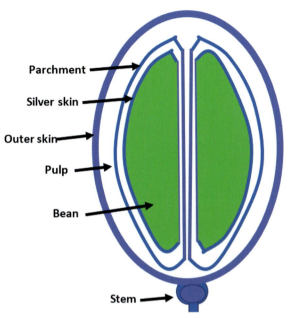

Figure 13.1 The cross-sectional view of the coffee cherry. (*Source*: Oliveira, L.S., Franca, A.S., 2015. Chapter 31—an overview of the potential uses for coffee husks. In: Preedy, V.R. (Ed.), Coffee in Health and Disease Prevention. Academic Press, San Diego, pp. 283–291. 10.1016/B978-0-12-409517-5.00031-0).

The coffee waste mostly consists of four forms such as coffee husks, coffee silver skin, coffee bean skins and SCGs. These wastes are shown in Fig. 13.2. The amount of coffee waste depends upon the processing technique and the roasting and brewing process. Coffee husk, pulp and peel consist of around 45% of the cherry (Murthy and Madhava Naidu, 2012). The coffee waste is considered as a lignocellulosic biomass. This is mainly composed of cellulose (59.2–62.94 weight%), hemicellulose (5–10 weight%) and lignin (19.8–26.5 weight%) (Mussatto et al., 2011a,b). Apart from these compounds, the coffee waste also has inorganic micronutrients such as calcium,

Figure 13.2 The major coffee waste a—Spent coffee ground, b—Husk, c—Pulp, d—Silver skin.
(*Source*: Campos-Vega, R., Loarca-Piña, G., Vergara-Castañeda, H.A., Oomah, B.D., 2015. Spent coffee grounds: a review on current research and future prospects. Trends in Food Science and Technology 45 (1), 24–36. 10.1016/j.tifs.2015.04.012).

magnesium or sodium but these compounds only contribute around 5% of the dry mass of waste.

SCG is accumulated coffee waste at the consumer end after the coffee brewing. SCGs typically contain 45.3% polysaccharides attached to cellulose and hemicellulose complexes and 14% phenol of the dry weight (Vítězová et al., 2019). Mannans are the principal component of polysaccharides in SCGs, whereas galactose, glucose and arabinose are also present as minor components (Vítězová et al., 2019). It is estimated that nearly six million tonnes of SCGs are accumulated every year (Hardgrove and Livesley, 2016). The huge waste of SCG mostly moves to the landfill and provides a mammoth task for the waste disposal. It also causes serious environmental problems as the decomposition of SCGs requires high biological oxygen demand and releases methane gas into the atmosphere on decomposition (Lessa et al., 2018).

The well-established concept of reduce, reuse and recycle becomes the three major pillars of the circular economy. So, the circular economy aims to form a close loop among manufacturing system with less resource consumption and causing minimum environmental pollution by transforming waste into input material for the next stage of production including coffee waste. The effective usage of waste to create new values is the new mantra of a circular economy instead of recycling (What Goes, 2018). Hence, the discarded waste must be reused efficiently for the value addition of the product as well as to reduce the burden of waste management. SCGs are such waste having components like phenolic compounds, terpenes, caffeine, etc. These compounds make the SCG to be used for special purposes mostly for odour control and to stop the growth of microorganisms. The SCG extract can be used as a natural dye for cotton and wool fabrics (Bae and Hong, 2019). The coffee oil extracted from SCG can be used as cosmetics, active ingredient for pharmaceutical as well as can be utilized to produce biodiesel (Oliveira et al., 2008). Typically, SCG contains around 11−20 weight% coffee oils (Al-Hamamre et al., 2012). The typical presence of fatty acids in the coffee oil is shown in Table 13.1. Also, the different parameters of the extracted coffee oil from SPG are listed in Table 13.2. Coffee oil has been identified as a suitable raw material for biodiesel production (Oliveira et al., 2008); however, the high content of free fatty acids in coffee oil might negatively influence the process of biodiesel production (Kwon et al., 2013). The extract of SPG can also be used for the

Table 13.1 Composition of coffee oil derived from SPG.

Fatty acid	Content (%)
Palmitic acid (C16:0)	35.7
Stearic acid (C18:0)	7.1
Oleic acid (C18:1)	9.4
Linoleic acid (C18:2n6c)	43.7
Arachidic acid (C20:0)	2.2
α-Linolenic acid (C18:3n3)	1.1
cis-11-Eikosenoic acid (C20:1)	0.3
Behenic acid (C22:0)	0.4

Table 13.2 The extracted coffee oil parameters from SPG.

Oil parameter	Value
Saponification value	166.1
Acid value	7.1
Iodine value	70.3
Ester value	158.9

dyeing and finishing of cotton and wool fabrics. The fabrics treated with the spent coffee extract displayed a good antioxidant capacity and significant antibacterial activity, especially towards Gram-positive bacteria (Koh and Hong, 2017). Therefore, the above discussion concludes that SPG (SCG) is a valuable resource and can be used in various ways for value addition instead of disposal. This valuable resource should not be dumped in a landfill which can become a potential pollutant to the environment.

13.3 Recycled PET (rPET)

The global demand all fibres in 2019 exceeded 11 million tonnes out of which the Polyester fibre tops lists as shown in Fig. 13.3 (Luppino, 2020).

Textile manufacturers are using polyester fibre extensively as this fibre is more affordable and can be assessed easily. Before polyester was discovered in the 1940s, textile production was limited as natural fibre availability is limited (Why Exactly, 2021). Hence, polyester is one of the major factors behind the exponential growth of textile articles. Apart from textile, PET (Polyethylene Terephthalate) is also extensively used for beverage bottles, packaging and electrical insulation. The global production of PET bottles reached 13.1 million tonnes alone (Global PET, 2021) The manufacturing of PET bottles needs input raw material derived from petroleum, fresh water for processing, fossil fuel consumption to run the machines, hence causes the GHG emissions apart from causing a huge burden on the disposal of these

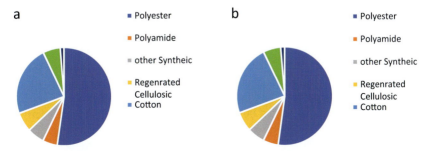

Figure 13.3 The production of different types of fiber; a:market share in %, b: production tons. *Source*: (Luppino, n.d.).

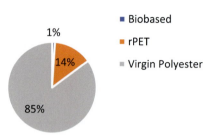

Figure 13.4 The market share of different types of polyester fiber.

bottles. So, PET bottles become one of the biggest threats to the environment. The use of recycled PET plastic is one of the solutions to reduce the plastic pollution menace. The recycled PET popularly known as rPET is a sustainable replacement for polyester fabrics. The rPET is produced by melting the discarded plastic bottles and containers and then doing the respinning them into new polyester fibre. The market share of the different types of polyester in the year 2019 is shown in Fig. 13.4. The big brands of apparel and retail companies like Adidas, H&M, Gap, Ikea, etc. are determined to use rPET up to 36% in their brand articles by 2030. It is also predicted that rPET share of all polyester-used fibres will increase to 20% by 2030.

Recycling of PET starts with collection of postconsumer waste material, transportation of the materials to the recycling plant, material soring and ultimately reprocessing of the material waste into pellets or chips. The rPET pellets are converted into plastic bottles using the injection and blow moulding process. Benavides et al. (2018) stated the requirement of material and to produce 1 kg of rPET pellets from PET that was recovered from waste plastic. Their report is shown in Table 13.3 (Benavides et al., 2018). The energy required for the conversion 1 kg of rPET pellets into bottles by injection and blow moulding is 2.9 and 6.2 MJ, respectively (Benavides et al., 2018).

13.3.1 Limitations of rPET

The rPET mostly produced from the PET bottles. It is really very challenging to recycle the textile products and it is almost impossible to recycle the polyester from blended fabrics. The blended yarns of polyester with natural fibres are very common in apparel for the optimization of comfort and mechanical properties to improve the moisture regain of the yarn. The recycling challenge becomes more complicated when the fabric is further finished for imparting value addition to the garments. It is also very difficult to recycle 100% polyester several times. The recycling of polyester is carried out either mechanically or chemically. The mechanical recycling of PET bottles follows the traditional fibre-making process after converting into polyester chips of the properly washed and shredded bottles. Whereas original monomers are derived in the chemical recycling process after applying suitable chemical processing techniques for plastic waste. The rPET typically needs 59% less energy to produce than virgin polyester. But the rPET derived from the mechanical means lacks mechanical properties as compared to the virgin polyester. Hence, rPET needs to be blended with virgin

Table 13.3 Requirement of material and energy for the transformation of 1 kg rPET pellets.

Material	Amount in kg
Mixed plastic waste	2.4
Sodium hydroxide	0.01
Sulphuric acid	0.02
Water	3.1
Energy	4.1 MJ

(*Source*: Benavides, P.T., Dunn, J.B., Han, J., Biddy, M., Markham, J. 2018. Exploring comparative energy and environmental benefits of virgin, recycled, and bio-derived PET bottles. ACS Sustainable Chemistry and Engineering 6 (8), 9725–9733. 10.1021/acssuschemeng.8b00750).

polyester for optimization of properties. So, rPET is a comparatively sustainable option for textiles.

13.4 Coffee fibres

The continuous research and development in the textile and fashion industry for the development of new sustainable fibres which can fulfil the aspirations of social needs and also be eco-friendly. The food biomass can be used to produce sustainable fibres that have a low carbon footprint. They are potential solutions for a more circular and sustainable fashion industry, which helps to achieve a better future. The coffee fibre is an optimal solution of new generation sustainable fibres which is created by the recycled polyester with SCG. The Taiwanese company SINGTEX introduced coffee fibre as S.Café brand in 2019. The S.Café yarn is created from recycled polyester and embedded SCGs on it. This fibre offers excellent natural antiodour qualities, protection against UV rays and fast drying (Textiles made, 2021). The S.Cafe brand was developed with an investment of 4 years of research and $1.7 million dollars. The fabrics woven or knitted from S.Café yarn contains around 5% upcycled SCGs crushed to nanoscale (sourced from places in Taiwan like Starbucks and 7–11) and 95% rPET. Jason Cheng, the promoter of SINGTEX estimated that a typical piece of garment requires five PET bottles and SCGs derived from three cups of coffee. The SCGs mixed with rPET enhance the technical capabilities. The majority component of S.Café is rPET, and this makes this yarn more sustainable than the fabrics made up of virgin polyester. Hence, the S.Café or the coffee fibre attracts major fashion brands like Rumi X, American Eagle (they made coffee jeans), Timberland, Sundried, Five12 Apparel, and Ecoalf (Brones, 2021).

13.4.1 Products of coffee fibre

Singtex is one of the leading manufacturers of coffee fibres. This company is also the leading fabric manufacturers in Taiwan and constantly developing sustainable textile

13.4.1.1 Coffee yarns

S.Café yarn is touted to have quick-drying capability with UV resistance and superb odour control. It is marketed in various variants such as Polyester filament, Polyester fibre, S café spun yarn spun with compact spinning system and, S.Café ICE-CAFÉ nylon filament yarn. The ICE-Café Yarn, S.Café ICE-CAFÉ is an eco-friendly yarn which incorporates a special functional powder to create an icy texture (SINGTEX-Singtex, 2021a). The manufacturer of S Café claims that the butane activity of the café yarn is 2.26, whereas the butane value of the typical polyester and cotton yarn is 1.28 and 1, respectively. Butane activity indicates the ability of activated carbon to absorb butane present in dry air, under certain specific conditions as per ASTM D5742 (ASTM D5742, 2020). It is the ratio (in percentage) between the mass of butane absorbed by a sample of activated carbon and the total mass of the same sample. The interest in measuring the butane activity is to allow the control and evaluation of the quality of granular activated carbons. The higher value refers to better odour control.

The excessive ultraviolet radiation is one of the major causes of skin ageing and can be effectively protected by clothing. The UPF (ultraviolet protection factor) value is tested as AATCC 183 (AATCC 183, 2021). It is stated that the S café has a UPF value of 5.66, whereas the UPF value of the typical polyester and cotton yarn is 4.16 and 1, respectively. The UPF (ultraviolet protection factor) is calculated as per Eq. (13.1) for the estimation of protection from UV rays. The greater values of UPF will provide more protection.

$$UPF = \frac{\sum_{290}^{400} E_\lambda S_\lambda \Delta_\lambda}{\sum_{290}^{400} E_\lambda T_\lambda S_\lambda \Delta_\lambda} \tag{13.1}$$

Source (Louris et al., 2018).

E_λ is the solar UVR spectral irradiance in Wm-2nm-1.

S_λ is the relative erythemal effectiveness according to the CIE (Humblet, 2006; Jkaybay, 2020; Gies et al., 2003).

Δ_λ is the wavelength interval of the measurements.

T_λ is the spectral transmittance at wavelength λ

λ is the wavelength (in nm).

The S Café yarns also have fast drying ability as the tested (GB/T 21,655.1-2008) value of the café yarn is 1.9, whereas this value of the typical polyester and cotton yarn is 1.45 and one, respectively (GB/T 21655.1-2008, 2021).

The ALAMBETA value of the ICE-Café Yarn, S.Café ICE-CAFÉ is 1.6 whereas the typical polyester and cotton yarn are 0.75 and one, respectively. The higher ALAMBETA value indicates the warm-cooling sensation for the S.Café ICE-CAFÉ. ICE Café, fabric is touted to cool the body temperature down by 10−20C and can be utilized during workouts. The thermal insulation, water vapour permeability, effective absorption of sweat and its fast in-plane distribution in large area are the major comfort aspects of the garments, especially for under-garments. Cotton and other cellulosic fabrics have good moisture absorption capability but lack in distribution of sweat into large areas due to very high adhesion forces. Hence, specially designed polyester fibres like COOLMAX are utilized for expensive under-garments (Consumer, 2021). ALAMBETA instrument is used for the determination of warm-cool feeling. Warm-cool feeling is the immediate sensation when an object like textiles comes into contact with human skin (Hes and Dolezal, 2018).

13.4.1.2 P4DRY

P4DRY is a patented fabric printing technology from Singtex and can be considered as a sustainable technology. This product is the derivative of the eco-friendly brand S.Café. This technology imparts a layer of SCG mixed with other chemicals on the fabric by a patented printing technology. The printed fabric layer is touted to have four key attributes such as dry touch, odour control, dropping condensation rate and sustainable too. P4Dry transfers the absorbed moisture quickly to the fabric surface when the sweat enters the fabric from the skin. This quick transfer of moisture to fabric surface leads to quick drying. Hence, P4DRY will be more comfortable as compared to the other type of fabrics. The odour control of the P4Dry is due to the presence of S Cafe yarns. These yarns will absorb odour-forming molecules constantly. The SCG might help to control the odour-forming molecules. P4Dry technology is helping to speed up the moisture transfer from the skin surface to the fabric surface. This allows the moisture condensation rate to drop significantly, that is, the less number of water vapour molecules is changing the phase from gas to liquid each second. The vapour pressure in the space above the liquid surface is the major factor which affects the condensation rate. Hence, the significant drop in condensation rate will improve the comfortability of the fabric (P4DRY, 2021).

13.4.1.3 S.Café AIRNEST

This is another innovative product of Singtex. This is a sustainable bio-foam having 25 percent of the material comprised of coffee oil extracted from SCGs. This membrane is a sustainable alternative to the petroleum-based Polyurethane (PU) foams. The AIRNEST applies to a wide range of functionality and end-user markets like shoes, home textiles, sports gloves and fabrics. This. S.Café's innovation continues with its cutting-edge fabric printing technology, the P4DRY. Repurposed coffee grounds give the print layer four key attributes: it dries quickly, controls odour, reduces the rate of condensation and is tremendously sustainable (S.Café-Product, 2021a).

13.4.1.4 Sefía

Sefia fabric has 36% bio-based material (Lyocell Filament) and can be considered as a sustainable material. Sefia fabric is woven from S cafe and Lyocell filament. Lyocell is a regenerated fibre mostly produced from wood pulp. The wood pulp is often extracted chemically from eucalyptus trees. The eucalyptus trees are fast-growing, don't need irrigation or pesticides, and are grown on land no longer fit for food and require very less water to grow. This makes lyocell more sustainable than cotton. LENZING Lyocell fibre offers excellent moisture regain value (approximately11%), elongation of around 13%, tenacity of 37 cN/tex and causes lesser wrinkle than cotton (Kilic and Okur, 2011). Hence, the Sefia fabric is also touted to have excellent odour control property combined with other properties like high drape coefficient, high tenacity and as well as pilling resistant (SINGTEX-Singtex, 2021a).

13.4.1.5 eco²sy

This is a patented multilayer insulation technology by Singtex. This product is also produced from recycled plastic bottles and used coffee grounds as like S Café. The site claims that its "breakthrough technology produces a layer of breathable, warm, windproof, waterproof, and odour control material for a variety of uses" (S.Café-Product, 2021b).

13.4.1.6 S.Café mylithe

This product is developed from the input S Café yarn by the air-jet texturing process. Air-jet texturing process is a purely mechanical method for filaments and is considered to be the most versatile among all types of texturing processes. This method uses the cold airstream to impart bulkiness into filaments. The introduced bulk will empower the yarn to feel like cotton touch and provides warmth due to the significant increase in air pockets in the yarn. The air-textured yarns appearance and physical characteristics resemble like spun yarn (Koo et al., 2014). This product can be used for active sports wears.

13.4.1.7 AIRMEM

Singtex claims that the AIRMEM is the "first bio-based coffee membrane" (S.Café-Product, 2021b) and is considered to be an eco-friendly product. The manufacturing of this product consumes 26% less energy than the traditional membrane. AIRMEM has 25% coffee oil extracted from the SCG. This membrane is more sustainable than the traditional membranes which are sourced from petroleum. This manufacturer of this product claims to have four key attributes which make it unique for the textiles. The four key attributes are odour control, windproofing, water-resistant and breathable. This product has also acclaimed the ISPO Award in 2015 (SINGTEX-Singtex, 2021b).

13.4.2 Coffee fibre in footwear

The presence of coffee fibre can also be found in footwear. The coffee fibre is mostly found in the upper part of the shoes.

13.4.2.1 Xpressole Panto

The tag line of the Xpressole Panto is "The World's First Outdoor Boots Made From Coffee" (Xpresole Panto, 2021). The majority components of these boots have some SCG. The main features of the boot are touted to be waterproof, lightweight, antiodour, quick dry, antimicrobial, cool to touch and slip resistant. The insole of the boot is made from coffee yarn and 5 mm flexible ortholite foam. The ortholite foam is a sustainable product as its product significantly uses the recycled rubber (Sustainability, 2021a). The lining is made of coffee-infused fabric, neoprene and lycra. The shell is consisting of SCG and Ethylene-Vinyl Acetate (EVA). The outsole of the boot has SCG as well as EVAN and rubber pads. The developer of this boot claims that the boot consumes 35% less petroleum derivatives than the traditional boots. The producer also upcycles 15 cups of SCGs for each pair of shoes.

13.4.2.2 Coffee Sneakers from RENS

Coffee Sneakers is a footwear from Rens. This Finland-based footwear brand developed this waterproof sneaker using recycled plastic bottles and SCGs. The manufacturer upcycles six recycled plastic bottles and 150 g of coffee waste for every pair of Rens (Sustainability, 2021b).

13.5 Sustainable products from coffee waste

The coffee waste other than SCG can also be used for value addition to reduce the stress on the environment. Some of such products derived from coffee waste are discussed hereafter.

13.5.1 Composite from coffee husk

Composite materials are the product derived from combining two or more materials for specific applications. The composite materials possess unique properties like high strength, high modulus-to-weight ratio than the traditional engineering materials. The lightweight composite materials help to consume less energy leading to an extent of 20%−30% and become an alternative sustainable material (Knight and Curliss, 2003). These materials also have key functionalities like corrosion resistance, shape flexibility and durability. Hence, these materials are extensively used in many areas like construction, aerospace, medical appliances, wind turbines and many more. Typically, composite materials have three components consists of the matrix, the reinforcements and the interface(Ngo, 2020). The appropriate selection of matrix, the reinforcement and the manufacturing process can make the composite materials suitable for customized applications. The increasing demand for the fibre-reinforced composite materials is causing the recycling problem and ultimately move to the landfills (Oliveux et al., 2015). Hence, biocomposites are evolving as a subproduct of composite material with the use of natural fibres combined with polymers which are aimed to achieve some degree of reinforcement from the fibres to the polymer (Hidalgo-Salazar et al., 2018). The most used natural fibres for composite material formation are hemp, flax, jute and sisal. Whereas polypropylene, polystyrene, epoxy resin, natural rubber and recycled polypropylene are the most used polymer matixes for composites materials (Arbelaiz et al., 2005). Natural fibres and biomass waste can be an alternative and sustainable choices for reinforcing materials (Christian, 2016). Coffee husk is one of the major wastes produced during the transformation of coffee bean. The average diameter of a coffee husk is around 1.2 mm. This natural plant fibre contains high holocellulose as well as a significant level of lignin (de Carvalho Oliveira et al., 2018). These products are more sustainable than the traditional composite materials due to the lower environmental burden.

Hidalgo-Salazar et al. developed biocomposites using recycled polypropylene and coffee husk. This biocomposite was prepared using extrusion and injection moulding processes (Hidalgo-Salazar et al., 2018). Maleated polypropylene was added as a coupling agent for the formation of this biocomposite. They found that the addition of coffee husk fibre leads to improvement of the flexural modulus and thermal properties of the composites but reduced impact properties.

Huang et al. also utilized the coffee husk fibre from the coffee industry as a reinforcing filler in the preparation of a cost-effective thermoplastic-based composite. They found that coffee husk fibre exhibited thermal behaviour like wood fibre (Huang et al., 2018). They prepared high-density polyethylene composites using melt processing and extrusion with 40%−70% coffee husk fibre content loadings of from 40% to 70%. They found that the increasing coffee husk fibre content of the high-density polyethylene matrix resulted in an increase in the modulus and thermal properties of the composites. The increase in coffee husk fibre content causes reduction in water resistance (Huang et al., 2018).

Llangovan et al. fabricated a bio-composite by compression moulding with coffee husk fibre as a reinforcement for polypropylene. They studied the effect of varying the weight ratio of coffee husk fibre to the polypropylene (70:30, 80:20, 90:10) and densities (0.5, 0.75 and 1.0 g cm^{-3}) on mechanical properties (Ilangovan et al., 2019). They found that the 1.0 g cm-3 composite at 80:20 ratio had the highest tensile of 24.5 MPa and flexural strength of 21.5 Mpa. They also observed that with an increase in density, the water stability and flame retardancy of the composites improved. Again, the highest sound absorption coefficient of 0.9 and thermal insulation coefficient of 51.8 mW mK^{-1} were observed 1.0 g cm^{-3} at 80:20 ratio CH/PP composites. They concluded that these composites can be used in false ceiling and insulation panels (Ilangovan et al., 2019).

Lule and Kim in 2021 fabricated an eco-friendly and economical PBAT (polybutylene adipate terephthalate) composite using different proportions of coffee husk fibre as a reinforcing interface (Lule and Kim, 2021). They observed that the silane-treated coffee husk fibre (T-CH) has better contact angle which makes it more compatible with PBAT. The composite fabricated with 40% T-CH shows greater tensile strength and Young's modulus. The use of T-CH reduces the composite cost by 32% and is more eco-friendly.

13.5.2 Extract of spent coffee ground as a natural dye

The SCGs can be used as a potential renewable source of the natural dye as enough pigments can be extracted from these discarded SCGs. The extracted pigment can become the potential resource for the dyeing of textiles.

The research observation of Hong (2018) shows that fabrics treated with SCG extract had superior antioxidant ability and showed antibacterial ability, particularly to Gram-positive bacteria. The fibres with amino groups show excellent colour and colour fastness properties when dyed with extract of SCG. Mussatto et al. observed that the extract of SCGs contains more valuable ingredients and also eco-friendly when are extracted by the usual espresso method than the methanolic extraction method (Mussatto et al., 2011a,b).

Koh and Hong (2017) applied the extract of SCG for the functional textile finishing. The cotton fabric was finished by pad-dry-cure process from the SCG extract. They found that finished cotton fabric inhibited the growth of Gram-positive *Staphylococcus aureus* but did not significantly affect the growth of Gram-negative *Klebsiella pneumoniae*. This finished cotton fabric shows slightly yellowness (Koh and Hong, 2017).

Mongkholrattanasit et al. (2021) used the biocolourant from the SCG to dye the cotton fabric. They also applied low molecular weight chitosan with citric acid as a cross-linker (Mongkholrattanasit et al., 2021). They applied low molecular weight chitosan to cotton at various concentrations by pad-dry-cure technique, followed by dyeing with SCG extract using an exhaustion process. The proper cross-linking behaviour was confirmed by the Fourier transform infrared spectrometry analysis. They have tested the colour properties using CIELAB (L*, a* and b*) and K/S values, colour fastness and whiteness index apart from the physical properties of the dyed fabric. Higher colour strength and wrinkle resistance were observed for these dyed fabric samples compared with untreated and alum-mordanted samples.

13.6 Sustainability of coffee fabric manufacturing

The process of fabric produced from the coffee fibre follows similar processes to other fibres. The process sequence of fabric is solely dependent on the end use of the product. The traditional textile industry mostly works for the bulk production of apparel and home furnishings and the production of these textile goods starts with raw material production followed by fibre preparation (like ginning for cotton, crimping or texturizing or cutting for synthetic fibres), yarn formation (spinning), weaving/knitting and wet processing processes.

13.6.1 Fibre production

The coffee fibre is mostly produced from the rPET with SCG. The fibre production is the first and basic phase of textiles. Energy, water, fertilizer and pesticide are the key requirements for growing up the crops for the production of natural fibre and become the key concern in this phase. Natural plant fibres are further processed for making the

Table 13.4 Typical requirement of energy (MJ) to produce 1 kg of fibre.

Fiber	Energy in MJ for 1 kg
rPET	66
Viscose	100
Polypropylene	115
Polyester virgin	125
Acrylic	175
Nylon	250

(*Source* Mukherjee, A. 2017. Clothes from Plastic Bottles).

fibre suitable for textile operation like ginning for cotton. The production of the animal fibres like wool and silk requires natural resources like water and animal feed. The production of regenerated fibre like viscose, lyocell, etc. is derived from the wood pulp and consumes reasonable chemicals and water for its spinning. Whereas the production of synthetic fibres like coffee fibre, polyester, nylon, acrylic, etc. begins with crude oil extraction and refining, followed by the production of the respective polymer. The polymers are then spun (melt spinning for polyester), crimped and cut for textile operations. The typical requirements of energy consumption in MJ to produce 1 kg of different kinds of synthetic fibre are shown in Table 13.4. The graphical representation of the process sequence is shown in Fig. 13.5.

13.6.2 Yarn manufacturing

The yarn manufacturing (spinning) is the process of the conversion of fibres to yarn, and this transformation of yarn from fibre requires a series of steps and machinery. The spinning process begins with fibre management where all the fibres are opened and blended if required. The first stage of yarn manufacturing starts with the opening and separation of trash or foreign materials from the fibres in the blow room line. Then the feed material passed through other machines like carding, breaker draw frame, finisher draw frame, combing (an optional machine), roving frame and ring frame. Ring spinning technique is the most used spinning technology for yarn production. Overall 80% of the yarns are produced by this technology. The ring spinning system is also extensively used for the formation of core-spun yarns. The core-spun yarns are produced by twisting a filament (spandex) or staple spun yarn around an existing yarn (Patra and Pattanayak, 2015). These core spun yarns are mostly used to induce stretching capacity in the fabric The rest yarns are produced by using open-end spinning system like rotor, air-jet, etc. The rotor yarns are mostly utilized for the manufacturing of the coarse yarns like yarns for denim. The spinning process requires a very small amount of antistatic chemicals for the smooth running of synthetic materials (Schönberger and Schäfer, 2003).

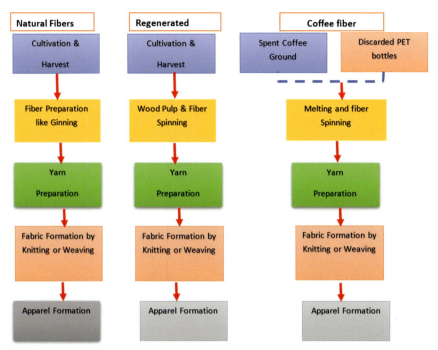

Figure 13.5 Graphical summary of key process for apparel formation for different types of fibres.

13.6.3 Fabric manufacturing

Fabric is the major component of all types of textiles goods and an important intermediate product of the textile value chain. Textile fabrics are suitable for apparel due to their unique properties like lightweight, flexibility, preamble and requisite strength. The fabrics are considered as two-dimensional materials rather than three-dimensional products as the thickness is negligible with respect to the length and width. The fabrics can be manufactured by different techniques such as weaving, knitting, nonwoven, braiding and tufting. The woven and knitted fabrics are mostly used for apparel and home furnishings (Pattanayak, 2020). Woven fabric manufacturing requires more processes than knitted fabrics and ultimately consumes more energy than knitted fabric. The woven fabrics production requires warping, sizing (optional), drawing and weaving, whereas the weft knitted fabrics can be produced directly from the cones from the winding process. The sizing process for woven fabrics requires adhesives (like natural starch, poly vinyl alcohol) and other chemical additives. This process also consumes water in the form of steam for drying wet warp sheet. Hence, the woven fabrics consume lots of water and energy for production as compared to knitted fabrics.

Table 13.5 Energy (MJ) and water (litters) consumption to produce 1 kg fabric.

Process	Cotton		Lyocell		Polyester	
Fibre production	48.7	576–4377	66–102	263	77.3–112	17.2
Fibre preparation[a]	18.4	0		0	13.6	0
Fabric formation[b]	5–30	0	5–30		5–30	
Dyeing and finishing	17.9–60.9	105–145	17.9–60.9	105–145	40	65–148

[a]Fibre preparation is the process needed to make the fibre ready for the spinning process.
[b]This stage includes the yarn formation.

13.6.4 Wet processing

The greige fabrics (the unfinished fabric) are not suitable for the apparel without proper wet processing. The greige fabrics require a series of wet processing treatments to make the fabrics suitable for the apparel and other applications. The wet processing of fabrics involves scouring, bleaching, dyeing and finishing. The treated fabrics are washed several times after the end of each of these processes. The washing of fabrics requires lots of water and chemicals to improve their tactile properties. The discharged wastewater from the textile processing plants can be a major source of pollution if untreated. The wastewater consists of organic, inorganic, elemental and polymeric substances (Sanmuga Priya and Senthamil Selvan, 2017). Hence, the chemical oxygen demand for the effluent treatment is estimated to 40−80 g per kilogram of fibre (Turley et al., 2011). Apart from all these processes, energy also required other essential activities like illumination, humidification, etc. Table 13.5 shows the typical energy and water consumption to produce 1 kg of fabric.

13.7 Conclusions and futuristic trends

There is already a trend to use more sustainable fibres other than the most widely used fibres such as cotton and polyester for garments. Coffee fibre is one of the sustainable fibres having excellent antibacterial and fast drying capacity. The innovation of the products using coffee fibre will be a sustainable option for the textile sector and has the potential to replace many unsustainable products. The conversion of the waste for the value addition of different products will be the key focus of the circular economy. The more water and energy-efficient fabric manufacturing processes will be the key challenges for the development of sustainable products. Some brands have started using coffee fibre in their product line. Others are yet to include these fibres in their production line. The use of coffee fibres, not only makes the fabrics sustainable but also solves the problems of coffee waste management in the supply chain.

References

AATCC 183: Test Method for Transmittance or Blocking of Erythemally Weighted Ultraviolet Radiation through Fabrics, 2021. https://global.ihs.com/doc_detail.cfm?document_name=AATCC%20183&item_s_key=00326499. (Accessed 23 June 2021).

Al-Hamamre, Z., Foerster, S., Hartmann, F., Kröger, M., Kaltschmitt, M., Jun. 2012. Oil extracted from spent coffee grounds as a renewable source for fatty acid methyl ester manufacturing. Fuel 96, 70−76. https://doi.org/10.1016/j.fuel.2012.01.023.

Arbelaiz, A., Fernández, B., Cantero, G., Llano-Ponte, R., Valea, A., Mondragon, I., Dec. 2005. Mechanical properties of flax fibre/polypropylene composites. Influence of fibre/matrix modification and glass fibre hybridization. Composites Part A: Applied Science and Manufacturing 36 (12), 1637−1644. https://doi.org/10.1016/j.compositesa.2005.03.021.

ASTM D5742: Determination of Butane Activity in the Laboratory, Mar. 2020. Analytice. Accessed 23 Jun 2021. https://www.analytice.com/en/astm-d5742-determination-of-butane-activity-in-the-laboratory/.

Bae, J., Hong, K.H., Apr. 2019. Optimized dyeing process for enhancing the functionalities of spent coffee dyed wool fabrics using a facile extraction process. Polymers 11 (4). https://doi.org/10.3390/polym11040574. Art. no. 4.

Benavides, P.T., Dunn, J.B., Han, J., Biddy, M., Markham, J., Aug. 2018. Exploring comparative energy and environmental benefits of virgin, recycled, and bio-derived PET bottles. ACS Sustainable Chemistry and Engineering 6 (8), 9725−9733. https://doi.org/10.1021/acssuschemeng.8b00750.

Brasquet, C., Roussy, J., Subrenat, E., Cloirec, P.L., Nov. 1996. Adsorption and selectivity of activated carbon fibers application to organics. Environmental Technology 17 (11), 1245−1252. https://doi.org/10.1080/09593331708616494.

Brones, A., 2021. Upcycled Coffee Textiles: Out Of The Garbage Can And Into The Dyer. https://sprudge.com/can-you-wear-the-coffee-you-drink-125116.html. https://sprudge.com. (Accessed 21 June 2021).

Campos-Vega, R., Loarca-Piña, G., Vergara-Castañeda, H.A., Oomah, B.D., Sep. 2015. Spent coffee grounds: a review on current research and future prospects. Trends in Food Science and Technology 45 (1), 24−36. https://doi.org/10.1016/j.tifs.2015.04.012.

Christian, S.J., 2016. 5−natural fibre-reinforced noncementitious composites (biocomposites). In: Harries, K.A., Sharma, B. (Eds.), Nonconventional and Vernacular Construction Materials. Woodhead Publishing, pp. 111−126. https://doi.org/10.1016/B978-0-08-100038-0.00005-6.

Consumer|Coolmax, 2021. https://www.coolmax.com/en. (Accessed 23 June 2021).

de Carvalho Oliveira, F., et al., Jun. 2018. Characterization of coffee (Coffea arabica) husk lignin and degradation products obtained after oxygen and alkali addition. Bioresource Technology 257, 172−180. https://doi.org/10.1016/j.biortech.2018.01.041.

EU FUSIONS, 2021. https://www.eu-fusions.org/index.php. (Accessed 18 June 2021).

GB/T 21655.1-2008: PDF in English, 2021. https://www.chinesestandard.net/PDF.aspx/GBT21655.1-2008. (Accessed 25 June 2021).

Gies, P., et al., Jan. 2003. Ultraviolet protection factors for clothing: an intercomparison of measurement systems¶. Photochemistry and Photobiology 77 (1), 58−67. https://doi.org/10.1562/0031-8655(2003)0770058UPFFCA2.0.CO2.

Global PET Bottle Market (2021 to 2026)—Industry Trends, Share, Size, Growth, Opportunity and Forecasts - ResearchAndMarkets.Com, Mar. 05, 2021. https://www.businesswire.com/news/home/20210305005232/en/Global-PET-Bottle-Market-2021-to-2026—Industry-Trends-Share-Size-Growth-Opportunity-and-Forecasts—ResearchAndMarkets.com. (Accessed 21 June 2021).

Hardgrove, S.J., Livesley, S.J., Aug. 2016. Applying spent coffee grounds directly to urban agriculture soils greatly reduces plant growth. Urban Forestry and Urban Greening 18, 1−8. https://doi.org/10.1016/j.ufug.2016.02.015.

Hes, L., Dolezal, I., Aug. 2018. Indirect measurement of moisture absorptivity of functional textile fabrics. Journal of Physics: Conference Series 1065, 122026. https://doi.org/10.1088/1742-6596/1065/12/122026.

Hidalgo-Salazar, M.A., Correa-Aguirre, J.P., Montalvo-Navarrete, J.M., López-Rodríguez, D., Rojas-González, A.F., 2018. Recycled Polypropylene-Coffee Husk and Coir Coconut Biocomposites: Morphological, Mechanical, Thermal and Environmental Studies. https://doi.org/10.5772/INTECHOPEN.81635.

Hoffmann, U., Jul. 2015. Can Green Growth Really Work and what Are the True (Socio-) Economics of Climate Change? Working paper United Nations Conference on Trade and Development (UNCTAD), CH-Geneva [Online]. Available: https://orgprints.org/id/eprint/29947/. (Accessed 3 July 2021).

Huang, L., Mu, B., Yi, X., Li, S., Wang, Q., Jan. 2018. Sustainable use of coffee husks for reinforcing polyethylene composites. Journal of Polymers and the Environment 26 (1), 48−58. https://doi.org/10.1007/s10924-016-0917-x.

Humblet, C., 2006. A System Dynamics Analysis of a Capilene Supply Loop. Thesis. Massachusetts Institute of Technology [Online]. Available: https://dspace.mit.edu/handle/1721.1/37215. (Accessed 3 July 2021).

Ilangovan, M., Guna, V., Hu, C., Takemura, A., Leman, Z., Reddy, N., Sep. 2019. Dehulled coffee husk-based biocomposites for green building materials. Journal of Thermoplastic Composite Materials. https://doi.org/10.1177/0892705719876308, p. 0892705719876308.

Jkaybay, ∼, Nov. 21, 2020. Ethical Investing—Carbios Recycles Polyester Clothing. The Green Stars Project. https://greenstarsproject.org/2020/11/20/ethical-investing-carbios-recycled-polyester-pet/. (Accessed 18 June 2021).

Kilic, M., Okur, A., Jan. 2011. The properties of cotton-Tencel and cotton-Promodal blended yarns spun in different spinning systems. Textile Research Journal 81 (2), 156−172. https://doi.org/10.1177/0040517510377828.

Knight, M., Curliss, D., 2003. Composite materials. In: Meyers, R.A. (Ed.), Encyclopedia of Physical Science and Technology, third ed. Academic Press, New York, pp. 455−468. https://doi.org/10.1016/B0-12-227410-5/00128-9.

Koh, E., Hong, K.H., Nov. 2017. Preparation and properties of cotton fabrics finished with spent coffee extract. Cellulose 24 (11), 5225−5232. https://doi.org/10.1007/s10570-017-1466-8.

Koo, B.-M., Kim, J.-H.J., Kim, S.-B., Mun, S., Aug. 2014. Material and structural performance evaluations of hwangtoh admixtures and recycled PET fiber-added eco-friendly concrete for CO_2 emission reduction. Materials 7 (8). https://doi.org/10.3390/ma7085959. Art. no. 8.

Kwon, E.E., Yi, H., Jeon, Y.J., May 2013. Sequential co-production of biodiesel and bioethanol with spent coffee grounds. Bioresource Technology 136, 475−480. https://doi.org/10.1016/j.biortech.2013.03.052.

Lessa, E.F., Nunes, M.L., Fajardo, A.R., Jun. 2018. Chitosan/waste coffee-grounds composite: an efficient and eco-friendly adsorbent for removal of pharmaceutical contaminants from water. Carbohydrate Polymers 189, 257−266. https://doi.org/10.1016/j.carbpol.2018.02.018.

Lou, C.-W., et al., Oct. 2007. PET/PP blend with bamboo charcoal to produce functional composites. Journal of Materials Processing Technology 192−193, 428−433. https://doi.org/10.1016/j.jmatprotec.2007.04.018.

Louris, E., et al., Dec. 2018. Evaluating the ultraviolet protection factor (UPF) of various knit fabric structures. IOP Conference Series: Materials Science and Engineering 459, 012051. https://doi.org/10.1088/1757-899X/459/1/012051.

Lule, Z.C., Kim, J., Jan. 2021. Properties of economical and eco-friendly polybutylene adipate terephthalate composites loaded with surface treated coffee husk. Composites Part A: Applied Science and Manufacturing 140, 106154. https://doi.org/10.1016/j.compositesa.2020.106154.

Luppino, R., 2020. Preferred Fiber and Materials Market Report (PFMR) Released! Textile Exchange. https://textileexchange.org/2020-preferred-fiber-and-materials-market-report-pfmr-released/. (Accessed 4 July 2021).

Mongkholrattanasit, R., Nakpathom, M., Vuthiganond, N., May 2021. Eco-dyeing with biocolorant from spent coffee ground on low molecular weight chitosan crosslinked cotton. Sustainable Chemistry and Pharmacy 20, 100389. https://doi.org/10.1016/j.scp.2021.100389.

Mukherjee, A., 2017. Clothes from Plastic Bottles!.

Murthy, P.S., Madhava Naidu, M., Sep. 2012. Sustainable management of coffee industry by-products and value addition—a review. Resources, Conservation and Recycling 66, 45—58. https://doi.org/10.1016/j.resconrec.2012.06.005.

Mussatto, S.I., Ballesteros, L.F., Martins, S., Teixeira, J.A., 2011a. Extraction of antioxidant phenolic compounds from spent coffee grounds. Separation and Purification Technology 83, 173—179. https://doi.org/10.1016/j.seppur.2011.09.036.

Mussatto, S.I., Carneiro, L.M., Silva, J.P.A., Roberto, I.C., Teixeira, J.A., 2011b. A study on chemical constituents and sugars extraction from spent coffee grounds. Carbohydrate Polymers 83 (2), 368—374. https://doi.org/10.1016/j.carbpol.2010.07.063.

Ngo, T.-D., 2020. Introduction to Composite Materials. IntechOpen. https://doi.org/10.5772/intechopen.91285.

Oliveira, L.S., Franca, A.S., 2015. Chapter 31—an overview of the potential uses for coffee husks. In: Preedy, V.R. (Ed.), Coffee in Health and Disease Prevention. Academic Press, San Diego, pp. 283—291. https://doi.org/10.1016/B978-0-12-409517-5.00031-0.

Oliveira, L.S., Franca, A.S., Camargos, R.R.S., Ferraz, V.P., 2008. Coffee oil as a potential feedstock for biodiesel production. Bioresource Technology 99 (8), 3244—3250. https://doi.org/10.1016/j.biortech.2007.05.074.

Oliveux, G., Dandy, L.O., Leeke, G.A., 2015. Current status of recycling of fibre reinforced polymers: review of technologies, reuse and resulting properties. Progress in Materials Science 72, 61—99. https://doi.org/10.1016/j.pmatsci.2015.01.004.

P4DRY—COFFEE PRINTING TECHNOLOGY|Steemhunt, 2021. https://steemhunt.com/tag/coffee/@jenina619/p4dry-coffee-printing-technology. (Accessed 19 June 2021).

Patra, A.K., Pattanayak, A.K., 2015. In: Denim, R.P. (Ed.), 16—Novel Varieties of Denim Fabrics. Woodhead Publishing, pp. 483—506. https://doi.org/10.1016/B978-0-85709-843-6.00016-0.

Pattanayak, A.K., 2020. 3—sustainability in fabric manufacturing. In: Nayak, R. (Ed.), Sustainable Technologies for Fashion and Textiles. Woodhead Publishing, pp. 57—72. https://doi.org/10.1016/B978-0-08-102867-4.00003-7.

Ross, G., Aug. 27, 2019. Australia Recycles Paper and Plastics. So Why Does Clothing End up in Landfill? | Graham Ross. the Guardian. http://www.theguardian.com/commentisfree/2019/aug/27/australia-recycles-paper-and-plastics-so-why-does-clothing-end-up-in-landfill. (Accessed 17 June 2021).

S.Café®-Product, 2021. https://scafefabrics.com/index.php?/en-global/product/airnest. (Accessed 19 June 2021).

S.Café®-Product, 2021. https://scafefabrics.com/index.php?/en-global/product/eco2sy. (Accessed 24 June 2021).

Sanmuga Priya, E., Senthamil Selvan, P., May 2017. Water hyacinth (*Eichhornia crassipes*)—an efficient and economic adsorbent for textile effluent treatment—a review. Arabian Journal of Chemistry 10, S3548—S3558. https://doi.org/10.1016/j.arabjc.2014.03.002.

Schönberger, H., Schäfer, T., 2003. Best Available Techniques in Textile Industry. Federal Environmental Agency, Berlin.

Severini, C., Caporizzi, R., Fiore, A.G., Ricci, I., Onur, O.M., Derossi, A., Feb. 2020. Reuse of spent espresso coffee as sustainable source of fibre and antioxidants. A map on functional, microstructure and sensory effects of novel enriched muffins. Lebensmittel-Wissenschaft and Technologie 119, 108877. https://doi.org/10.1016/j.lwt.2019.108877.

SINGTEX-Singtex, 2021. http://www.singtex.com/en-global/technology/fabrics_info/p4dry. (Accessed 19 June 2021).

SINGTEX-Singtex, 2021. http://www.singtex.com/en-global/technology/fabrics_info/airmem. (Accessed 25 June 2021).

Subic, A., Mouritz, A., Troynikov, O., 2009. Sustainable design and environmental impact of materials in sports products. Sports Technology 2 (3–4), 67–79. https://doi.org/10.1002/jst.117.

Sustainability, 2021. OrthoLite. https://www.ortholite.com/sustainability/. (Accessed 26 June 2021).

Sustainability, 2021b. Rens Original US. https://rensoriginal.com/pages/sustainability. (Accessed 26 June 2021).

Textile Market Size|Industry Analysis Report, 2021–2028, 2021. https://www.grandviewresearch.com/industry-analysis/textile-market. (Accessed 18 June 2021).

Textiles made from coffee grounds?—MTIX International, 2021. https://mtixinternational.com/2019/09/textiles-made-from-coffee-grounds/. (Accessed 4 July 2021).

Tournier, V., et al., Apr. 2020. An engineered PET depolymerase to break down and recycle plastic bottles. Nature 580 (7802), 216–219. https://doi.org/10.1038/s41586-020-2149-4.

Turley, D.B., et al., 2011. The Role and Business Case for Existing and Emerging Fibres in Sustainable Clothing: Final Report to the Department for Environment, First. London U.K: Food and Rural Affairs (DEFRA) Department for Environment.

Vítězová, M., et al., Jan. 2019. The possibility of using spent coffee grounds to improve wastewater treatment due to respiration activity of microorganisms. Applied Sciences 9 (15). https://doi.org/10.3390/app9153155. Art. no. 15.

What Goes, Oct. 09, 2018. Around: How Coffee Waste Is Fueling a Circular Economy. Daily Coffee News by Roast Magazine. https://dailycoffeenews.com/2018/10/09/what-goes-around-how-coffee-waste-is-fueling-a-circular-economy/. (Accessed 4 July 2021).

Why, Exactly, Is Polyester So Bad for the Environment?, Jan. 19, 2021. Ecocult. https://ecocult.com/exactly-polyester-bad-environment/. (Accessed 17 June 2021).

Xpresole Panto, 2021. Xpresole. https://www.xpresolepanto.com. (Accessed 26 June 2021).

Recycled fibres from polyester and nylon waste

14

Sanat Kumar Sahoo and Ashwini Kumar Dash
Department of Textile Engineering, Odisha University of Technology and Research, Bhubaneswar, Odisha, India

14.1 Introduction

The increase in population increases consumption and creates waste generation in the garment sector. The use of textile products has become a key focus of worldwide property efforts. The event of recycled technology has allowed the textile business to supply huge amounts of products that exhaust natural resources. Textile use techniques are developed to deal with this increase in textile waste and new solutions are still being researched. Beyond recycled, the textiles business is seeking bio-based polymer alternatives that would finish up being a lot of property than recycled choices (Carneiro et al., 2016).

There are several reasons for the waste production from the textile industry to lead the recycling processes. They embody reduction of the landfills would like, resources conservation, paying the connected tipping fee and facility of low-priced raw materials for creating products. Even then, in actuality, the textile's use rate isn't terribly high. Moreover, the often-endorsed reason of inadequate public can participate in use; economic science is often the explanation for acceptance of alternative waste disposal modes.

Besides, the textile and clothing business is the most polluting business in the world. Textiles have a big impact on the surroundings throughout their lifecycle (Blackburn, 2009). A great deal of water, energy and chemicals are used in textile production. Textile use is important to cut back on pollution. So as to cut back the gas emissions, efforts are created to extend textile use.

As we tend to take into account the case of textile and attire use it becomes apparent that the method impacts several entities and contributes considerably, in an exceedingly broader sense, to the social responsibility of up-to-date culture. Environmental problems also are related to the sector—embody high energy and water usage and use of toxic chemicals. Synthetic fibre products won't decompose within landfills. Taking 100's of years to decompose such waste discarded in lowland has no merchandizing price and it fouls the atmosphere, if not degraded they get accumulated and unfold infectious diseases and foul smells.

By 2025, the Recycled Polyester Challenge is a crucial catalyst for amendment within the attire and textile business. We tend to be difficult for the fashion business

---

Sustainable Fibres for Fashion and Textile Manufacturing. https://doi.org/10.1016/B978-0-12-824052-6.00008-1
Copyright © 2023 Elsevier Ltd. All rights reserved.

to arrange transportation for the proportion of recycled polyester up from 14% to 45% at 17.1 million metric tons by 2025 (Wani et al., 2020).

The Nylon Waste Recycling Market report provides revenue for the worldwide polymer or Nylon Waste usage Market between 2015 and 2026, with 2019 serving because the base year and 2020–26 serving because of the forecast year. Additionally, the study includes the compound annual growth rate (CAGR) for the worldwide market trend over the forecast amount. The worldwide polymer or Nylon Waste usage market investigates historical and gift growth patterns and prospects so as to get perceptive insights into these trade metrics over the market forecast amount of 2020–26 (Martínez Silva and Nanny, 2020).

The elimination of microfibers released by synthetic textiles, such as polyester and nylon to make them as a useful product for society and also sustainable for the ecosystem is reported by Ellen Macarthur Foundation (Liu et al., 2021). By using emerging new materials and methods will reduce microfiber shedding and emission of poisonous gas to the environment. For the future generation and ecosystem, it's good to avoid the plastic-based fabrics from polyester and nylon. Otherwise, it will be very difficult to collect these pre-production and post-production waste and reuse or reproduce them into a new product.

14.2 Textile recycling

Recycling is the process of reusing the existed one. In the textile sector, recycling means reusing or reproducing the same or different products from the previous waste products. These past waste materials may be fibres, fabric, garments, plastic, carpet etc. There are two types of recycling processes adopted worldwide, and they are mechanical and chemical processes. Also, two types of waste are collected from the environment names pre-consumer and post-consumer wastes. The mechanical method involves mainly five steps sorting, cutting, shredding, carding and winding to make a new product (Goyal, 2021). And in the chemical process, some chemicals are used to dissolve the waste into its original position and then new products are formed. By using melt-processing, plastic can be formed into several forms from its original one with the help of recycling, hydrolysis, burning the solid waste and utilizing the fibrous product by heat generation (Wang, 2010) (Fig. 14.1).

14.3 Polyester

14.3.1 Virgin polyester

vPolyester is the global leader in fibre in its popularity and end-use. The chemical name of polyester is polyethylene terephthalate (PET). Polyester comes under the synthetic fibre category and was patented in 1941 (Houck et al., 2001). Since then, it became the most popular material in the textile and fashion industry. Today's data

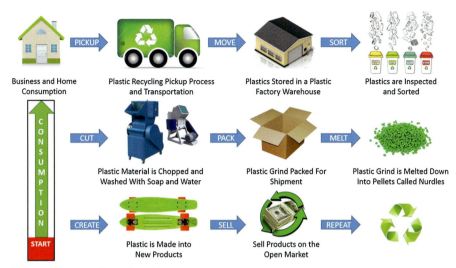

Figure 14.1 Recycling of textiles.

says that the worldwide production of polyester is approximately 77 million tons and it holds around 55% of the market share in the global fibre market (Becerra and Alberto Valderrama, 2000).

PET belongs to the polyester family of polymers and it is a general-purpose thermoplastic polymer. Thermoplastics can be heated, melted and cooled into different shapes and sizes. Polyester resins are having an excellent combination of properties like mechanical, thermal, chemical resistance and dimensional stability. PET is often a transparent plastic with strong, lightweight in its property.

We are using many types of fibres in our daily life but PET is the most used among them. The application of polyester are fabrics, packaging, moulded parts for automotive, films, electronics and many more. This transparent clear plastic is mainly used as a water bottle or soda bottle container, trays and domes holding fruits and salads, peanut butter containers and cleaning product bottles as packaging made from PET plastic.

Polyester is popular because it is a durable, cheap and easily available fabric. But it is highly unsustainable. It is having low toxicity and it makes the material for single-use drinks bottles:

- In its virgin form, PET is widely considered safe to wear as it is free from known endocrine disruptors such as Bisphenol A (BPA) and bisphenol S (BPS) (Lusher et al., 2017).
- PET is made with the use of antimony (a potential carcinogen), but scientific studies have shown that antimony only leaches from PET in temperatures above 65°C for 38 days (Vermaas, 2011).

14.3.2 Recycled polyester

Recycled polyester, conjointly called RPET, is obtained by melting down discarded plastics and re-spinning it into new polyester fibre. Polyethylene terephthalate (PET) will be recycled from each post-industrial and post-consumer material. By mistreatment of plastic bottles and containers discarded by the customers, RPET is created. The PET scraps are taken and broken into fibre elements which will then be spun into the recycled yarns. Like ancient polyester, recycled polyester may be a manmade fibre from synthetic fibres. However, recycled polyester makes use of existing plastic, rather than utilizing new materials to craft the material (i.e., petroleum) (Thiounn and Smith, 2020). In several cases, those existing plastics are your recent water bottles, which are then processed and as if by magic reworked into that awe-inspiring, rip-resistant Bluff Utility backpack sitting next to your table.

PET products are 100% reusable and are the foremost recycled plastic worldwide. PET will be simply known by its exercise method. Low diffusion constant makes PET way more appropriate than different plastic materials to be used as a recovered, recycled material. PET and RPET plastic will be known by the image related to its which may simply be noticed on the packaging material, typically at very cheap. Several brands and retailers will create this disclaimer on their packaging typically stating what proportion recycled content has been used (e.g., 100% or 30%−70%) (Tua et al., 2019). Some firms value more highly to use the nomenclature, 'postconsumer' recycled PET. Although they share constant chemical compounds, there's a transparent distinction between PET and RPET.

The 'R' in RPET stands for 'recycled'. Virgin PET plastic that is created from oil and petrochemicals will be recycled into RPET. Within the usage method, plastic is collected, clean and remade into products. Technically, something made of PET will be replaced with RPET. It's good for the environment to divert plastic from our landfills. The assembly of recycled polyester needs so much fewer resources than that of recent fibres and generates fewer dioxide emissions. New polyester fibres are produced by recycling polyester which is obtained from existing plastic by melting down and re-spinning it. Study shows that 5 water bottles are enough to prepare fibre for one jersey. Although usage of plastic looks like an associate indisputable sensible plan, recycled polyester is way from being the most effective property fashion world. To preserve the system and therefore the atmosphere for future generations, RPET is the only solution. RPET is the innovative recycled polyester yarn factory-made from discarded PET bottles. Victimization environment-friendly usage processes, RPET doesn't rely on fossil fuel and therefore expeditiously preserves natural resources and reduces environmental strain while not compromising quality and producing eco-friendly yarns. The main concern with recycled PET is the risk of alternative, more toxic, plastics being accidently mixed in throughout the usage method.

Despite the widespread health issues, polycarbonate plastic remains factory-made with the employment of BPA and bits per second. So, there's a clear stage that polycarbonate plastic accidently gets mixed in with PET throughout the usage method. This risk of contamination from these chemicals is low, not least as a result of these contaminants worsen the standard of recycled PET, thus lowering the merchandizing price—not one thing makers are going to be dashing to do! Today, automatically recycled polyester from plastic water bottles makes up the overwhelming majority of recycled polyester; but, chemical usage and, additional specifically, textile-to-textile usage is going to be a necessary part of reaching our goal. One year past, the non-profit organization Textile Exchange challenged over 50 textile, fashion and retail corporations (including giants like Adidas, H&M, Gap and Ikea) to extend their use of recycled polyester by 25% by 2020 (Black, 2015).

14.3.3 Sources

Most materials employed in textile use may be split into two categories: pre-consumer and post-consumer waste. Pre-consumer or post-industrial waste consists of textile waste created at the commercial stage of the assembly of textile material. Typically, these by-products are created by the textile, garment, cotton, and fibre industries and are repurposed by the article of furniture, home building, automotive, and alternative industries. Post-consumer waste consists of discarded clothes or house articles made of factory-made textiles. Some post-consumer waste is directed towards second-hand retailers to be oversubscribed once more. A number of this waste is collected in municipal assortment bins; however, the bulk of this waste is found in landfills.

RPET is recycled PET material sourced from post-consumer waste plastic drink bottles, and rHDPE uses HDPE material (high density polyethylene) sourced from post-consumer recycled milk jugs and house cleaner bottles. PET is the most recycled plastic in the world and HDPE is the second-most recycled plastic in the world. At an equivalent time, there's an associate imperative ought to cut back the number of wastes generated by the fashion product, which is calculable at 92 million tonnes a year, the report says (Duan et al., 2008).

Indeed, recycled fibres are derived from garments and wastes generated from textile are attracting high business interest, and a variety of the fashion industry's key players, together with H&M cluster, Kering and Patagonia, have endowed start-ups innovating in this field. Alternative firms, like Adidas, Bestseller, Levi Strauss and Co, PVH and Wrangler, have established partnership agreements through that they're exploring the employment of such fibres within the manufacture of an innovative new product. However, it ought to be noted, the report points out, that start-ups getting into the market are competitive with some business heavyweights, together with Asahi Kasei, Birla Cellulose, Lenzing, Sateri, and Tangshan Sanyou (Hurley, 2017).

14.3.4 Manufacturing process

14.3.4.1 Methods of textile recycling

Now technology has grown to a high level but it can't recycle the 100% polyester clothes. There are two types of recycling process of PET and these are mechanical and chemical (Fig. 14.2).

14.3.4.2 Mechanical

Mechanical usage is taking a plastic bottle, washing it, shredding it then turning it back into a polyester chip that then goes through the normal fibre-creating method. With this

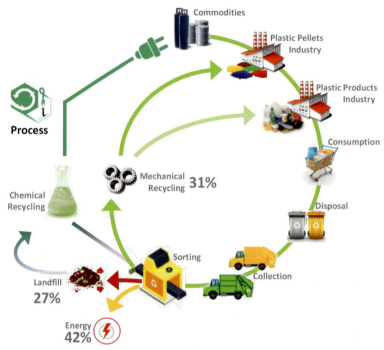

Figure 14.2 Manufacturing process of recycled polyester.

method, the recycled materials aren't polyester textiles, but plastic bottles. Each recycled fibre is manufactured from identical material referred to as PET. Once the materials are sent to the industry, first they're sorted by colour and category. The same as cotton, then PET plastic is cut into small pieces and washed to get rid of contaminants. The dried, sliced plastic is wrought into PET pellets and undergoes extrusion to form new fibres. However, through the mechanical method, the fibre will lose its strength and so has to be mixed with virgin fibre (Park and Kim, 2014) (Fig. 14.3).

14.3.4.3 Chemical

The chemical process is typically used on synthetic fibres like Polyethylene terephthalate (PET) as these fibres will bear a breaking down and recreation method. This method isn't used widely, however, there are industries that are researching and implementing chemical use. Within the case of PET, the starting materials are initially broken down into monomers. This act of de-polymerization additionally removes contaminants from the beginning material like dyes and unwanted fibres. From here, the fabric is polymerized to be accustomed to turn-out a textile product. Not like the mechanical methodology of use, chemical use produces high-quality fibres just like virgin fibre used. Therefore, no new fibre square measure is required to support the merchandise of the activity. Completely different chemicals and pathways are used for alternative materials like nylon and cellulose-based fibres; however, the structure of the method is the same (Debowski et al., 2019) (Fig. 14.4).

Recycled fibres from polyester and nylon waste 315

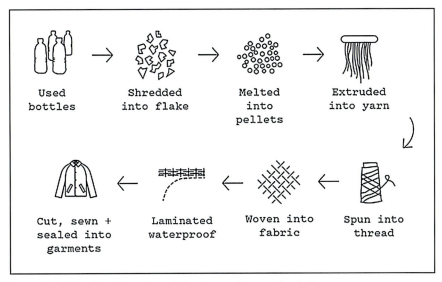

Figure 14.3 Manufacturing of recycled polyester by a mechanical process.

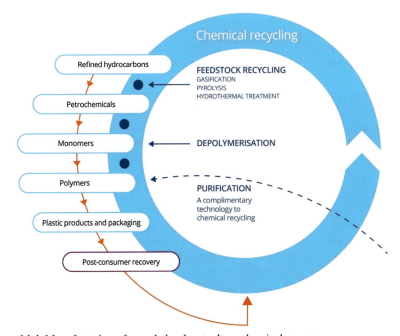

Figure 14.4 Manufacturing of recycled polyester by a chemical process.

14.3.4.4 Recycled polyester yarn process

Sorting

The sorting method is described as a pyramid model in terms of the quantity of fabric when recycling is done for post-consumer textile waste. The pyramid is built in order from the bottom with their largest volume are as follows: crude sorting, followed by the exportation of used clothes, conversion to new material, wiping and shinning them into cloths, lowland burning for energy, and finally diamonds. Usually, among the pyramid model, it's found that the quantity of textile things is reciprocally proportional to its value, furthermore which means that despite diamonds creating up the littlest sector (1%–2%) of the sorting method they have an inclination to be the foremost profitable.

Once the bales of PET come to recycling industries, they are first staged. Then, they're placed onto a conveyor that takes them through a bale breaker. This breaks the bales, and also the bottles then become sorted individually. It's important that each bottle are properly washed and the labels are removed. Then their labels are removed. Steam is employed, yet as varied chemicals. To do this, the bottles are placed into a hot air trommel, which is understood because of the 'pre-wash stage' (Vadicherla et al., 2015). If any PVC bottles have accidently created their method into the bale, they're going to flip slightly brown throughout this method. This then allows them to be removed.

Crude sorting

Within crude sorting, waste things are usually separated manually into different classes while conjointly removing bulkier things, like coats and blankets. The classes of textile waste are also divided primarily based upon components like material, condition, quality, or covering items like shirts. On the crude sorting method, recycled textiles are appointed unconditional grades representing their industrial worth primarily based upon varied fibre characteristics like length, colour, and also the homogeneity of its chemical composition (Hawley, 2014).

Cleaning

PET bottles are collected and sorted into different classes of packages at recycling centres and then sent to the specific PET recycling unit. Then caps and labels should be removed from the bottles and containers and cleaned properly. Further, they're steamed and washed to get rid of any non-PET elements (Moazzem et al., 2021).

Once all the steps are completed and therefore the facility is certain that no contaminants stay, the materials are rinsed once again. Once it's dried, it becomes a brand-new producing material by being reintroduced. Actually, producing new things out of the RPET is mostly done at alternative facilities.

Shredding

The bottles are chopped into small chips using a mechanical bale breaker. Shredding permits the unwanted liquid (water, soda, cold drinks etc.) to drain out. In order that it doesn't have an effect on the standard of Plastic. This waste is effective, primarily for the style trade. Later, the clear plastics are different from the coloured plastics. The cleared plastics are wont to build white garments. It is unreal therefore, it's notably valuable (Hussain et al., 2021).

Melting

The chips are then melted passing through a spinner. Then a cooling arrangement is there to cool it instantly and it turns into fine polyester filaments. These are then placed in a smoothed place, after that stretching occurs and stored into reels to roll as a bundle.

It is then ready for the manufacturing of recycled yarn for the using in weaving or knitting or in different industrial use. Polyester Waste utilization has constant aesthetics and operative performance as regular polyester yarn. Recycled polyester offers a brand-new life to a fabric that is non-biodegradable and would otherwise find you within the ocean and landfills that is harmful to the water bodies and alternative animals.

Sometimes, the material is required to be purified even more. This can be done by the melt-filtering process. In this process, extra contaminants that still remain are melted and taken out of the RPET. Then it is passed through screens, forming pellets after drying again and known as pelletized plastic, which is more efficient for transportation and re-manufacturing, as well as provides more uniform sizing (Tapia-Picazo et al., 2014).

14.3.5 Properties

To counterpart the conventional polyester, this is soft yet tough fibre with a sustainable option. It has all the properties same as virgin polyester. It is good to prefer the materials having high-performance durability; those are made from the wastes that are collected from the landfills. It also helps to reduce greenhouse gas emissions, which is a good symbol for society.

RPET is having properties like light-weight, toughness, shatterproof and resistance to micro-organisms and is generally used for transparent applications (carbonated and non-carbonated bottles, etc). Also, it is inexpensive, stretchable, resealable and, most important of all, recyclable (Kumar and Janet Joshiba, 2020).

14.3.6 Applications

Many wear retailers across the planet have enforced recycled textiles in their products. Industries like Patagonia, Everlane, Lindex, Pure Waste and Heavy Eco sell produces sustainable garments. These firms incorporate materials derived from textile recycled waste likewise as recycled plastics into the garments they sell (Fig. 14.5).

In Sweden, firms like Lindex and H&M are using pre- and post-consumer waste fibres inside their new wear lines. Equally, in a European country, Pure waste could be used to make t-shirts from recycled fibres in their 95% wind-powered factories.

Aside from wear, a Danish carpet-producing company named Egetæpper creates its carpets from recycled fishnets fibres.

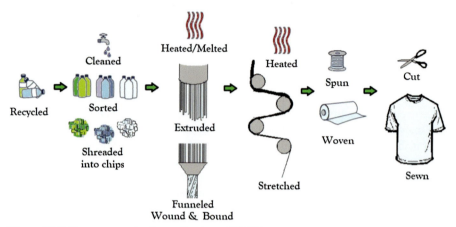

Figure 14.5 From sources to the application of RPET.

Glen Raven, Inc., a garment manufacturer, designed a program that collects unused items throughout production and recycles them into a replacement cloth. They additionally established a program for purchasers to recycle their previous materials with the corporate.

RPET can be used in a huge variety of ways. These include (Leonas, 2017):

- Tee shirt fabric
- Polyester carpet fibres
- Athletic shoes
- Long underwear
- Fibrefill for winter coats, sleeping bags and sweaters
- Upholstery
- Luggage
- Sheet and film
- Industrial strapping
- New PET containers
- Automotive parts
- Sportswear
- Outdoor and casual wear
- Uniforms
- Home textile

14.3.7 Advantages

By limiting the utilization of virgin materials, recycled polyester dramatically lowers its environmental impact versus ancient polyester (Duarte et al., 2016).

   i. Recycled polyester
  ii. Reduces reliance on virgin fossil fuel as a staple
 iii. Diverts used plastic from landfills

iv. Prevents used plastic from ending up in our oceans and harming marine life (more thereon here)

v. Decreases gas emissions from making and processing virgin polyester

vi. Can be unceasingly recycled once more and once more while not quality degradation

14.3.7.1 Keeping plastics from attending to lowland and also the ocean

Recycled polyester provides second life to a cloth that's not perishable and would otherwise find you in the lowland or the ocean. Consistent with the organization Ocean Conservancy, eight million metric loads of plastic enter the ocean per annum, on high of the calculable a 150 million metric tons that presently flow into marine environments. If we tend to keep this pace, by 2050 there'll be additional plastic within the ocean than fish. Plastic has been found in 60% of all seabirds and 100% of all marine turtle species, as a result of the mistake of plastic for food (Elias, 2018). Taking plastic waste and turning it into a helpful material is incredibly vital for humans and the environment.

14.3.7.2 RPET is simply nearly as good as virgin polyester, however, takes fewer resources to form

Recycled polyester is sort of identical to virgin polyester in terms of quality; however, its production needs 59% less energy compared to virgin polyester, consistent with a 2017 study by the Swiss people federal department for the atmosphere. 'If you inspect life cycle assessments, RPET scores considerably higher than virgin PET', adds Magruder. In addition, recycled polyester will contribute to cutting back the extraction of fossil oil and gas from the planet to form additional plastic. The exploitation of recycled polyester lessens our dependence on fossil fuel as a supply of raw materials (Matthes et al., 2021).

14.3.8 Disadvantages

Even if recycled, polyester continues to be plastic. Once washed, it releases microplastics. Microplastics are little plastic items and fibres that are about 5 mm long. They're too little to urge caught within the filters and typically finish into the water systems and, consequently, within the oceans. As per Fashion Revolution International, the textiles sector is the only largest supply of micro-plastics, making up 34.8% of worldwide micro-plastic pollution. This has devastating consequences on marine life and therefore the system.

First of all, recycled polyester clothes are typically made of recycled PET bottles, not recent clothes. And whereas it's nice we have a tendency to have found a use for all those non-reusable plastic bottles we consume. It additionally implies that there is still a lot of polyester clothes out there, in landfills around the world, that don't biodegrade (Vadicherla et al., 2015).

Finally, the utilization method of polyester still features a vital environmental impact. It's going to be not up to creation of the new plastic, however except for greenhouse emissions being released; the method uses chemicals to realize the colour consistency, as wishing on the mechanical procedure won't guarantee this. Sadly, the handling and disposal of the chemical material aren't continuously clear.

14.3.9 Sustainability

Both PET and RPET plastic will be recycled. Thanks to the physics qualities of PET, the fabric will be heated, molten and remodelled into new materials once more and once more. For this method to require a place, adequate native infrastructure is required to confirm that the plastic material will be transported to utilization plants World Health Organization have the correct machinery to method them. However, all of this starts with customers disposing of recyclables within the applicable bin. In keeping with the Ellen McArthur Foundation, the worldwide utilization rate for PET bottle variety was approximated at 55th in 2012 with a possible yield of 70%−78%. RPET plastic is having extreme properties, because it will be 100% utterly recycled: bottle, label and cap (Muthu and Senthilkannan, 2020).

Recycled polyester is unquestionably a property choice for our wardrobe. However, we'd like to remember that it's still non-biodegradable and takes years to disappear once thrown away. It additionally still releases plastic microfibers.

14.3.9.1 Brands promoting RPET (Segran, 2019)

In 2017, Textile Exchange's Recycled Polyester (RPET) spherical Table created associate RPET Commitment to encourage brands and retailers to public plan to fast their use of recycled polyester by 25th by 2020. As per studies 59 textile, and retail companies—including major brands like Adidas, DiBella, Eileen Fisher, Gap Inc., H&M, IKEA, Lindex, Meta Wear, Target and Timberland—committed to or square measure supporting a rise in their use of RPET by a minimum of 25th by 2020.

Patagonia
Patagonia recycles used plastic bottles, unusable producing waste and worn-out clothes into polyester fibres to supply consumer goods.

Everlane
In October 2018, Everlane proclaimed plans to eliminate all virgin plastics from its offer chain by 2021. 75% of the plastics used by the corporate primarily come from polyester, nylon and elastane utilized in clothing, undergarment and a few sweaters were recycled.

IKEA
IKEA is committed to finishing the dependency on virgin fossil materials and using solely renewable or recycled materials by 2030. Nowadays 50% of all polyester textile

products made by IKEA square measure made from recycled polyester, and through innovations and new styles, all products are by next year.

NIKE
Yarn, soles and basketball courts square measure a number of samples of the various products Nike creates by remodelling plastic bottles, producing scraps and used products into new materials. In fact, 75% of all Nike shoes and accessories currently contain some recycled material.

Recron Green Gold
Reliance is one in every of the biggest polyester yarn and fibre producers in the world with a capability of 2.3 million tons each year. Green Gold is a fibre that has the bottom carbon footprint globally with a 25% reduction in carbon footprint. It additionally follows zero waste conception i.e., all the waste generated is employed.

14.4 Nylon

14.4.1 Conventional nylon

Nylon was the primary fibre created entirely in an exceedingly laboratory. Nylon (or polyamide) could be an artificial compound, a sort of plastic that was fictional in 1935 by Wallace Carothers, an American chemist, at the chemical-producing DuPont Company (Heeger, 2001). The primary aim use of nylon was the nylon-bristled toothbrush created in 1938. However, the invention that created nylon's success was women's stockings, back in 1940. These quickly became a staple in women's wardrobes and were thought of as a good replacement for silk in hose. Throughout war II, nylon production was pleased to supply parachutes, fuel tanks, ropes, and alternative military instrumentation. Nylon is currently widely utilized in the style business to form swimsuits, raincoats, tights, socks, and athletic wear. It represents around 12% of all artificial fibres created worldwide (Bellussi and Perego, 2000).

There are different types of nylons. However, the foremost common one is named nylon 6, 6 (because every of the two beginning molecules has six carbon atoms). Adipic acid and hexamethylene diamine are the two main molecules found in petroleum that are used for the nylon 6, 6 productions. These molecules are heated and once they reach an exact temperature and pressure, they fuse releasing water and making a giant compound molecule. This method is named condensation polymerization. To form garments, the tiny nylon bits are fusible, drawn through a spinneret, and loaded onto a spool. This creates nylon fibres that, when being stretched, are spun into a yarn from which it can be used to form nylon garments.

14.4.2 Recycled nylon

Since Nylon is produced from petroleum products it'll not biodegrade. Nylon doesn't break down simply and accounts for concerning 100% of the rubbish within the ocean. In line with the World Society for the Protection of Animals, over 0.6 million tons of fishing nets are dropped into oceans each year, together with nylon nets. Fishermen

322 Sustainable Fibres for Fashion and Textile Manufacturing

usually discard the nets as a result the choice is far costlier—paying somebody to lose them properly.

Though tons of experiments were conducted an intensive analysis on how nylon may well be regenerate to its recycled perishable type was disbursed, it absolutely was solely in 2013 onward that it truly created desired results. In any case, incorporating the maximum amount of recycled nylon will lessen our dependence on petroleum products as a material supply. It curbs discards, thereby prolonging lowland life and reducing poisonous emissions from incinerators. It helps promote new exercise streams for nylon products that are not any longer useable. And it causes less air, water, and soil contamination compared to victimization non-recycled nylon (Leonas, 2017).

Nylon is molten at a lower temperature which means some contaminants—nonrecyclable materials and microbes or—will survive. This is often why all nylons have to be compelled to be cleansed totally before the exercise method. 'When you've dragged fishnet through a ship, on the ocean bottom, and where else, its tons more durable to scrub before you'll be able to recycle it', Johnston says (Labunska et al., 2013).

Since we have a tendency to be reusing used nylon and turning it into new material, we have a tendency to be reducing our demand for brand new nylon to be created. This reduces our want for additional oil to be extracted from the world. Because it is usually created from previous fishing nets that are abandoned within the ocean, selecting recycled nylon conjointly implies that we have a tendency to be fun existing nylon from aiming to the ocean or landfills. It reduces the footprint of the 'new' item. The main drawback, however, is that recycled nylon continues to be plastic, thus it's not biodegradable. This implies that micro-plastics are still being free in water streams, ending in our oceans. That's why we have to require correct care of recycled nylon and use a Guppy Bag when laundry it.

14.4.3 Sources

Recycled Nylon has the same profit as recycled polyester. It diverts waste from landfills and its production uses abundant fewer resources than virgin nylon (including water, energy and fossil fuel). Some of the recycled nylon we tend to use comes from post-industrial waste fibre, yarn collected from spinning works, and waste from the weaving mills that may be processed into reusable nylon fibre. Another recycled nylon fibre we tend to area unit experimenting with is made from discarded industrial fishing nets. A large part of the recycled nylon made comes from recent fishing nets (Vallés-Lluch et al., 2002). This is often an excellent resolution to divert garbage from the ocean. It conjointly comes from nylon carpets, tights, etc. (Fig. 14.6).

One of Econyl's main inputs is ghost fishing nets forced from the ocean, thus victimization of this fibre created sense for Outer known's environmental and communications goals, as a result clean water could be a cause that resonates with its ocean-minded client base.

Recycled fibres from polyester and nylon waste   323

Figure 14.6 Sources for recycled nylon.

14.4.4 Manufacturing process

14.4.4.1 Recycled nylon

For the recycling, waste products like saved fishing nets are taken to pre-treatment facilities wherever they're sorted and cut into items sufficiently little to be placed through the producing method. The cut material is then moved to a recycling plant wherever they're placed into vast chemical reactors that, through a method of de- and re-polymerization break down the elements of the fabric and re-generate the nylon (Mihut et al., 2001). The ultimate product is then processed into yarn (Fig. 14.7).

Figure 14.7 Stages of recycled nylon products.

14.4.4.2 Recycling of nylon baggage

Nylon baggage's are difficult to recycle unless you buy one from an organization that gives a take-back program. San Francisco-based Timbuk2 is one such company. Once your nylon traveller or camera bag is wiped out, merely stick it in a box and mail it to the corporate at the address provided on its website. Timbuk2 can utilize or recycle several of the materials as potential.

There isn't any charge for the company's employment services (other than the price of postage), and customers that send out the product to be recycled can receive a 20% discount on a future purchase (unnatisilks).

There might also be artistic ways in which to utilize unwanted nylon baggage. If you've got a backpack that's in fine condition that you simply now not wish, contemplate donating it to a thrift look or a program that helps youngsters get college provides.

If you've got an oversized poke with a hole, cut it apart and use the nice nylon to form a smaller storage bag.

14.4.4.3 Recycle or utilize nylon fabric

Leftover nylon material from a stitching project could be a nice material to utilize. See if your community has a corporation that gives material and provides to artists and faculties. Materials for the humanities in NY City and also the Scrap Exchange in Durham, NC, are a number of examples. If you've got nylon fabric you wish to recycle, and you bought that fabric from in style outside gear manufacturer Patagonia, you'll come it to the corporate for employment (unnatisilks). Get a lot of data concerning Patagonia's employment program on its website.

14.4.4.4 Recycle and utilize nylons or tights

No Nonsense, which makes nylons, tights and alternative styles of leggings, offers an employment program for customers. The primary step is visiting their tights employment page and printing a post-paid mailing label. Next, place all of your unwanted nylon leggings in a box and placed them on the shipping label. Drop it at your nearest post-workplace or alternative mailing location, and your recent nylons area unit on their thanks to an employment facility. No Nonsense sends the fabric to a plant that recycles it into things like playground instrumentation, toys and vehicle insulation. There are various ways followed by the industries in which they utilize recycled nylons further (unnatisilks).

Put a bar of soap within the toe of clean nylon (make certain there's no run in this section). Tie off the open finish and suspend the sock by the sink. Once you move to wash your hands, get them lots wet then roll the sock between your hands. This works very well in potting sheds, barns or alternative places wherever a fixture may not be sensible.

- Use nylons to holdup tomatoes or alternative plants that require support as they grow.
- Fill clean nylon with potpourri or lavender.
- Use it as a bag in your drawers, automobile or the other space you wish to smell contemporary.

14.4.4.5 Recycle nylon carpet

If your community incorporates a carpet recycler, it's going to be able to take your recent nylon carpet. Shaw Floors could be a nationwide company that recycles a definite variety of nylon carpets. It's locations are in more than 30 cities, together including San Jose, CA; Tulsa, OK; Tallahassee, FL; and Boston (Mihut et al., 2001).

14.4.4.6 Recycling nylon is sweet for the earth—a lot of has to be done

A few years ago, an organization fashioned by three people determined that it'd be creating skateboards and glasses from recycled nylon. They were planning to collect the 'trash' floating within the ocean. 'When we want to recycle ocean waste, we should learn that there's a continuing process of nylon fishing nets being dropped by the fishermen into the ocean per annum (Singh et al., 2017). Nets that are placed in the ocean simply planned to be there for generations. These items don't break down'. Today, the companies pay fishermen in Chile to collect recent nylon fishing nets, that are then recycled into skateboards and glasses.

14.4.5 Properties

Quality and sturdiness are essential for technical, superior end products that are why nylon is usually used. Rather than evaluating property exclusively from a production viewpoint, contemplate it over the complete lifecycle of a garment. This would possibly mean creating style changes to make sure a product made from recycled materials stands the check of your time or searching for ways in which to increase the amount it may be used—for instance, through repair programs further as used ones. It is light-weight nevertheless sturdy, and it's typically touted for its quick-drying capabilities. Garment makers love it as a result it holding dye well. It is more cost-effective to supply than silk and doesn't get broken as simply (Metteb et al., 2020).

Thanks to their low permeable ness, nylon garments dry quicker than natural materials like cotton, and they don't want ironing. It's conjointly waterproof, which makes it appropriate to supply raincoats or umbrellas. Nylon is elastic and stretchable; therefore, it is a nice candidate material to make athletic wear. This fabric conjointly takes dye well, which may be a bonus for the fashion designers. All those characteristics of nylon supply an outsized spectrum of prospects once it involves its business applications. That's why it's widely used these days!

"A recycled fibre that pills or abrades easier or has poor tear strength doesn't build a lot of property product"', Boren said.

- Recycled Nylon fibres deliver similar quality and performance to straightforward yarns.
- It is offered in high vary from 22 to 78 dtex.

14.4.6 Applications

Nylon is currently widely employed in the style business to create swimsuits, raincoats, tights, socks, parachutes, fuel tanks, ropes, and alternative military instrumentation and dress. It represents around 12% of all artificial fibres created worldwide. This material is additionally employed in alternative industrial sectors like automotive and astronautics, still within the production of packaging and varied home goods (Chang et al., 1999).

A California-based company named Patagonia has additionally been adding a lot of recycled nylon to its line-up. Currently, the corporate has over 50 outputs that contain recycled nylon in varying percentages. The Torrent shell jackets, for example, have outer layer textile created with 100% with chemicals recycled nylon. Econyl has products having all the properties same as historically made by nylon 6. H&M and La Perla, and others have already begun to use Econyl in an exceeding variety of garments together with swimwear, underwear, and athletic wear (Perella). It additionally has applications among the house, with firms like Forbo Flooring and carpet idea mistreatment it in their production processes.

14.4.7 Advantages

The cost of recycled nylon is more expensive than new nylon, but it can be avoidable due to its advantages. To improve the quality and reduce the costs of the recycling process, a lot of research is currently being conducted.

The production of 1 ton of recycled Nylon reduces the emissions of an equivalent to (Sotayo et al., 2015):

- CO_2 produced by **265.451 cell phones** being charged.
- CO_2 produced a **car driving 8.191 km.**
- CO_2 is absorbed by **101.000 m^2 of forest** in a year.

A company named LIBOLON "esigns for the environment"—increases good ideas among the society for the use of the eco textile product with eco-friendly practices. One of the Taiwan manufacturers manufactured its first dope-dyed fibre in 2000. Its speciality is after weaving no dyeing process is required. This process reduces the water and energy consumption; also, it saves around 70%—80% in the emission of CO_2 and wastewater (Yee et al., 2020).

The target of the recycling process is to collect waste from landfills and the ocean. It is the same as recycled polyester. For the production of recycled nylon fewer resources are required than virgin nylon (including energy, water and fossil fuel). The main raw material for the recycling of nylon comes from old fishing nets. This is a great idea to clean garbage from the ocean. Some other resources are nylon carpets, tights, etc.

Recycled textiles allow designers to access the functionality of nylon, and contribute to a good environmental outcome by using used fishing lines and other post-consumer waste that have been collected from the oceans.

Since the invention of nylon, it is popular in the production of a variety of household and clothing items. There are different reasons for that. The main properties of nylon which makes it more resistant to wear and tear are strong and durable.

14.4.8 Disadvantages

Despite these benefits, there are drawbacks to recycled nylon that we tend to cannot ignore. The main issue thereupon is that this material isn't biodegradable: it can't be naturally counteracted by microorganisms and in a very means that's not harmful to the ecosystem. Scientists estimate that nylon takes between 30 and 40 years to decompose. Throughout that point, life risks intake nylon bits or obtaining at bay in nylon fishing nets, one among the largest sources of ocean pollution (Singh et al., 2017).

We even have to say the countless microplastics shed by nylon fabric once washed in our laundry machines that find themselves within the oceans. In total, nylon accounts for 100% of the junk within the oceans. Another downside is that nylon comes from fossil fuels that may be non-renewable energy. Making things out of nylon, therefore contributes to the depletion of Earth's natural resources. To not mention, the oil industry is one among the foremost harmful and polluting ones for our planet.

In addition to being tons a lot of energy-intensive than cotton production, manufacturing recycled nylon emits high levels of carbon dioxide gas and nitrous oxide. The latter may be a gas that's 310 times stronger than carbonic acid gas, contributory even a lot of too heating globally (Karthik and Rathinamoorthy, 2017).

The nylon used as clothing is additionally heavily treated with harmful chemicals, artificial dyes, and bleaching agents. They contribute to pollution as they're typically discharged in water streams. These poisonous chemicals also are coupled with enhanced risks of skin allergies, immune problems and cancer.

To prime it all, garment products of nylon aren't breathable. Therefore, carrying them, particularly throughout a physical exertion, creates a piece of ground for micro-organism to grow as sweat is at bay against the skin. This can be not ideal in terms of hygiene and will cause skin issues.

14.4.9 Sustainability

If you're on the lookout for a brand-new piece of covering, consider choosing one thing made from nylon. The material has some positive characteristics; however, it has a footprint on the ecosystem that weighs more. Sustainability is the main aim of any product and we should look at it seriously (Patti et al., 2021).

As we studied, nylon is a plastic material then research says that plastic can be recycled. It means recycled nylon can even be recycled. There are many brands and accreditations which will facilitate customers to realize additional property nylon items. After all, simply because you would like to avoid wasting the world, doesn't mean you would like your stockings all baggy.

Scientists are researching polymers to be utilized in nylon production that do not come back from oil and gas extraction. These new bio-based polymers come back from metabolic engineering of microorganisms to provide an associate degree increasing the range of chemicals, materials, and fuels from low-priced renewable resources. Whereas presently there's not a viable replacement for crude monomers, extremely promising biological blocks of polyamides are found. Because the value of crude continues to fluctuate, and awareness of the climate crisis will increase, it's seemingly that alternatives to these parts of nylon are going to be developed.

Swedish Stockings, as an example, produces lovely pantyhose from recycled yarn. Their factories conjointly use eco-friendly dyes, post-dyeing water treatment and alternative energy for a lot of the energy required within the producing method (Patti et al., 2021).

Three things to contemplate once coming up with sustainability property collections:

1. Think about operating with recycled alternatives of your chosen materials, the larger the demand the larger the interest and investment within the use business
2. Strive operating with mono instead of merging materials, as merging materials square measure presently tough to recycle.
3. Think about the dyes and chemical finishes that get into your cloth as this adds another layer of complexness to the use method.

14.4.9.1 Brands promoting recycled nylon

ECONYL

There are various brand of recycled nylon but the well-known recycled nylon product is Econyl, the primary post-consumer recycled nylon to hit the market from Italian manufacturer Aquafil. Econyl is created of nylon waste from landfills and oceans during a closed-loop method and is infinitely reusable. In line with Aquafil, Econyl avoids 50% of dioxide emissions and uses 50% less energy compared to virgin nylon yarns. Econyl received certification from Oeko-Tex standard 100. It guarantees that the fabric doesn't contain any harmful products to our health (Luo and Deng, 2021). Econyl is used by several sustainable brands to produce nylon wear, as well as Reformation, Peony, Outer known, and Arkitaip.

EcoRib

EcoRib could be a stretch-ribbed material made up of nylon fibre scraps. It is having property of light-weight and breathable. It creates a spread of vitamin A throughout the swimsuits in EcoRib also as in EcoLux material, another kind of recycled nylon. Since most swimsuits are made up of artificial materials, EcoRib could be a lot of property choices to strive for next time you would like new swimwear! Conjointly, once thinking of the word nylon, stockings could fir comes back to mind. Two brands promoting and creating tights out of recycled nylon: Swedish Stockings and Organic Basics.

Bio-nylon

Bio-nylon is the most typical one. It's a yarn that's made up of plant-based renewable ingredients, like sugarcane or corn flour. As bio-nylon is created from plants, there's no plastic within the final product. Therefore, there's no risk to shed microplastics in water streams. Producing bio-nylon conjointly doesn't contribute to increasing the demand for petroleum. And as mentioned on top of, it's perishable underneath the correct environmental conditions.

A great example of a bio-nylon is EVO by the corporate Fulgar. This material comes from petroleum oil, a natural resource that isn't terribly water-intensive.

EVO material is lighter than most artificial materials and encompasses a high physical property. It conjointly dries double as quickly as typical nylon and is thermo-insulated.

Another plant-based nylon is Bio Sculpt material. It's created with plant-based fibre created from castor beans. Again, this brand creates a number of vitamin A within its swimsuits in Bio Sculpt material.

Biodegradable synthetic nylon: Amni Soul Eco (Saxena, 2018)

Amni Soul Eco material could be a polymer, an artificial material like nylon. However, the shocking issue is that it's biodegradable! This material biodegrades in 5 years once disposed of during a lowland, which is regarding 10 times faster than most alternative artificial materials. It's breathable, reusable and Oeko-Tex commonplace 100 certified. However, it's still a lot of eco-friendly various to standard nylon. Different brands have created wear in Amni Soul Eco, as well as Florita and daring Swim. Bio-based nylon uses renewable feed stocks, like Fulgar's Evo made up of 100% castor oil rather than crude.

14.5 Conclusion

Recycling synthetic fibre definitely helps in decreasing the manufacturing of virgin fibres; which results in less dependency on petroleum as raw material and also decreases the number of waste PET bottles and fishing nets that goes into the landfill each year. But even after recycling, polyester and nylon still remain non-biodegradable fibre and fabric. Recycled products will also end up in the landfill after a few years of usage like a closed loop. Overall, recycling is a great step towards sustainability as it helps in decreasing the number of wastes from the environment that are thrown away. There are some brands that are promoting circular economy; through the use of RPET and the brands, that have created a whole new line, especially for the recycled products. Same as RPET, nylon has certain characteristics that make it more suitable for specific purposes. Though it has some negative impacts on the earth and ecosystems by wide-spreading its use. More sustainable materials should be chosen and create a demand for alternative fabrics to help save our planet. Supporting businesses and manufacturers that are concrete in the manner towards the sustainability of the apparel industry is vital in creating a distinction.

14.6 Sources of further information

Further information can be obtained from the following sources:

1. Muthu, Subramanian Senthilkannan, ed. Recycled Polyester: Manufacturing, Properties, Test Methods, and Identification. Springer Nature, 2019.
2. Thomas, Sabu, et al. eds Recycling of PET bottles. William Andrew, 2018.
3. Kasserra, H. Peter. "Recycling of Polyamide 66 and 6'." *Science and Technology of Polymers and Advanced Materials.* Springer, Boston, MA, 1998.
4. La Mantia, Francesco. Handbook of plastics recycling. iSmithers Rapra Publishing, 2002.
5. Tobler-Rohr, Marion I. Handbook of sustainable textile production. Elsevier, 2011.

References

Becerra, Alberto Valderrama, C., 2000. World cotton demand in the future: issues on competitiveness. In: 25th International Cotton Conference, Bremen, Germany.

Bellussi, G., Perego, C., 2000. Industrial catalytic aspects of the synthesis of monomers for nylon production. Cattech 4 (1), 4–16.

Black, K., 2015. Magnifeco: Your Head-To-Toe Guide to Ethical Fashion and Non-toxic Beauty. New Society Publishers.

Blackburn, R., 2009. In: Sustainable Textiles: Life Cycle and Environmental Impact. Elsevier.

Carneiro, N., Refosco, E., Soares, G., 2016. Contribution to an Efficient Transmission of Information to the Textile Fashion Consumer and the Influence in Sustainable Attitudes.

Chang, Y., Chen, H.-L., Francis, S., 1999. Market applications for recycled postconsumer fibers. Family and Consumer Sciences Research Journal 27 (3), 320–340.

Debowski, M., et al., 2019. Chemical recycling of polyesters. Polimery 64.

Duan, H., et al., 2008. Hazardous waste generation and management in China: a review. Journal of Hazardous Materials 158 (2–3), 221–227.

Duarte, I.S., et al., 2016. Chain extension of virgin and recycled poly (ethylene terephthalate): effect of processing conditions and reprocessing. Polymer Degradation and Stability 124, 26–34.

Elias, S.A., 2018. Plastics in the ocean. Encycl. Anthropocene 1, 133–149.

Goyal, A., 2021. "Management of Spinning and Weaving wastes." Waste Management in the Fashion and Textile Industries. Woodhead Publishing, pp. 61–82.

Hawley, J.M., 2014. "Textile recycling." Handbook of Recycling. Elsevier, pp. 211–217.

Heeger, A.J., 2001. Semiconducting and Metallic Polymers: The Fourth Generation of Polymeric Materials, pp. 8475–8491.

Houck, M.M., et al., 2001. Poly (trimethylene terephthalate): a "new" type of polyester fibre. Forensic Science Communications 3 (3), 217–221.

Hurley, N., 2017. Dirty Fashion: How H and M, Zara and Marks and Spencer Are Buying Viscose from Highly Polluting Factories in Asia, 1786. Guardian, Sydney, p. 12.

Hussain, A., et al., 2021. Circular economy approach to recycling technologies of postconsumer textile waste in Estonia: a review. Proceedings of the Estonian Academy of Sciences 70 (1), 82–92.

Karthik, T., Rathinamoorthy, R., 2017. "Sustainable Synthetic Fibre production." Sustainable Fibres and Textiles. Woodhead Publishing, pp. 191–240.

Kumar, P.S., Janet Joshiba, G., 2020. "Properties of Recycled polyester." Recycled Polyester. Springer, Singapore, pp. 1–14.

Labunska, I., et al., 2013. Levels and distribution of polybrominated diphenyl ethers in soil, sediment and dust samples collected from various electronic waste recycling sites within Guiyu town, southern China. Environmental Sciences: Processes & Impacts 15 (2), 503–511.

Leonas, K.K., 2017. The use of recycled fibers in fashion and home products. Textiles and clothing sustainability 55–77.

Liu, J., et al., 2021. Microfiber pollution: an ongoing major environmental issue related to the sustainable development of textile and clothing industry. Environment, Development and Sustainability 1–17.

Luo, Y., Deng, K., 2021. Sustainable fashion innovation design for marine litter. Journal of Physics: Conference Series 1790 (1). IOP Publishing.

Lusher, A., Peter, H., Mendoza-Hill, J., 2017. Microplastics in Fisheries and Aquaculture: Status of Knowledge on Their Occurrence and Implications for Aquatic Organisms and Food Safety. FAO.

Martínez Silva, P., Nanny, M.A., 2020. Impact of microplastic fibers from the degradation of nonwoven synthetic textiles to the Magdalena River water column and river sediments by the City of Neiva, Huila (Colombia). Water 12 (4), 1210.

Matthes, A., et al., 2021. Sustainable Textile and Fashion Value Chains. Springer.

Metteb, Z.W., Abbas Abdalla, F., Ehsan, S., Al-Ameen, 2020. Mechanical properties of recycled plastic waste with the polyester. AIP Conference Proceedings 2213 (1). AIP Publishing LLC.

Mihut, C., et al., 2001. Recycling of nylon from carpet waste. Polymer Engineering and Science 41 (9), 1457−1470.

Moazzem, S., et al., 2021. Environmental impact of discarded apparel landfilling and recycling. Resources, Conservation and Recycling 166, 105338.

Muthu, Senthilkannan, S., 2020. In: Environmental Footprints of Recycled Polyester. Springer Singapore.

Park, S.H., Kim, S.H., 2014. Poly (ethylene terephthalate) recycling for high value-added textiles. Fashion and Textiles 1 (1), 1−17.

Patti, A., Cicala, G., Acierno, D., 2021. Eco-sustainability of the textile production: waste recovery and current recycling in the composites world. Polymers 13 (1), 134.

Perella, M. "New fabrics make recycling possible, but are they suitable for high street." The Guardian. http://www.theguardian.com/sustainable-business/sustainable-fashion-blog/2015/jan/22/fabric-recycling-closed-loop-process-high-street-fashion. Accessed Apr (2016).

Saxena, S., 2018. SUSTAINABLE FASHION FABRICS.

Segran, E.L.I.Z.A.B.E.T.H., 2019. H & M, Zara, and other fashion brands are tricking shoppers with vague sustainability claims. Fast Company 8.

Singh, N., et al., 2017. Recycling of plastic solid waste: a state of art review and future applications. Composites Part B: Engineering 115, 409−422.

Sotayo, A., Green, S., Turvey, G., 2015. Carpet recycling: a review of recycled carpets for structural composites. Environmental Technology and Innovation 3, 97−107.

Tapia-Picazo, J.C., et al., 2014. Polyester fiber production using virgin and recycled PET. Fibers and Polymers 15 (3), 547−552.

Thiounn, T., Smith, R.C., 2020. Advances and approaches for chemical recycling of plastic waste. Journal of Polymer Science 58 (10), 1347−1364.

Tua, C., et al., 2019. Life cycle assessment of reusable plastic crates (RPCs). Resources 8 (2), 110.

unnatisilks https://www.unnatisilks.com/blog/recycling-nylon-is-good-for-the-planet-more-needs-to-be-done/.

Vadicherla, T., Saravanan, D., Senthil Kannan Muthu, S., 2015. "Polyester Recycling—Technologies, Characterisation, and applications." Environmental Implications of Recycling and Recycled Products. Springer, Singapore, pp. 149−165.

Vallés-Lluch, A., et al., 2002. Influence of water on the viscoelastic behavior of recycled nylon 6, 6. Journal of Applied Polymer Science 85 (10), 2211−2218.

Vermaas, J.F., 2011. Evaluation of the Antimicrobial Effect and Strength Properties of Polyester, Polyester/cotton and Cotton Treated with Anolyte. Diss. University of the Free State.

Wang, Y., 2010. Fiber and textile waste utilization. Waste and biomass valorization 1 (1), 135−143.

Wani, Ahmad, K., et al., 2020. "Conversion of waste into different by-products of economic value in India." Innovative waste management technologies for sustainable development. IGI Global 259—272.

Yee, F.Y., et al., 2020. Save environment by Replace plastic, the case of PepsiCola. International Journal of Asian Business and Information Management 5 (1), 51—59.

Composites derived from biodegradable Textile wastes: A pathway to the future

Saniyat Islam
School of Fashion and Textiles, RMIT University, Melbourne, VIC, Australia

15.1 Introduction

The worth of the global fashion industry approximately is three trillion dollars with 2% of global Gross Domestic Product (GDP) (Fashion United Group, 2021). Researchers are looking into synthetic fibres and resin systems alternatives due to the diminishing reserves of non-renewable petroleum resources and their non-biodegradability. Some of the techniques used to improve the performance of natural composites have been successful, however, they are still below the performance of the synthetic fibres. This chapter will investigate some commercially used natural fibres, and natural resins for their application and their relevance for the fabrication of composites.

Textile composites are fibre-reinforced composite materials, the reinforcement can stem from the form of a textile fabric that could either be woven, knitted or braided. The main advantage of using the natural fibre and resin system is it is cheaper than synthetic composites, bio-degradable and abundantly available. Chemical treatment and mechanical loading methods are used commonly to improve the tensile strength attributes of natural fibres. Natural Fibre-reinforced polymer matrix composites have gained commercial success in semi-structural and structural applications such as aviation, automotive, sporting equipment, computer electronics, and appliances (Thyavihalli Girijappa et al., 2019). By using long, aligned natural fibres in conjunction with naturally derived resins, 100% bio-based composites with improved mechanical properties can be achieved and can potentially be employed for more structural purposes. This could be a pathway to the future utilization of the material streams in the fashion and textile industry's waste, which now predominantly depends on the synthetics counterparts. The increase in awareness of the damage caused by synthetic materials to the environment has led to the development of eco-friendly materials. Recent research have shown a lot of interest in developing bio-based materials, which can replace synthetic materials in many applications.

15.2 Textile waste

Textile manufacturers undertake a range of waste-generating activities such as washing/drying, warp preparation, weaving, dyeing, printing, finishing, quality and process control, and warehousing. Each operational step that a raw textile material passes through from the moment it enters a textile or apparel establishment until it becomes a finished article is a potential source of pollution. 55% of the global production of clothes and textiles is based on synthetic fibres, mainly polyester, and the remainder is natural materials, mainly cotton (Zamani, 2014). The management of textile waste is a formidable problem. However, the overall guiding principle, agreed by everyone, to protect the environment is to 'reduce, re-use, repair or recycle', and actual disposal of waste should be the last option or phased out. Different types of textile wastes are shown in Table 15.1.

The major wastes generated by this sector are fibre waste. These include soft fibre wastes, yarn spinning (hard fibre) wastes, beaming wastes, off-cuts, packaging, spools and creels. Wet finishing processes use up to 200 litres of water per kilogram of fibre,

Table 15.1 Type of textile wastes in the fashion and textiles industry (Islam, 2021).

Type of textile waste			
Pre-consumer	**Post-consumer**	**Special waste**	**Medical waste**
Fibre, yarn, fabric, water, dyes, chemicals and finishes, packaging, seconds	Garments, packaging, buttons, zips, plastic and metal accessories, shoes, bags, sewing tags	Uniforms, specialized garments, specialized shoes, accessories, automotive textiles, upholstery	Biomedical waste, surgical gowns, dressings, sutures
Pathway	Pathway	Pathway	Pathway
Recyclable, reusable	Recyclable, reusable	Recyclable but challenging due to standards and conformity of materials	Incineration, landfill (special protective covering needed)
Suitable for reuse/ repurpose	Suitable for reuse/ repurpose and damaged or end of life materials can be utilized for composites	Suitable for possible application in composites	Hygiene, sterilization, cleaning and collection are challenging, further R&D needed

making wastewater the largest waste in this sector by volume (Morrissey and Browne, 2004). Consumer awareness and understanding better consumption decisions for clothing and textiles and manufacturers are liable to make that initiative by better designing the composites that are derived from textiles.

15.3 Material thinking

The environmental and social implications of fashion production and disposal make it of high relevance to integrate the targets of the 17 Sustainable Development Goals (United Nations, 2015) into this industry, to make it fully sustainable and positive for people and the ecosystem. The aim of the commitment is to integrate the UN SDGs in this industry, to tackle the environmental and social issues related to clothing production and consumption: Implementation of sustainable design strategies, the promotion of the use of sustainable technologies and appropriate resource management throughout the textile supply chain, proposing new business models and engaging with consumers for better consumption habits. Knowledge of the SDGs and their relation to the fashion industry is pivotal to initiating the change in product development (Islam, 2020).

15.4 Designing out waste with a material circularity approach

Circular Economy (CE) is a new approach that alters the production process and could reduce the use of resources for the fashion and textiles industry. In 2017, (MacArthur, 2017) paved steps to promote a circular economy and inspire people and fashion businesses to think in new ways and start shifting from linearity to circularity in the way they do business. In the context of fashion, the economy is still characterized by a linear model of consumption: companies extract materials, add labour and energy, and sell it to consumers who dispose of them after a short while when it no longer serves their purpose or has just gone out of trend. This can be termed as a take-make-dispose model as illustrated in Fig. 15.1. In general, circularity requires 12 holistic approaches (MacArthur, 2014) to define and apply to raw materials, energy and the use of chemicals as well, and hence it also involves renewable energy (burning of fossil fuels is a linear activity since the recreation of oil, coal, and gas is basically nonexistent). Chemicals should preferably be nontoxic and within closed technical loops.

To design out waste, it is pivotal to overcome the caveats of textiles should not end up in landfills and textiles should be designed for recycling/reuse and the loop can only be fulfilled if raw materials are remanufactured to replace virgin materials. One of the avenues to reduce landfills is to utilize the waste stream into textile composites.

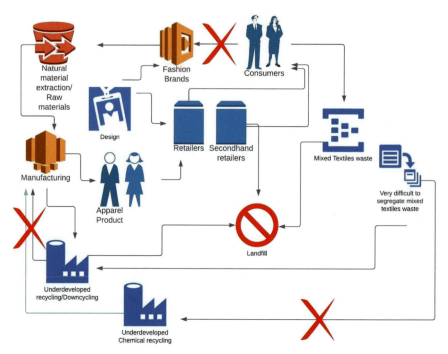

Figure 15.1 Current material flow of fashion and textiles industry.

15.5 What are textile composites?

Textile composite materials consist of a polymer matrix (thermoplastic or thermoset) combined with textile reinforcement. Materials of interest to the group include commingled glass/polypropylene fabrics. In addtion, dry fabrics (woven, warp-knitted or braided) combined with thermoset resins via liquid moulding (Misnon et al., 2014). Textile Composite materials could fulfil the demands of sustainability. They are preferred for the construction of many built environments, assisting in producing high energy-efficient, lightweight and durable products without affecting their performance. Composites are offering superior properties compared to metals like high strength to weight ratios, high flexural and impact strengths, corrosion and weather resistance, durability, dimensional stability, design flexibility and aesthetics (Pastore, 2000). Biocomposites could be an answer to reducing textile landfills to a significant degree.

15.6 What are biocomposites?

Biocomposites combine using natural fibres such as flax, hemp, sisal, bamboo, jute, kenaf, etc for reinforcement with polymer matrices either from renewable or non-renewable resources. They offer numerous advantageous attributes such as biodegradability, low density, high specific strength, reduced dermal and respiratory irritation, low cost, better insulation, ecofriendly, etc (Mishra et al., 2020). Green composites are bio-based resins derived from renewable resources such as starch, sugar, vegetable oils and soy oil reinforced using natural fibres. Polylactic acid (PLA) is the common type used (Bhat et al., 2021).

Recently these materials have expanded attention as a possible alternative to synthetic-based polymers for diverse categories of industrial applications such as automotive, films, construction, paper coating, packaging and biomedical applications (Bhat et al., 2021). Synthetic polymers pose many negatives towards the environment in avenues such as the number of greenhouse gases released after incineration and inappropriate disposal. Research needs to focus on novel green biopolymeric materials and their effective utilization in sustainable composite applications (To et al., 2019).

15.7 Aspects of biodegradability of natural cellulose-based fibres

'Biodegradable' denotes the capacity of materials to become disintegrated (decomposed) by the influence of biological micro-organisms such as bacteria or fungi (with or without oxygen) while being integrated into the natural environment (Eskander and Saleh, 2017). It should not cause any ecological harm during this irreversible disintegration-integration process. In general, natural fibres are prone to decompose and biodegrade into the natural environment. However, the key issue here is blending, mixing and other forms of additives and finishes inhibit the biodegradability aspect (Islam, 2021). This is one of the key challenges in how future composites are designed and produced. Plant-based textile fibres are attained from the hair, leaf, bast and fruit of a plant and are cellulosic polymers listed in Table 15.2.

Some of these fibres which are more common in applications in the fashion and textile industries and are potential candidates to be incorporated in the composite fabrication process are discussed in further detail in Section 15.7.

Table 15.2 Common natural fibres and their origin (Mwaikambo, 2006).

Fibre name	Botanical name	Fibre origin
Abaca	*Musa textilis*	Leaf
Bagasse	*Saccharum officinarum* L	Stem
Banana	*Musa ulugurensis* warb.	Leaf
Bamboo	*Gigantochloa scortechinii Dendrocalamus apus*	Stem (bast)
Coir	*Cocos nucifera* L.	Fruit
Cotton	*Gossypium* spp.	Seed
Flax	*Linum usitatissimum*	Stem (bast)
Hemp	*Cannabis sativa* L.	Stem (bast)
Jute	*Corchorus capsularis, Corchorus olitorius*	Stem (bast)
Kapok	*Ceiba pentandra*	Seed
Kenaf	*Hibiscus cannabinus*	Stem (bast)
Pineapple	*Ananas cosmosus* merr	Leaf
Ramie	*Boehmeria nivea*	Stem (bast)
Sisal	*Agave sisilana*	Leaf

15.7.1 Cotton

Cotton fibre is one of the most used fibres in the fashion and textiles industry that belongs to the plant sub-tribe Hibisceae and the family of Malvaceae (Elmogahzy and Farag, 2018). An annual crop, cotton is generally harvested mechanically in the form of large modules and later the modules are transported to a further processing stage known as the cotton ginning. The ginning process separates the seeds and other vegetable matters from the cotton fibres and is ready for yarn manufacturing as bales weighing between 180 and 225 Kgs. Cotton fibre is used widely in textile industries as raw materials for apparel, and recent research (Alomayri and Shaikh, 2013; Dobircau et al., 2009; Graupner, 2008; Kamble and Behera, 2020; Sathishkumar et al., 2017) showed developments on utilizing cotton as a component for composites for industrial and technical applications.

15.7.2 Kenaf

Kenaf is known as one of the bast fibres with superior tensile strength and has the potential to be used as a component of a composite (Saba et al., 2015). Current applications of Kenaf are in paper and rope manufacturing (Giwa Ibrahim et al., 2019; Hamidon et al., 2019). With a stiff and strong tensile attribute, Kenaf has been cultivated abundantly in many parts of the world such as Asia, some regions of Europe, the USA and Africa (Saba et al., 2015). The fibres are mainly extracted from the outer fibre, and inner core of the stem. The outer fibre is known as bast which makes 40% of the stalk's dry weight and the inner core comprises 60% of the stalk's dry weight. After harvesting the stem is further processed by using a mechanical fibre separator and the whole stalk is used in pulping (Thyavihalli Girijappa et al., 2019). These fibres can be woven into making good quality fabrics however, in recent times Kenaf has found its application in composites in conjunction with other materials such as in electronics, automotive, construction, packaging, furniture, textiles, mats, paper pulp, etc (Atiqah et al., 2014; Jaafar et al. 2018; Nishino et al., 2003; Nordin et al., 2013; Ochi, 2008; Rowell et al., 1999; Salleh et al., 2013; Serizawa et al., 2006).

15.7.3 Hemp

Similar to Kenaf, Hemp is a taproot herbaceous bast fibre plant with an erect stem that can grow up to 4 m in height (Amaducci and Gusovius, 2010). Its benefits such as suppressing weeds, being free from diseases, improving soil structure and zero consumption of pesticides and more than one harvest a year make hemp an attractive crop for sustainable fibre production in comparison to cotton. Hemp has great adaptability to a variety of climatic conditions, and it does not require chemical pesticides or water for irrigation (Islam, 2020). The consumption of fertilisers is modest and hemp as a plant suppresses weed growth and some soil-borne diseases, which means that at the end of its cultivation, soil condition is improved and healthier (González-García et al., 2007). The hemp plants are harvested, and the woody core from bast fibres is separated by a sequence of mechanical processes. The woody core also known as hurd is cleaned to obtain the required core content and sometimes they are cut to the desired size. The

Composites derived from biodegradable Textile wastes: A pathway to the future 339

separated bast fibres are then further processed to form yarns or bundles of fibres (Shahzad, 2012). Traditional applications of hemp have been making ropes, textiles, garden mulch, an assortment of building materials and animal beddings. In recent developments, it is used to fabricate different bio-based composites (Hepworth et al., 2000; Pickering et al., 2007; Sepe et al., 2018; Stevulova et al., 2013).

15.7.4 Jute

The jute fibre also known as the golden fibre of the Indian subcontinent, grows in heavy monsoon regions of Asia such as India, Bangladesh, China, and Myanmar (Khan et al., 2014). The jute plant grows up to $15-20$ cm in a quarter of a year and is harvested. The retting process involves treatments with chemicals ($N_2H_8C_2O_4$, Na_2SO_3, etc.) or biologically by keeping the fibres in stagnant monsoon water. The biological retting process is popular in south-east Asia. The process involves the harvested stalks arranged in bundles and allowed to soak in water for about $2-3$ weeks (Banik et al., 2003). This eliminates the pectin between the bast and the wood core which assists in the separation of the fibre strands. Subsequently, these fibres are dried and made into different end applications. The jute fibre has significant tensile properties (Berhanu et al. 2014), however, the lignin content makes the fibre not useable for apparel. The fibre has been extensively used as packaging materials, carpet backing, reusable bags, ropes, twines etc.

15.7.5 Flax

Another bast fibre is flax which has long been used as a material for a variety of applications (Quillien, 2014). Canada is the largest producer of flaxseed (40%) in the world. Other producing countries include China, the United States (US), and India together for 40% of global production (Duguid, 2009). The fibres from the flax plant are separated from the stems of the plant and are mainly used to produce linen. The fibre extraction involves the retting and scorching both these processes will make some alterations in the properties of the fibres (Van de Velde and Baetens, 2001). The retting involves the enzymes which degrade the pectin around the flax fibres which results in the separation of fibres. The linen fibres can measure up to 90 cm in length and diameter of $12-16$ μm (Goutianos et al., 2006). Linen fibres have applications in furniture materials, textiles bed sheets, linen, interior decoration accessories, etc (Bos, 2004; Goudenhooft et al., 2019). With favourable tensile properties, flax also becomes a worthy candidate to be used as a composite material (Le Duigou et al., 2011; Zafeiropoulos et al., 2001).

15.7.6 Ramie

Ramie is again one of the bast fibres cultivated extensively in the region native to China, Japan, and Malaysia where it has been used for over a century as one of the raw materials for textile fabrics (Nam and Netravali, 2006; Rehman et al., 2019; Yang et al., 2021). Ramie as a plant grows very fast (non-branching) up to $100-200$ cm in height. The fibres extracted from the stem are the strongest and longest of the natural bast fibres (Lu et al., 2006). The processing of the ramie fibres is comparable to linen from flax (Angelini and Tavarini, 2013). Ramie is blended with other cellulosic fibres to make apparel and in addition, also has applications in upholstery (Cengiz and Babalık, 2009), bulletproof panels (Marsyahyo et al., 2009), fishing nets, marine packings, etc (Sen and Reddy, 2011). In addition to this attempt has been made for developing bio-based products by utilizing them in the field of automotive, furniture, construction, etc. The ramie fibre has been extensively used to fabricate a wide range of textiles, pulp, and paper, agrochemicals, composites, etc (He et al., 2015; Kishi and Fujita, 2008; Romanzini et al., 2013; Yu et al. 2010, 2014).

15.7.7 Nettle

The nettle plant provides strong fibres, which are a high fixed carbon source and have beneficial properties such as moisture absorption, breathability and high tensile strength (Bajpai et al., 2013; Guo et al., 2005). Nettle plants are protected with fine hairs, particularly in the leaves and stems. When touched, it releases chemicals, stings and triggers inflammation that causes redness, itching, bumps and irritation to the human skin (Bacci et al., 2009). Processing Nettle Fibre involves the collection of nettles and subsequent removal of the leaves. This follows by retting the nettle stalks for at least a week to break down the cellulose surrounding the fibres so the fibres can be extracted. The fibres can also be extracted through mechanical decortication. The extracted fibres are then dried either in greenhouses, sauna or by the sun (Guo et al., 2005). The typical applications of nettle fibres are in the textile industry, biofuel, oil absorption, housing, etc (Bajpai et al., 2013; Viju et al., 2019). In recent years, the application of nettle fibres as composite materials for various applications attempts have been made to use the nettle fibres on an industrial scale (Bacci et al., 2009; Bajpai et al., 2013; Mortazavi and Moghaddam, 2010).

15.7.8 Pineapple leaf

The pineapple is a multiple fruit-bearing plant, herbaceous perennial that grows in countries near the equator. The tropical countries such as The Philippines, Thailand and the Indian subcontinent have the right climate for pineapple plant growth. The leaves of the pineapple fruit can be treated as agricultural waste and hence utilization of these became popular in many applications in textiles and other sectors (Kengkhetkit and Amornsakchai, 2012). Pineapple leaf fibres are long and are blended with other fibres instead of jute and other bast fibres. They are used to make heavier fabrics for upholstery and furnishings. The fibres are used in industrial applications like tires and conveyor belts. The sturdy and lustrous fibre finds wide applications from apparel to technical textiles (Lopattananon et al., 2006). They are among the latest entrants in the paper manufacturing industry. With its various vitamins and compounds, it finds use in the pharmaceutical and cosmetic industries. Other applications include automobiles, technical textiles, mats, construction, leather footwear etc. Surface-modified pineapple fibres have found their utilization in the fabrication of conveyor belt cords, air-bag, advanced composites, etc. (Rajeshkumar et al., 2020; Threepopnatkul et al. 2009)

15.7.9 Sisal

Brazil is at the forefront of the production of sisal, which is one of the most used natural fibres. Another leaf fibre that provides opportunity beyond its traditional applications such as ropes, carpets, mats, etc. to aviation and automobile sectors (Naveen et al., 2019; Thyavihalli Girijappa et al., 2019). The sisal plant crops about 200−250 useable leaves in the life span of 6−7 years (Devaraju and Harikumar, 2019). A favourable range of mechanical properties such as high tensile strength (Kim and Netravali, 2010) provides avenues to be used in the automotive and shipping industry, property constructions, twines etc(Thyavihalli Girijappa et al., 2019).

15.7.10 Coconut fibre

The coconut fibre is sourced from the shell of the coconut fruit. The major share of the commercially produced coconut fibre comes from tropical and coastal areas of India, Sri Lanka, Indonesia, the Philippines, and Malaysia (Pham, 2016). Among the different natural fibres, coconut fibre is the thickest with high lignin content than cellulose and hemicellulose (Arsyad et al., 2015). The fibres are sourced from the shell which can be counted as food waste and provides sustainable options to utilize this waste in other applications such as fabrication of ropes, mats, mattresses and brushes, in the upholstery industry, agriculture, construction, etc. (Pham, 2016; Thyavihalli Girijappa et al., 2019)

15.7.11 Kapok

Kapok fibre is a standout amongst natural fibres, as it caters for unique properties such as buoyancy, hydrophobicity with bulk and lightweight (Chen et al., 2013). Traditionally, kapok fibre is used as buoyancy material (filling of life jackets), pillows and mattresses, oil-absorbing material(Quek et al., 2020; Wang et al., 2012), reinforcement material(Venkata Reddy et al., 2008), adsorption material (Zheng et al., 2015), biofuel (Putri et al., 2012), etc.

15.7.12 Bamboo

Bamboo, due to the alignment of fibres in the longitudinal directions is also acknowledged as a natural glass fibre (Zakikhani et al., 2017). Bamboo grows abundantly in dense forests particularly in China, about 40 families, and 400 species have been found (Fan and Fu, 2016). Bamboo fibre has applications as reinforcement in polymeric materials (Liu et al., 2012) due to its lightweight, cost-effective, high tensile strength, and stiffness properties that are unparallel in the natural fibre domain. Bamboo has been traditionally used for building materials, bridges, traditional boat structures, etc. The fibres extracted from bamboo can be used as reinforcement for making advanced composites in various industries (Thyavihalli Girijappa et al., 2019; Zakikhani et al., 2017).

15.8 Natural fibres as reinforcement for composites materials

The key advantages of designing composites from natural fibres are biodegradability, lightweight, tensile properties, versatility, plant-based and in many instances can be sourced as agricultural waste. The dependence of the natural fibres on using landmass, irrigation and other chemicals is always going to be challenged in the future as the population grows (United Nations, 2019) and creates more pressure on food production for cultivable landmass. Demand for the newer materials which have better properties than the existing one's will rise, and research and development will continue to utilize different types of natural fibrous materials sourced from fruits, seeds, leaves, stem, animals, etc. As discussed in Section 15.7, the natural fibres have on many occasions been modified by using different chemical treatments thus adjusting and enhancing the properties of natural fibre composites. In addition, polymers and other synthetic materials have been used along with the natural fibres to enhance the properties of the natural fibres and these ideas have led to the development of several hybrid composites reinforced with natural fibres, and filler materials (Thyavihalli Girijappa et al., 2019).

15.9 Opportunities and challenges around natural fibres reinforced polymers

In the current context of fashion and textiles, there is an upsurge in awareness regarding environmental pollution due to waste (pre and post-consumer) which has led to looking into alternatives of potentially harmful synthetic materials with more environmentally friendly materials. After the advent of synthetic fibres, they have

Table 15.3 Properties of common plant-based cellulose fibres.

Fibre	Density (g/m³)	Tensile strength (MPa)	Young's modulus (GPa)	Source
Cotton	1.51	400	12	Brouwer (2000)
Hemp	1.48	514	24.8	Beckermann and Pickering (2008) and Brouwer (2000)
Jute	1.3–1.45	393–773	13–26.5	Mohanty et al. (2000)
Flax	1.50	345–1100	27.6	Mohanty et al. (2000)
Sisal	1.45	468–640	9.4–22	Mohanty et al. (2000)
Ramie	1.51	500	44	Brouwer (2000)
Coir	1.15	131–175	4–6	Mohanty et al. (2000)

become mainstream in terms of their applications in many households and commercial use. The use of synthetic materials from non-renewable sources such as petroleum has led to the accumulation of non-biodegradable wastes, which are a hazard to the ecological system. Therefore, extensive research is being conducted on the biodegradation of plastics and building materials such as composites. Natural fibres together with synthetic biodegradable materials (bioplastics) can be used to fabricate Biocomposites that have beneficial effects on the environment such as biodegradability, renewability of the base material, and decrease in emission of greenhouse gasses. Table 15.3 shows the density (g/m^3), tensile strength (MPa) and Young's modulus (GPa) of common plant-based fibres with a potential to be used as Biocomposites.

As can be seen in Table 15.3 that flax and jute have the highest tensile strength. Although the values are lower than the tensile strength of their synthetic counterparts, natural fibres offer a lower density and competitive Young's modulus. Biodegradation and renewability offer a lot of advantages such as the reduction of plastic waste and mitigation in the cost of waste management and impact the landfill (Jia et al., 2014). Degradation of the manufactured composite material occurs with the breakdown of the composite materials which comes with the loss of mechanical properties. In the open-air environment, the degradation of natural fibre-reinforced composites can be impacted by ambient moisture level, temperature, exposure to ultraviolet light and microbes. The degradation transpires from the breakdown of hemicelluloses, lignin, and cellulose components of the fibre matrix, thereby deteriorating the bonding between fibres and polymer matrix (de Melo et al., 2017). It is pivotal for the industry to move towards opportunities to utilize and promote the use of natural fibres as reinforcement in the polymer so that the aspect of biodegradation comes into play. In the current context, synthetic materials which are compostable or degradable could be a new opportunity for the next generation of composite materials and lead to a plethora of design innovations.

15.10 Design innovations

One of the key areas to adopt circular thinking (MacArthur, 2014, 2017) in the fabrication of composites are in design. In the scenario of utilizing waste mitigation and utilization of cellulose-based composites, design direction must follow the material streamlining. This will facilitate the reduction of waste ending up in landfills, which is a key feature of the current way the fashion and textiles industry is operating. Fig. 15.2 proposes the design consideration and illustrates how waste generation pathways can be mitigated through adopting circular approaches in designing composites in the future. One of the challenges in utilization of waste textiles is the absence of labelling due to wear and tear and traceability of the supply chain of the textiles in question.

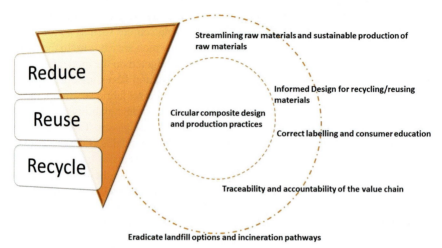

Figure 15.2 The key aspects to introduce circular practices into composites design.

Emphasis on consumer education (Montagna and Carvalho, 2018) and awareness of how to care for the garment is pivotal in achieving circularity in material flow in the fashion and textile industry (McCorkill, 2021).

15.11 Streamlining waste

Fig. 15.2 proposed streamlining of materials for product design. One of the systemic issues in waste generated from fashion and textiles is the mixing and blending of fibres (Laitala et al., 2018). Fibres are blended and mixed together at the yarn manufacturing stage to minimize cost and achieve specific functionality of the intended end uses. This has been unavoidable as the industry intrinsically never designed textiles for recycling or other avenues of reuse of materials.

With the current consumption and amount of textile waste that ends up in landfills (Desore and Narula, 2018), the fashion and textile industry needs to react radically to change its current practice of how it utilises materials (Braungart et al., 2007; Earley and Goldsworthy, 2015; Islam, 2021). Fig. 15.3 speculates how the biobased composites can be remanufactured from waste collection and material streamlining segmenting natural and synthetic waste material. Design-led product innovation in combination with technological advancement and automation can lead to new business opportunities and models in the future.

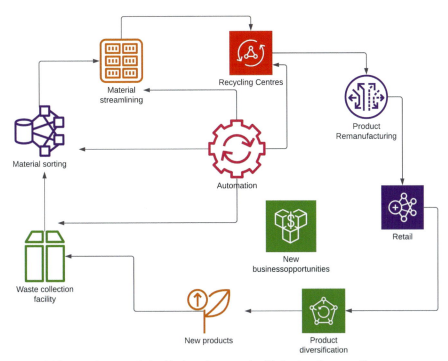

Figure 15.3 New framework for bio-based composite fabrication from textile waste.

15.12 Way forward

Waste mitigation in the fashion and textile industry has become one of the focal points in recent years. There are a lot of conversations happening to embed the UNSDGs (United Nations, 2015) in relation to Circular Economy (MacArthur, 2017) adoption between the stakeholders. As the waste sorting technology and framework develop within the next decade, we will experience a lot of research and development into the utilization of recovered waste materials. The textile recycling industry is still realising and improving the mechanical recycling of fibres as chemical pathways are either very expensive or underdeveloped. In the current business scenario, natural fibres reinforced composites are experiencing comprehensive growth with good prospects in the automotive and construction industries (Bhat et al., 2021). Bast fibres such as hemp, kenaf, flax, etc., are preferred for automotive applications (Mishra et al., 2020). On the other hand, wood plastic composite is the material of choice for construction industries (Mishra et al., 2020). Looking at the developments of the current trends, Europe is projected to remain the largest consumer and the driving market for natural fibre-reinforced composites due to the high acceptance level of environmentally friendly composite materials. This has been possible due to the stakeholder industries, government agencies, and advances in small-scale environmentally friendly industry initiatives. The improvement in materials performance will drive the growth of natural fibre-reinforced polymer composites in new potential areas. Natural fibre-based composites are still in a nascent stage in the electronics and sports goods segments; however, we may see the potential to capture a good market share in the future.

15.13 Conclusion

The world is moving towards sustainable material choices to better cater for the future generations' consumption. The fashion textile industry's current focus is to design and provide superior sustainable products. There is an increase in demand for commercial use of natural fibre-based composites in recent years for various industrial sectors. As discussed in this chapter, natural fibres are biodegradable materials with advantages like low cost, lightweight, renewability, and high specific properties. Increased environmental awareness has resulted in the mitigation of waste generated by the fashion and textile industry and the recovered waste then can be utilized to streamline the natural fibres as an effective reinforcement material in composite polymer matrices. In many applications, natural fibres could be deemed suitable to replace the existing synthetic polymers and there are potential for new research and development of product ranges. Soon, natural fibres are envisaged to become one of the sustainable and renewable resources in the composite field which can replace synthetic fibres in many applications.

References

Alomayri, T., Shaikh, F.L., 2013. Characterisation of cotton fibre-reinforced geopolymer composites. Composites Part B: Engineering 50, 1—6.

Amaducci, S., Gusovius, H.-J., 2010. 'Hemp—cultivation, Extraction and Processing', Industrial Applications of Natural Fibres Structure, Properties and Technical Applications. John Wiley & Sons, Ltd, Chichester, West Sussex, United Kingdom, pp. 109—134.

Angelini, L.G., Tavarini, S., 2013. Ramie [Boehmeria nivea (L.) Gaud.] as a potential new fibre crop for the Mediterranean region: growth, crop yield and fibre quality in a long-term field experiment in Central Italy. Industrial Crops and Products 51, 138—144.

Arsyad, M., Wardana, I., Irawan, Y.S., 2015. The morphology of coconut fiber surface under chemical treatment. Matèria. Revista Internacional d'Art 20, 169—177.

Atiqah, A., Maleque, M., Jawaid, M., Iqbal, M., 2014. Development of kenaf-glass reinforced unsaturated polyester hybrid composite for structural applications. Composites Part B: Engineering 56, 68—73.

Bacci, L., Baronti, S., Predieri, S., di Virgilio, N., 2009. Fiber yield and quality of fiber nettle (Urtica dioica L.) cultivated in Italy. Industrial Crops and Products 29 (2—3), 480—484.

Bajpai, P.K., Meena, D., Vatsa, S., Singh, I., 2013. Tensile behavior of nettle fiber composites exposed to various environments. Journal of Natural Fibers 10 (3), 244—256.

Banik, S., Basak, M., Paul, D., Nayak, P., Sardar, D., Sil, S., Sanpui, B., Ghosh, A., 2003. Ribbon retting of jute—a prospective and eco-friendly method for improvement of fibre quality. Industrial Crops and Products 17 (3), 183—190.

Beckermann, G., Pickering, K.L., 2008. Engineering and evaluation of hemp fibre reinforced polypropylene composites: fibre treatment and matrix modification. Composites Part A: Applied Science and Manufacturing 39 (6), 979—988.

Berhanu, T., Kumar, P., Singh, I., 2014. Mechanical behaviour of jute fibre reinforced polypropylene composites. Proceedings of AIMTDR 1—6, 289 2014, 12-14 December.

Bhat, K.M., Rajagopalan, J., Mallikarjunaiah, R., Rao, N.N., Sharma, A., 2021. 'Eco-Friendly and Biodegradable Green Composites', intechOpen Book Series.

Bos, H.L., 2004. The Potential of Flax Fibres as Reinforcement for Composite Materials. Citeseer.

Braungart, M., McDonough, W., Bollinger, A., 2007. Cradle-to-cradle Design: Creating Healthy Emissions Ea Strategy for Eco-Effective Product and System Design.

Brouwer D, W., 2000. Natural Fiber Composites in Structural Components: Alternative Applications for Sisal. Seminar, Commond Fund for Commodities-Alternative Applications for Sisal and Henecuen.

Cengiz, T.G., Babalık, F., 2009. The effects of ramie blended car seat covers on thermal comfort during road trials. International Journal of Industrial Ergonomics 39 (2), 287—294.

Chen, Q., Zhao, T., Wang, M., Wang, J., 2013. Studies of the fibre structure and dyeing properties of C alotropis gigantea, kapok and cotton fibres. Coloration Technology 129 (6), 448—453.

de Melo, R.P., Marques, M.F., Navard, P., Duque, N.P., 2017. Degradation studies and mechanical properties of treated curaua fibers and microcrystalline cellulose in composites with polyamide 6. Journal of Composite Materials 51 (25), 3481—3489.

Desore, A., Narula, S.A., 2018. An overview on corporate response towards sustainability issues in textile industry', Environment. Development and Sustainability 20 (4), 1439—1459.

Devaraju, A., Harikumar, R., 2019. Life Cycle Assessment of Sisal Fiber. Reference Module in Materials Science and Materials Engineering Book Chapter. Elsevier Publication.

Dobircau, L., Sreekumar, P., Saiah, R., Leblanc, N., Terrié, C., Gattin, R., Saiter, J., 2009. Wheat flour thermoplastic matrix reinforced by waste cotton fibre: Agro-green-composites. Composites Part A: Applied Science and Manufacturing 40 (4), 329—334.

Duguid, S.D., 2009. Flax. Oil Crops. Springer, pp. 233—255.

Earley, R., Goldsworthy, K., 2015. Designing for Fast and Slow Circular Fashion Systems: Exploring Strategies for Multiple and Extended Product Cycles', Paper Presented to PLATE Conference. Nottingham Trent University, 17-19 June.

Elmogahzy, Y., Farag, R., 2018. Tensile properties of cotton fibers: importance, research, and limitations. In: Handbook of Properties of Textile and Technical Fibres. Elsevier, pp. 223—273.

Eskander, S., Saleh, H.E.D., 2017. Biodegradation: process mechanism. Environ. Sci. & Eng 8 (8), 1—31.

Fan, M., Fu, F., 2016. Advanced High Strength Natural Fibre Composites in Construction. Woodhead Publishing.

Fashion United Group, 2021. Global Fashion Industry Statistics—International Apparel. Fashion United Group viewed 2/10/2021. https://fashionunited.com/global-fashion-industry-statistics/.

Giwa Ibrahim, S., Karim, R., Saari, N., Wan Abdullah, W.Z., Zawawi, N., Ab Razak, A.F., Hamim, N.A., Umar, R.A., 2019. Kenaf (Hibiscus cannabinus L.) seed and its potential food applications: a review. Journal of Food Science 84 (8), 2015—2023.

González-García, S., Hospido, A., Moreira, M., Feijoo, G., 2007. Life Cycle Environmental Analysis of Hemp Production for Non-wood Pulp.

Goudenhooft, C., Bourmaud, A., Baley, C., 2019. Flax (Linum usitatissimum L.) fibers for composite reinforcement: exploring the Link between plant growth, Cell Walls development, and fiber properties. Frontiers in Science : Plant Biophysics and Modeling 10 (411).

Goutianos, S., Peijs, T., Nystrom, B., Skrifvars, M., 2006. Development of flax fibre based textile reinforcements for composite applications. Applied Composite Materials 13 (4), 199–215.

Graupner, N., 2008. Application of lignin as natural adhesion promoter in cotton fibre-reinforced poly (lactic acid)(PLA) composites. Journal of Materials Science 43 (15), 5222–5229.

Guo, Y., Wu, H-l, Sun, X-y, Qian, X., Shen, Y-q, 2005. Analysis on the properties of nettle fiber. Journal of Textile Research 26 (4), 27.

Hamidon, M.H., Sultan, M.T., Ariffin, A.H., Shah, A.U., 2019. Effects of fibre treatment on mechanical properties of kenaf fibre reinforced composites: a review. Journal of Materials Research and Technology 8 (3), 3327–3337.

He, L., Li, W., Chen, D., Zhou, D., Lu, G., Yuan, J.J.M., Design, 2015. Effects of Amino Silicone Oil Modification on Properties of Ramie Fiber and Ramie Fiber/polypropylene Composites, 77, pp. 142–148.

Hepworth, D., Hobson, R., Bruce, D., Farrent, J., 2000. The use of unretted hemp fibre in composite manufacture. Composites Part A: Applied Science and Manufacturing 31 (11), 1279–1283.

Islam, S., 2020. Sustainable raw materials: 50 shades of sustainability. In: Nayak, Rajkishore (Ed.), Sustainable Technologies for Fashion and Textiles. Elsevier, pp. 343–357. https://doi.org/10.1016/B978-0-08-102867-4.00015-3.

Islam, S., 2021. 15 - waste management strategies in fashion and textiles industry: challenges are in governance, materials culture and design-centric. In: Nayak, R., Patnaik, A. (Eds.), Waste Management in the Fashion and Textile Industries. Woodhead Publishing, pp. 275–293. https://www.sciencedirect.com/science/article/pii/B9780128187586000156.

Jaafar, C.N.A., Rizal, M.A.M., Zainol, I., 2018. Effect of kenaf alkalization treatment on morphological and mechanical properties of epoxy/silica/kenaf composite. International Journal of Engineering and Technology 7, 258–263.

Jia, W., Gong, R.H., Hogg, P.J., 2014. Poly (lactic acid) fibre reinforced biodegradable composites. Composites Part B: Engineering 62, 104–112.

Kamble, Z., Behera, B.K.J.J.E.F., 2020. Mechanical properties and water absorption characteristics of composites reinforced with cotton fibres recovered from textile waste. Journal of Engineered Fibers and Fabrics 15, p. 1558925020901530.

Kengkhetkit, N., Amornsakchai, T., 2012. Utilisation of pineapple leaf waste for plastic reinforcement: 1. A novel extraction method for short pineapple leaf fiber. Industrial Crops and Products 40, 55–61.

Khan, J.A., Khan, M.A., Islam, M.R., 2014. A study on mechanical, thermal and environmental degradation characteristics of N, N-dimethylaniline treated jute fabric-reinforced polypropylene composites. Fibers and Polymers 15 (4), 823–830.

Kim, J.T., Netravali, A.N., 2010. Mercerization of sisal fibers: effect of tension on mechanical properties of sisal fiber and fiber-reinforced composites. Composites Part A: Applied Science and Manufacturing 41 (9), 1245–1252.

Kishi, H., Fujita, A., 2008. Wood-based epoxy resins and the ramie fiber reinforced composites. Environmental Engineering and Management Journal 7 (5), 517–523.

Laitala, K., Ingun Grimstad, K., Henry, B., 2018. Does use matter? Comparison of environmental impacts of clothing based on fiber type. Sustainability 10 (7), 2524.

Le Duigou, A., Davies, P., Baley, C., 2011. Environmental impact analysis of the production of flax fibres to be used as composite material reinforcement. Journal of Biobased Materials and Bioenergy 5 (1), 153–165.

Liu, D., Song, J., Anderson, D.P., Chang, P.R., Hua, Y., 2012. Bamboo fiber and its reinforced composites: structure and properties. Cellulose 19 (5), 1449–1480.

Lopattananon, N., Panawarangkul, K., Sahakaro, K., Ellis, B., 2006. Performance of pineapple leaf fiber–natural rubber composites: the effect of fiber surface treatments. Journal of Applied Polymer Science 102 (2), 1974–1984.

Lu, Y., Weng, L., Cao, X.J.C., 2006. Morphological, Thermal and Mechanical Properties of Ramie Crystallites—Reinforced Plasticized Starch Biocomposites, 63, pp. 198–204 no. 2.

MacArthur, E., 2014. Towards the Circular Economy: Accelerating the Scale-Up across Global Supply Chains, 3. Ellen MacArthur Foundation, United Kingdom. https://ellenmacarthurfoundation.org/towards-the-circular-economy-vol-3-accelerating-the-scale-up-across-global.

MacArthur, E., 2017. A New Textiles Economy: Redesigning Fashion's Future. Ellen MacArthur Foundation, United Kingdom, pp. 1–150. https://ellenmacarthurfoundation.org/a-new-textiles-economy.

Marsyahyo, E., Jamasri, Rochardjo, H.S.B., Textiles, S.J.J.I., 2009. Preliminary Investigation on Bulletproof Panels Made from Ramie Fiber Reinforced Composites for NIJ Level II, IIA, and IV, 39, pp. 13–26 no. 1.

McCorkill, G., 2021. 'Fashion Fix: Exploring Garment Repair from a Critical Fashion Practice Perspective', Paper Presented to 4th Conference on Product Lifetimes and the Environment (PLATE), Virtual, pp. 26–28. https://ulir.ul.ie/handle/10344/10230.

Mishra, R., Wiener, J., Militky, J., Petru, M., Tomkova, B., Novotna, J., 2020. Bio-composites reinforced with natural fibers: comparative analysis of thermal, static and dynamic-mechanical properties. Fibers and Polymers 21 (3), 619–627.

Misnon, M.I., Islam, M.M., Epaarachchi, J.A., Lau, K.T., 2014. Potentiality of utilising natural textile materials for engineering composites applications. Materials & Design 59, 359–368.

Mohanty, A., Misra, M., Hinrichsen, G., 2000. Biofibres, biodegradable polymers and bio-composites: an overview. Macromolecular Materials and Engineering 276 (1), 1–24.

Montagna, G., Carvalho, C., 2018. Textiles, Identity and Innovation: Design the Future. Chapman and Hall/CRC, Milton, United Kingdom. http://ebookcentral.proquest.com/lib/rmit/detail.action?docID=5509964.

Morrissey J, Anne, Browne, J, 2004. Waste management models and their application to sustainable waste management. Journal of Waste Management 24 (3), 297–308.

Mortazavi, S., Moghaddam, M.K., 2010. An analysis of structure and properties of a natural cellulosic fiber (Leafiran). Fibers and Polymers 11 (6), 877–882.

Mwaikambo, L., 2006. Review of the history, properties and application of plant fibres. African Journal of Science and Technology 7 (2), 121.

Nam, S., Netravali, A.N., 2006. Green composites. I. Physical properties of ramie fibers for environment-friendly green composites. Fibers and Polymers 7 (4), 372–379.

Naveen, J., Jawaid, M., Amuthakkannan, P., Chandrasekar, M., 2019. 21—mechanical and physical properties of sisal and hybrid sisal fiber-reinforced polymer composites. In: Jawaid, M., Thariq, M., Saba, N. (Eds.), Mechanical and Physical Testing of Biocomposites, Fibre-Reinforced Composites and Hybrid Composites. Woodhead Publishing, pp. 427–440. https://www.sciencedirect.com/science/article/pii/B9780081022924000217.

Nishino, T., Hirao, K., Kotera, M., Nakamae, K., Inagaki, H., 2003. Kenaf reinforced biodegradable composite. Composites Science and Technology 63 (9), 1281–1286.

Nordin, N.A., Yussof, F.M., Kasolang, S., Salleh, Z., Ahmad, M.A., 2013. Wear rate of natural fibre: long kenaf composite. Procedia Engineering 68, 145–151.

Ochi, S.J.M., 2008. Mechanical Properties of Kenaf Fibers and Kenaf/PLA Composites, 40, pp. 446–452 no. 4–5.

Pastore, C.M, 2000. Opportunities and challenges for textile reinforced composites. Mechanics of composite materials 36 (2), 97–116. https://link.springer.com/article/10.1007/BF02681827.

Pham, L.J., 2016. Coconut (cocos nucifera). In: Industrial Oil Crops. Elsevier, pp. 231–242.

Pickering, K.L., Beckermann, G., Alam, S., Foreman, N.J., 2007. Optimising industrial hemp fibre for composites. Composites Part A: Applied Science and Manufacturing 38 (2), 461–468.

Putri, E.M.M., Rachimoellah, M., Santoso, N., Pradana, F., 2012. Biodiesel production from kapok seed oil (Ceiba pentandra) through the transesterification process by using cao as catalyst. Global Journal of Researches in Engineering 12 (2-D).

Quek, C.S., Ngadi, N., Ahmad Zaini, M.A., 2020. The oil-absorbing properties of kapok fibre—a commentary. Journal of Taibah University for Science 14 (1), 507–512.

Quillien, L., 2014. Flax and linen in the first millennium Babylonia BC: the origins, craft industry and uses of a remarkable textile. Prehistoric, Ancient Near Eastern Aegean Textiles Dress 18, 271–296.

Rajeshkumar, G., Ramakrishnan, S., Pugalenthi, T., Ravikumar, P., 2020. Performance of surface modified pineapple leaf fiber and its applications. In: Pineapple Leaf Fibers. Springer, pp. 309–321.

Rehman, M., Gang, D., Liu, Q., Chen, Y., Wang, B., Peng, D., Liu, L., 2019. Ramie, a multipurpose crop: potential applications, constraints and improvement strategies. Industrial Crops and Products 137, 300–307.

Romanzini, D., Lavoratti, A., Ornaghi Jr., H.L., Amico, S.C., Zattera, A.J., 2013. Influence of fiber content on the mechanical and dynamic mechanical properties of glass/ramie polymer composites. Materials and Design 47, 9–15.

Rowell, R.M., Sanadi, A., Jacobson, R., Caulfield, D., 1999. Properties of kenaf/polypropylene composites. In: Ag & Bio Engineering, pp. 381–392.

Saba, N., Paridah, M., Jawaid, M., 2015. Mechanical properties of kenaf fibre reinforced polymer composite: a review. Construction and Building Materials 76, 87–96.

Salleh, Z., Berhan, M., Hyie, K.M., Taib, Y., Kalam, A., Roselina, N.N., 2013. Open hole tensile properties of kenaf composite and kenaf/fibreglass hybrid composite laminates. Procedia Engineering 68, 399–404.

Sathishkumar, T., Naveen, J., Navaneethakrishnan, P., Satheeshkumar, S., Rajini, N., 2017. Characterization of sisal/cotton fibre woven mat reinforced polymer hybrid composites. Journal of Industrial Textiles 47 (4), 429–452.

Sen, T., Reddy, H.J., 2011. Various industrial applications of hemp, kinaf, flax and ramie natural fibres. International Journal of Innovation, Management and Technology 2 (3), 192.

Sepe, R., Bollino, F., Boccarusso, L., Caputo, F., 2018. Influence of chemical treatments on mechanical properties of hemp fiber reinforced composites. Composites Part B: Engineering 133, 210–217.

Serizawa, S., Inoue, K., Iji, M., 2006. Kenaf-fiber-reinforced poly (lactic acid) used for electronic products. Journal of Applied Polymer Science 100 (1), 618–624.

Shahzad, A., 2012. Hemp fiber and its composites—a review. Journal of Composite Materials 46 (8), 973–986.

Stevulova, N., Kidalova, L., Cigasova, J., Junak, J., Sicakova, A., Terpakova, E., 2013. Lightweight composites containing hemp hurds. Procedia Engineering 65, 69–74.

Threepopnatkul, P., Kaerkitcha, N., Athipongarporn, N., 2009. Effect of surface treatment on performance of pineapple leaf fiber–polycarbonate composites. Composites Part B: Engineering 40 (7), 628–632.

Thyavihalli Girijappa, Y.G., Mavinkere Rangappa, S., Parameswaranpillai, J., Siengchin, S., 2019. Natural fibers as sustainable and renewable resource for development of eco-friendly composites: a comprehensive review. Frontiers in Materials 6, 226.

To, M.H., Uisan, K., Ok, Y.S., Pleissner, D., Lin, C.S.K., 2019. Recent trends in green and sustainable chemistry: rethinking textile waste in a circular economy. Current Opinion in Green Sustainable Chemistry 20, 1–10.

United Nations, 2015. Transforming our world: the 2030 Agenda for Sustainable Development. United Nations. https://sustainabledevelopment.un.org/post2015/transformingourworld.

United Nations, 2019. World Population Prospects 2019, Department of Economic and Social Affairs and Population Dynamics. United Nations viewed 7th October. https://population.un.org/wpp/Download/Standard/CSV/.

Van de Velde, K., Baetens, E., 2001. Thermal and mechanical properties of flax fibres as potential composite reinforcement. Macromolecular Materials and Engineering 286 (6), 342–349.

Venkata Reddy, G., Venkata Naidu, S., Rani, S., 2008. Kapok/glass polyester hybrid composites: tensile and hardness properties. Journal of Reinforced Plastics and Composites 27 (16–17), 1775–1787.

Viju, S., Thilagavathi, G., Vignesh, B., Brindha, R.J.T.J.T.I., 2019. Oil sorption behavior of acetylated nettle fiber 110 (10), 1415–1423.

Wang, J., Zheng, Y., Wang, A., 2012. Effect of kapok fiber treated with various solvents on oil absorbency. Industrial Crops and Products 40, 178–184.

Yang, X., Fan, W., Ge, S., Gao, X., Wang, S., Zhang, Y., Foong, S.Y., Liew, R.K., Lam, S.S., Xia, C., 2021. Advanced textile technology for fabrication of ramie fiber PLA composites with enhanced mechanical properties. Industrial Crops and Products 162, 113312.

Yu, T., Jiang, N., Li, Y.J.C.P.A.A.S., Manufacturing, 2014. Study on Short Ramie Fiber/poly (Lactic Acid) Composites Compatibilized by Maleic Anhydride, 64, pp. 139–146.

Yu, T., Ren, J., Li, S., Yuan, H., Li, Y.J.C.P.A.A.S., Manufacturing, 2010. Effect of Fiber Surface-Treatments on the Properties of Poly (Lactic Acid)/ramie Composites, 41, pp. 499–505 no. 4.

Zafeiropoulos, N., Baillie, C., Matthews, F., 2001. A study of transcrystallinity and its effect on the interface in flax fibre reinforced composite materials. Composites Part A: Applied Science and Manufacturing 32 (3–4), 525–543.

Zakikhani, P., Zahari, R., Sultan, M.T.H.H., Majid, D.L.A.A., 2017. Morphological, mechanical, and physical properties of four bamboo species. Bioresources 12 (2), 2479–2495.

Zamani, B., 2014. Towards Understanding Sustainable Textile Waste Management: Environmental Impacts and Social Indicators.

Zheng, Y., Wang, J., Zhu, Y., Wang, A., 2015. Research and application of kapok fiber as an absorbing material: a mini review. Journal of Environmental Sciences 27, 21–32.

Part Five

Organizations, standards and challenges

Organizations and certifications relating to sustainable fibres

Kunal Singha[1], Subhankar Maity[2] and Pintu Pandit[1]
[1]Department of Textile Design, National Institute of Fashion Technology, Patna, Bihar, India; [2]Department of Textile Technology, Uttar Pradesh Textile Technology Institute, Kanpur, Uttar Pradesh, India

16.1 Introduction

Organizations are always playing a crucial role in the certifications and sustainable business in the textile and fashion industries. These certifications are very handy for promoting the fashion or garment brand and overall profit. These various standards also guarantee the general business policies, business supply chain and flow of materials from the raw material to end-products, wages policy, social and environmental effects and impact of a textile or fashion business to the society (Certifications; Unsustainable magazine; Futurelearn; Considerate). In the following sections, various sustainable organizations, certifications and standards are discussed.

16.2 Key sustainability organizations and certifications

16.2.1 American Forests

American Forests is a non-profit forest conservation organization founded in 1875 in the United States (Fig. 16.1). It focuses on the reforestation movement building and encompasses climatic change, wildlife balance, water health quality and social equity. This standard is also used in the textile and garment industries to show their prudence in sustainability lines (Certifications).

Figure 16.1 American Forests logo.

16.2.2 Clean Clothes Campaign

Clean Clothes Campaign (Fig. 16.2) is a mission to improve the working condition of the fibre and garment industry worldwide. This global network, founded in 1989, consists of more than 230 organizations to ensure worker fundamental rights and fight for their betterment. This mission fights for preserving sustainable fibre manufacturing and clothing processes and women's rights, consumer advocacy and poverty reduction (Certifications).

Figure 16.2 Clean Clothes Campaign logo.

16.2.3 ISO Certification

ISO Certification is the most followed standard in sustainable third-party auditors, managing the management system, manufacturing process, service or documentation procedures to meet all the requirements for sustainable standardization and quality assurance. ISO 9001 and ISO 14001 certification provides proof of creditability and deposits of well-read sustainability auditing and improvement. In textile and garment industries, ISO 9001 abides the management system that ensures that all the products meet all statutory regulations and are consistent in quality to complete customer satisfaction (Certifications; Unsustainable magazine).

16.2.4 Made-By

The Made-By is an environmental benchmark for textile fibres, and it is working as a not-for-profit organization that compares the environmental impact in the garment industry (Fig. 16.3). This uses to rank 28 fibres depending on greenhouse gas emissions, toxicity to human body, eco-toxicity, land, water, energy and water usage. It is a division of the Common Objective (CO), a global tech company providing sustainable fashion business solutions by providing various companies the rewards for maintaining the best practices and turning sustainability at a lower opportunity cost.

Figure 16.3 Made-By logo.

16.2.5 1% for the planet

One% for the Planet (Fig. 16.4) is a global organization created by Yvon Chouinard, founder of Patagonia, and Craig Matthews, founder of Blue Ribbon Flies, in 2002 and currently, they have more than 3000 members. It believes that businesses can join by donating 1% of their sales back to the environment, regardless of their profitability. It collectively engages itself to save resources across the globe to solve the world's problem. Individuals can also join the group by donating 1% of their salary.

Figure 16.4 1% for the planet.

16.2.6 *Centre for Advancement of Garment Making (CAGM)*

Centre for Advancement of Garment Making (CAGM) empowers fashion professionals to build a transparent, regenerative industry circularly and economically. It provides ample values in craft for human life and also for the advancement of the craft in the planet. They also educate the people on sustainable fashion who are involved in the fashion business (Certifications; Unsustainable magazine; Futurelearn; Considerate) (Fig. 16.5).

Figure 16.5 Centre for Advancement of Garment Making (CAGM) logo.

16.2.7 Fashion Forum Fellowship 500

Fashion Forum Fellowship 500 (Fig. 16.6), launched by the Ethical Fashion Forum in 2011, is an invitation that follows the fair-trade organizations to maintain sustainability and fashion for developing nations. It cares for pioneers and innovators working collaboratively towards sustainability.

Figure 16.6 Fashion Forum Fellowship 500.

16.2.8 Fashion Revolution

Fashion Revolution (Fig. 16.7) is a global non-profit organization established by The Fashion Revolution Foundation and Fashion Revolution CIC and expanded over 100 countries which performs various campaign on diverse issues like fashion supply chain and reforming of the fashion industries by using sustainable textile and fashion materials. They use #WhoMadeMyClothes as their call to arms and also became the number one global trend on twitter (Certifications; Unsustainable magazine; Futurelearn; Considerate).

Figure 16.7 Fashion Revolution.

16.2.9 Fashion Transparency Index

Fashion Transparency Index rates the 250 fashion brands based on their sustainability and social and environmental policies and impacts. It considered 220 indicators of any fashion, textile or garment industry based on their social and ecological aspects. Its thought process is like animal welfare, biodiversity, chemical usage, climate due to diligence, labour issues such as working conditions, living wages or forced labour, freedom of association and gender equality, purchasing practices, supplier disclosure, waste and recycling. They have published their fifth index in 2020 (Unsustainable magazine; Sustainability).

16.2.10 Good on You

Good on You publish brand ratings for sustainable and ethical clothing companies and also provides an app so that any customer can rank their fashion or textile buying. Their certificate is based on the company's action plan on sustainability treatment of the planet, people, and animals. They also use to publish articles on fashion trends and sustainability indexes.

16.2.11 Canopy

Canopy (Fig. 16.8) works to transform unsustainable product supply chains by encouraging executives to become conservation and sustainability champions. They have worked with publishers such as Penguin-Random House and Scholastic and leading clothing brands like H&M and Zara-Indeitex, Stella McCartney, Target and Uniqlo.

Figure 16.8 Canopy logo.

16.2.12 Carbon Neutral (CN)

Carbon Neutral (Fig. 16.9) standard founded in 2002 is working on carbon neutrality and reducing the carbon footprint close to zero by controlling the energy efficiency and external emission reduction or purchasing offsets. This certification helps the fashion and textile industries set goals, achieve them and advertise their efforts. CN-certification ensures the customer about the company's zero carbon emission and consists of a five-step process certification technique.

Figure 16.9 Carbon Neutral.

16.2.13 Climate Neutral

Climate Neutral (Fig. 16.10) certification is a standard given to the textile and garment industries to guarantee almost zero or less greenhouse gas emissions. This process measures all the accreditation for a company which involves measuring all of the carbon emissions from making and delivering products and services to customers and purchasing their carbon footprint by funding projects like reforestation or renewable energy. It also helps to find out the developing and future emission plan implementation. In 2019, almost 150 companies globally had joined this CNeutral-standard and already reduced an enormous amount of offsetting carbon by 228,314 tonnes (Certifications; Unsustainable magazine; Futurelearn; Considerate; Anandjiwala, 2007; Maia et al., 2013; Chavan, 2014).

Figure 16.10 Climate Neutral.

16.2.14 Fabscrap

Fabscrap (Fig. 16.11) is a non-profit recycling program from New York City and works on recycle fabric scraps from manufacturers. They used to pick up and design and re-distribute the fabric scraps to students, artists, crafters and other designers for reuse, while the smaller scraps are used in insulation. This certification for any textile and garment industries working in sustainable lines will undoubtedly prove the company is working on new fibre-to-fibre recycling technologies whenever possible. Eileen Fisher, Nautica and Mara Hoffman are some of the members of this certification.

Figure 16.11 Fabscrap.

16.2.15 Upcycling certification

Upcycling certification (Fig. 16.12) guarantees getting new and creative, reused and transforming products mainly made from waste materials or useless or unwanted products. This upcycling certification is essential in fashion and textiles to convince their customer of the concept of reusability-recyclability and sustainability. Some of the prominent textile and fashion brands who follow this certification are Econyl, Zara, H&M, Prada, Arvind Mill Ltd., Reliance fires, Herme, Louis Vuitton, Anthropology, Patagonia, Banana Republic, GAP, Tommy Hilfiger, C&K, Armani, Zara-Inditex, Versache and Levi's Strauss, etc.

Figure 16.12 Upcycling certification.

16.2.16 Bluesign

The bluesign (Fig. 16.13) standard is an independent citification used in the textile industry to ensure and focus on legal compliance regarding environmental health and safety. The certification standard comprises various aspects such as water and air emission, occupational health conditions, consumer safety and methodologies for reducing harmful emissions combined in the early production stages. The details of it can be found at their official website: www.bluesign.com.

Figure 16.13 Bluesign.

The independent bluesign programme focuses on ZDHC (Zero Discharge of Hazardous Waste) (Fig. 16.14), which is also used as a primary measurement tool for most chemical-based dyeing units of textile cotton or natural fibres. This standard helps the industries to manage and reduce their chemical usage. Bluesign companies have a holistic approach to sustainable processing and manufacturing. To be Bluesign certified, a manufacturing facility must not only reduce greenhouse gases, but must also make an active contribution to climate protection (Certifications; Unsustainable magazine; Futurelearn; Considerate; Wong and Ngai, 2021; Rani and Saha, 2021; Periyasamy and Militky, 2020; Seisl and Hengstmann, 2021).

Figure 16.14 ZDCH.

Bluesign Certified Manufacturing companies target to have the lowest impact on the environment during production, which claimed as 'guarantees the application of sustainable ingredients in a clean process resulting in a safely manufactured product'. These companies also need to compliance the toughest criteria to protect soil, air, water management to reduce the environmental hazards and by ensuring the maximum resources intact for future generations.

16.2.17 CMiA

Cotton Made in Africa (Fig. 16.15) is an initiative launched by the Aid by Trade Foundation that promotes environmental protection. Its mission and aims are to combat

Figure 16.15 Cotton made in Africa.

poverty by extending various on-site training modules or courses to small farmers and are to combat poverty by extending various on-site training modules or courses to small farmers and help them a higher rate of cotton crop yielding.

Plexus Cotton Group, Cooee Kids Ltd, and Jack Jones, Armani Exchange and Otto Group organizations are using Cotton Made in Africa. The details of it can be found at their official website: www.cotton-made-in-africa.org.

16.2.18 Cradle to Cradle

In 2002, Braungart and William McDonough in their book, Cradle to Cradle: Remaking the Way We Make Things, have discussed the cradle-to-cradle (C2C) as a statement or programme for product or process design which gives specific details of how to achieve C2C design (also referred to as 2CC2, C2C, cradle two cradle, or regenerative design).

C2C (Fig. 16.16) certification is a multi-attribute label programming system that comprises showing efforts in eco-intelligent product design. To achieve C2C certification, companies need to focus and achieve on multiple areas such as eco materials, recycling, renewable energy, water efficiency and social responsibility. They are being awarded for maintaining the C2C practices with various levels of certification levels such as basic, silver, gold and platinum level.

This basically comprises a biomimetic approach to the design of products and systems for any industry where human and manufacturing nature are very important to ensure long-term sustainability and materials or resources crunch. C2C-certified companies are better positioned to ensure minimum use of resources with the maximum output circulating in healthy, safe metabolisms. C2C comprises of primary, silver, gold and platinum-level certification. This applies to materials, sub-assemblies and

Figure 16.16 Cradle to cradle.

finished products. Textile and fashion brands and suppliers who obliged C2C certification are included Metawear Organic, Wolford, Pratibha and Pacific Jeans Ltd (Periyasamy and Militky, 2020; Tobler-Rohr, 2011).

C2C is a multi-attributed certification process that provides eco-label ways of demonstrating efforts to design a product eco-intelligent. C2C divides and analyses company efforts in multiple focus lines such as eco materials, recycling, renewable energy, water efficiency and social responsibility. They also use to give awards to successful companies by giving them various ranks or awards like Basic, Silver, Gold or Platinum-level certification. This process also included the business analysis in materials, sub-assemblies and finished products. The details of it can be found at their official website: www.mbdc.com/. Fig. 16.17 explains C2C process in fashion and textiles.

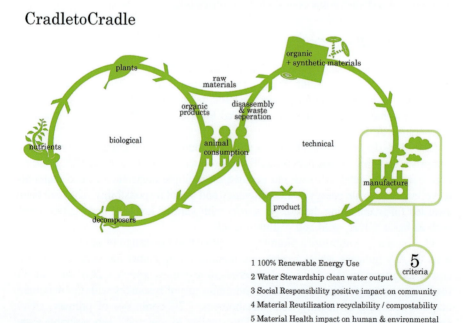

Figure 16.17 Cradle to cradle process.

16.2.19 EU Eco Label

The European Union (EU) Eco Label serving as a voluntary scheme since 1972 to encourage businesses to market products and services which involve environmental interaction and efforts. Ecolabel criteria are based on the product's impact on the product's environment and service and product lifecycle. The EU Eco label is one of the vital certifications for sustainable and organic fibre used worldwide. The details of it can be found at their official website: www.mbdc.com/.

16.2.20 Better Cotton Initiative (BCI)

The Better Cotton Initiative (BCI) (Fig. 16.18) is an organization made up of significant cotton consumers in the textile sector. It is the most commonly used cotton scheme. BCI is made to promote and implement the core production standard by using sustainability, social justice, economic development and applied supply chain management. The Better Cotton Standard system works in a holistic way to make the cotton fibre production most sustainable by following the 3BL (three bottom lines) of sustainability. They are the three pillars of sustainability: environmental, social and economic. BCI system is designed to guarantee the right exchange of good business practices and scaling up collective business action to establish a Better Cotton as a mainstream commodity. BCI is the world's largest cotton sustainability program and works as a nonprofit organization. They intend to make better cotton production for the farmers who produce them in a much better and sorted way and environment-friendly. BCI provides training on sustainable farming practices to more than 2.3 million cotton farmers in 23 countries, with more than 1840 members from farms to fashion to a textile brand (Certifications; Unsustainable magazine; Futurelearn; Considerate; Seisl and Hengstmann, 2021; Tobler-Rohr, 2011; Levering and Vos, 2019; Vadicherla and Saravanan, 2014; Annapoorani, 2018). The world's largest fashion brands are members of the BCI including ASOS, Puma, Zara, Levi Strauss & Co., Nike, American Eagle, H&M and C&A.

Figure 16.18 Better Cotton Initiative.

16.2.21 Sustainable Fair Trade Management System

Sustainable Fair Trade Management System (SFTMS) is a certification system operating after World War II. It had been launched from the house of World Fair Trade Organization. SFTMS is an all-new worldwide independent certification system and standard who demonstrates Fair Trade business practices by providing a fair and dynamic and integrated way for business production, trading and communication. The details of it can be found at their official website: www.wfto-europe.org/the-wfto-way.

16.2.22 Fair Trade

The Fair Trade movement (Fig. 16.19) certifies the various measurement and works towards the production, promotion and sale of commodities and artisanal items, which are made as per abiding the aspects of transparent supply chain, minimum prices, proper lead time management to ensure the security and self-economic sufficiency as well as sustainable production practices and fairer terms of business production.

Figure 16.19 Fair Trade movement.

16.2.23 Fairtrade certified cotton

The Fairtrade mark is an independent certification for consumer label. It appears in the product labelling on sustainable fibre, which actually ensures that the product has met the sustainable process and standard. The Fairtrade Textile Standard is one component of the greater Fairtrade Textile Programme to facilitate change in textile supply chains and related business practices (Fig. 16.20). That is a comprehensive approach to engage all the manufacturers and workers across the whole supply chain to ensure better wages, working environment and conditions required for a textile brand to commit to fair terms of trade. The consumer label on the textile fibre products certainly shows the certification of the products to signify that Fairtrade standards have been met. This standard relates to fulfilling the minimum expectation and demands are met for making the fibre and related products in terms of various aspects such as terms of trade, better prices, longer lead times to promote security, self-economic sufficiency and

Figure 16.20 Fairtrade certified cotton.

sustainable production best practices. Fairtrade cotton is also in use under a Fairtrade Sourcing Partnership that provides different on-product labelling to the familiar Mark. These standards are authorized and circulated by the Fair Labelling Organization and set under the ISEAL Code of Good Practice requirements in standards-setting. The details of it can be found at their official website: www.flo-cert.net and www.fairtrade.org.uk.

16.2.24 Global recycle standard

The Global Recycle Standard (GRS) (Fig. 16.21) has been developed to meet demands in the textile and fibre industry and the recycled parts in the import and export of the textile fibre. Control Union (CU) Certifications developed the GRS in 2008 and ownership was passed to the Textile Exchange (TE) on 1 January 2011. iIt is a third-party certification which acts as voluntary on recycled issues, chain of custody, social and environmental practices and chemical restrictions. The GRS is used in the textile and fashion industries to offer track and track and trace certification system for sustainable manufacturing of cotton fibre cultivation and growing process. GRS is made to insure the highest cotton-making process in ginning, spinning, weaving and knitting, dyeing and printing and stitching in more than 50 countries.

The Global Recycle Standard counts MUD Jeans, Sarvin, Contiential Clothing Co. and Been London amongst the organizations using its verification systems.

Figure 16.21 The Global Recycle Standard.

16.2.25 Sedex

Sedex (Fig. 16.22) is a supplier ethical data exchange-based non-profit organization founded by UK Retailers in 2001. They are in the standardization operation with a high-level social audit and monitoring supplier monitoring practices in all the businesses including sustainable textile fibre and fashion trade. They also openly transparently shared the audits of suppliers apart from their regular auditing job as their primary goal is to promote and verify ethical and responsible business practices.

Figure 16.22 Sedex.

16.2.26 SGS

SGS (Fig. 16.23) is the world's leading inspection, verification, testing and certification company. SGS offers specialized solutions to make businesses faster, simpler and more efficient in all class of business apart from textile and fashion. They offer several certification verifications, including ISO. They will provide independent testing to ensure that suppliers meet health and safety standards, and fabric content and quality also applicable for bio-cotton and sustainable fiber-based textile and fashion articles.

Figure 16.23 SGS.

16.2.27 The Global Organic Textile Standard (GOTS) certified

The Global Organic Textile Standard (GOTS) (Fig. 16.24) was developed by leading standard by eco-textile processing in the field of was created in 2006, by leading standard organisations to unify or combine all the existing textile or eco-friendly standard in the field of processing to explore worldwide recognised needs that ensures global

Figure 16.24 Global organic textile standard.

organic status of textiles i.e., from harvesting of the raw materials and responsible manufacturing of environmentally assured and credible end consumer. Since its introduction in 2006, GOTS has significantly achieved popularity due to its practical feasibility and unified processing criteria from the industry and retail sectors dealing with organic and sustainable fibre like cotton, viscose rayon (VR) and others fibre like soybean fibre (SOY), milk fibre, banana fibre and ramie ,etc. GOTS ensures the organic status of textiles from the harvesting of the raw materials through environmentally and socially responsible manufacturing to labelling, to provide credible assurances to the consumer. The standard covers the processing, manufacturing, packaging, labelling, trading and distribution of all textiles made from at least 70% certified natural organic fibres. The final fibre products may include, but are not limited to, yarns, fabrics, clothes and home textiles. However, this standard does not set criteria for leather products (Unsustainable magazine; Futurelearn).

National and regional organic standards bodies use GOTS rather than their individually built textile standards. GOTS is a better standard than any local standard, including the soil association and other national bodies' organic standards.

GOTS focuses on tracing certified organic fibres (cotton and certified wool and silk) through the supply chain from the agricultural field to the final supplier. GOTS has a broad scope covering environmental and social issues in textile supply chains. Brands and suppliers using the GOTS standard include People Tree, Nomads Clothing, Arthur & Henry, Dibella India and Mantis World. The details of it can be found at their official website: www.global-standard.org.

16.2.28 Oeko-Tex

The International Oeko-Tex (Fig. 16.25) Association has been a constant certification agency for testing harmful substances since 1992. They have two certification level as Okeo-Tex level and SA8000 certification level. Boy Wonder, Boo Surfwear, Plummy Fashion Ltd. and Kantala are just some brands using the Oeko-Tex 100 standard.

Figure 16.25 Oeko-Tex.

16.2.28.1 Oeko-Tex Standard 1000

The Oeko-Tex Standard 1000 (Fig. 16.26) is a testing and certification system for environmentally friendly production sites for the whole textile processing chain. Oeko-Tex Standard 1000 ensures the Green labels of the textile fibres and fashion products that is the best-known guarantees that textiles and leathers have been tested for harmful substances such as colorants, heavy metals, formaldehyde and azo dyes. Customers can be confident that a high product safety standard is met and that the garment was manufactured using sustainable processes under socially responsible working conditions. Their site provides a handy buying guide. The details of it can be found at their official website: www.oeko-tex.com.

Figure 16.26 Oeko-Tex standard 1000.

16.2.28.2 SA8000 certified

This is a part of Okeo-Tex standard which is serving as a voluntary social standard certification grounded on the principles of core International Labour Organization (ILO) conventions, United Nations (UN) Conventions and an ISO-style management system.

SA8000 Standard (Social Accountability International) is the world's leading social certification program that proves a company's commitment to social accountability and ethical treatment of employees, including the Universal Declaration of Human Rights. This standard also considers most importantly the continual management improvement and supply chain performance.

SA8000 (Fig. 16.27) is widely applied in all sectors of various industries. It consists of various aspects of Social Accountability International (SAI) aspects in a global format, interests of multi-stakeholder, human rights of the involved workers, and

Figure 16.27 SA 8000.

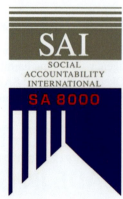

vision and mission of the textile ad fashion companies. The details of it can be found at their official website: www.sa-intl.org. Yabal, Boy Wonder, Darzah, and White Stuff and Ocean Bluu all work with SA8000 certified facilities.

16.2.29 PETA

PETA standard (Fig. 16.28) is related to the PETA-Approved Vegan label to more than 1000 companies in textile, fashion and leather manufacturing field that use vegan alternatives to animal-derived materials such as leather, fur, silk, feathers or bone. The certification label can be used for apparel, accessories, furniture or home decor.

Figure 16.28 PETA standard.

16.3 Fair labour schemes and initiatives

16.3.1 Business Social Compliance (BSCI)

Established in 2002 by the FTA (Foreign Trade Association), the Business Social Compliance Initiative works towards improving the working conditions in the global supply chain for any leading business-driven scheme. BSCI membership can assist retail, brand, trading and importing companies to meet their progress towards social compliance-related goals. More details can be found at their official website: www.bsci-intl.org/.

16.3.2 Fair Wear Foundation

Fair Wear Foundation is an independent verification initiative that works for the garment industries to improve labour conditions inside the company's existing supply chain. This also ensures FWF codes of conduct for labour practices. More details can be found at their official website: www.fairwear.org/.

16.3.3 Fair Labor Association (FLA)

The Fair Labor Association (FLA) is a nonprofit organization dedicated to labour issues in manufacturing factories worldwide. The FLA has its own FLA Workplace Code of Conduct and the textile and garment companies under that need to abide this code of conduct compliance. More details can be found at their official website: www.fairlabor.org/.

16.3.4 Textile Exchange (TE)

TE (Fig. 16.29) is also known as Organic Exchange (OE). It is a non-profit, member-based organization dedicated to expanding the organic cotton market with a recent strategic shift to include other sustainable textiles. TE closely monitors and works as a vital certification agency in various areas such as organic farming engagement and public education. TE provides exclusive benefits to its members who are engaged in the whole global textile supply chain to achieve better idea exchange inside the textile and fashion trade. More details can be found at their official website: www.textileexchange.org/.

The TE offers standard reference for textile and garment manufacturers about which materials are sustainable and maintain a database detailing the 17 United Nations Sustainable Development Goals (UNSDG) and tracing the progress toward reaching these

Figure 16.29 Textile Exchange.

goals (Futurelearn; Considerate; Poole et al., 2009; Gullingsrud, 2017; Lee, 2017). TE offers the following wide range of certifications such as

- Global Recycled Standard (GRS)
- Content Claim Standard (CCS)
- Organic Content Standard (OCS)
- Recycled Claim Standard (RCS)
- Responsible Down Standard (RDS)
- Responsible Mohair Standard (RMS)
- Responsible Wool Standard (RWS)

16.3.4.1 Organic Content Standard (OCS)

Based on the TE aided 'Content Claim Standard', the OCS (Fig. 16.30) is a sustainability-related standard that verifies the final product's organic content. Their report contains the accurate amount of a given organically grown material in the final end-customer product. It thoroughly analyses and thus along the organizational supply chain which takes sufficient steps to assure that the integrity and identity of the final product are ensured. The OCS does not cover the raw material certification, chemical processing inputs (e.g., chemicals), environmental management, health and safety, or labour rights issues. Therefore, it uses these inputs from a separately verified source or party. The various brands that use this standard includes Continental Clothing, Mango, Pact and C&A.

Figure 16.30 Organic content standard.

16.3.4.2 Responsible Wool Standard (RWS)

The RWS (Fig. 16.31) offers the chance to the farmers involved in sheep farming, to show their best practices to the public. This initiative aims to elevate the brands and

Figure 16.31 Responsible wool standard.

consumers and assure that the wool products they buy and sell are in line with best values.

The RWS is a voluntary global standard that illustrates the welfare of sheep and the land they graze on and it also provides verification of the practices that are happening at sheep farm level. Thus it helps to highlight the brands with claims about their wool sourcing with confidence. An international working group developed this standard through an open and transparent process. CU was very much involved in this RWS standard development at the early stages as the technical group. Then this standard became very favourite all over the world.

The RWS evolves from the TE, and this provides a voluntary global standard that addresses the welfare of sheep and the land they graze on. This also discusses the wool production to the marketing supply chain, that is, chain of custody from farm to final product. The final consumers can fully be confident in the RWS logo while pondering the aforesaid facts. The known RWS standard brands include Eileen Fisher, H&M, Patagonia and Esprit.

16.3.4.3 Responsible Down Standard (RDS)

The textile industry has always faced the biggest challenge in raw material sourcing, especially in sourcing natural fibres and materials such as feather and down, wool, angora, cashmere and leather. RDS (Fig. 16.32) serves as a leading standard

Figure 16.32 Responsible Down Standard.

for animal welfare in down, leather products and all animal products included in textiles.

RDS ensures the continuous improvement of best animal welfare practices in the down industry and the wide stakeholder review process, which involves worldwide down supply chain members and major animal welfare organizations. The scope of the RDS includes the full down supply chain from the farms and slaughter facilities (animal welfare) to the garment factories (traceability).

16.3.4.4 Recycled Claim Standard (RCS-100)

The RCS-100 (Recycled Claim Standard) (Fig. 16.33) is generally used for Materials Traceability Working Group as a part of OIA's Sustainability Working Group. This standard is also utilized to track recycled raw materials through the supply chain.

The RCS-100 verifies the content and quantity of recycled material in a final product by applying and analysing through input and chain-of-custody verification via a third party. This also verifies the transparency, consistency and comprehensiveness and evaluates the recycled raw materials content claims on the products. RCS-100 can also be utilized as a business-to-business tool to control the right selling quality products and get the right payment for that. It also ensures apt and honest communication with consumers. RCS-100 standard is widely utilized in textile sectors, including ginning, spinning, weaving and knitting, dyeing, printing and stitching.

Figure 16.33 Recycled Claim Standard.

16.3.5 Organic Content Standard (OCS-100)

The OCS (Fig. 16.34) is utilized for any non-food product containing 95%−100% organic material. It verifies the content of organic materials in the final end-consumer product and also traces out the production processes and system from the raw material sources to the making of the final product.

A third-party verification does the OCS 100 to verify and confirm the final product holds the right percentage of organic content or not. It is a transparent, comprehensive and consistent evaluation applied to textile and fashion materials, ranging from processing, manufacturing, packaging, labelling, trading and distribution of a product

Figure 16.34 Organic Content Standard.

having a minimum of 95% certified organic content. It can be used as a business tool to ensure the selling of organic product and getting a reasonable payment for them.

16.3.6 OCS blended – Organic Content Standard

The OCS blended standard (Fig. 16.35) is a third-party verification to confirm whether a final product is made from a minimum 5% of organic content or not. This is applied to any blended product apart from the food products. In textile and fashion industries, the standard is used to define the quality of the blended fibre and fabric while ensuring a minimum presence of 5% organic content. Like the OCS-100 standard, it is just a derivative of this standard and it is also a transparent, comprehensive and consistent evaluation and verification which covers various sectors like processing, manufacturing, packaging, labelling, trading and distribution of a textile and fashion garment containing at least 5% of certified organic content.

Figure 16.35 Organic blended content standard.

16.3.7 IVN – Naturtextil

The organisation IVN (Internationale Verband der Naturtextilwirtschaft) (Fig. 16.36) was founded in 1999 by two quality seals named as Naturtextil IVN certified BEST

Organizations and certifications relating to sustainable fibres

Figure 16.36 IVN standard.

and Naturleder IVN certified. These two seals are responsible for safeguarding and reviewing the overall textile production chain related to its social standards and ecological accountability.

The IVN standard assists in evaluating and verifying the ecological quality and encourages the eco-friendly textile and leather products. Certified textiles may carry the IVN certified or IVN certified BEST label. These certifications can certainly carry a greater sense of confidence to the end-customer and make the company feel more legally confident in their quality-related claims for their products. This certification is used as a premium certification ranking after the globally accepted GOTS. It also addresses all production stages from raw material to sale and the use of the finished leather (not the finished leather products), such as in maintaining the standard for eco-friendly leather products in Europe.

16.3.8 Content Claim Standard (CCS)

The CCS (Content Claim Standard) (Fig. 16.37) offers companies a tool to verify the amount and the quality of specific input materials. An independent third party checks every organization along with its supply chain and all the type of input materials have

Figure 16.37 Content Claim Standard.

been checked by this third party to verify all the claims declared by the company. The CCS is the foundation for all of TE's chain of custody standards.

This standard is widely introduced in most European countries-based textile and garment industries, including ginning, spinning, weaving and knitting, dyeing and printing and stitching. CCS can be used as a token of appreciation and foundation for content claim standards developed around specific raw materials in fashion and textile production.

16.4 Examples of sustainable textile fibres and fabric materials

16.4.1 Repreve

Repreveis the leading branded performance fibre manufactured from recycled materials like plastic bottles. This fibre has excellent wicking, adaptive warming and cooling, water repellency properties and it also emits less greenhouse gases during the manufacturing process.

16.4.2 Supima

Supima Cotton is a highly sustainable, recyclable, soft and durable long-staple cotton, and it is just less than 1% in production in the USA. It makes up less than 1% of all cotton grown in the United States.

16.4.3 Refibra

Refibra (Fig. 16.38) is an upcycled cotton scarps-made fibre used in garment production. It is a respuned cotton fibre material.

Figure 16.38 Refibra.

16.4.4 ZQ Merino Wool

ZQ Merino Wool (Fig. 16.39) is the globally leading sustainable wool brand made from recycled merino sheep from Australia.

Figure 16.39 Zq merino wool.

16.4.5 Cocona

Cocona (Fig. 16.40) is a natural fabric with enhanced performance made from activated carbon. This type of fibre is created from the waste coconut husks produced in the water filter industry. This fibre made fabric resist moisture, shield ultra-violet (UV) radiation and odour control.

Figure 16.40 Cocona.

16.4.6 Cupro

Cupro (cuprammonium rayon) (Fig. 16.41) is used in the cotton industry from the waste from the textile cotton industry such as cotton linter which regenerates cellulose fibre by reacting solution of copper and ammonia to dissolve bamboo wood cellulose. This type of cupro-fibre-based fabric is breathable and can control body temperature and along with that it also drapes finely like silk and has silky feel. This type of material can be dyed or strained and blends with other natural fibre very easily.

Figure 16.41 Cupro (cuprammonium rayon).

16.4.7 Econyl

Econyl (Fig. 16.42) fibre is made from synthetic wastes such as industrial plastic, waste fabric and fishing nets collected from oceans. They are then recycled and regenerated into new nylon yarn by using closed-loop process, which reduces water usage and waste.

Figure 16.42 Econyl.

16.4.8 EcoVero

EcoVero (Fig. 16.43) is another important sustainable fibre developed by Lenzing. This fibre is derived from wood sources with the help of highly environmental standards. During the making process, all the chemicals are used in -loop processes that take 50% less energy and water consumption than the manufacturing of standard VR (Certifications; Unsustainablemagazine; Futurelearn; Considerate).

Figure 16.43 EcoVero.

16.4.9 Filium

Filium is a patented process for making natural and sustainable fibres such as cotton, silk, wool, linen and also makes these fibres water-resistant, stain-resistant and odor-resistant. The fabric and cloths made of Filium need less attention in laundering. It also significantly reduces carbon emissions and pollution as it leeches almost zero nanoparticles, harmful emission and carbon emission and pollution to human skin and the surrounding environment.

16.4.10 Luxe

Luxe (Fig. 16.44) is sustainable and extra-long-staple cotton found in the USA.

Figure 16.44 Luxe.

16.4.11 Viscose

VR is a sustainable natural cellulosic fibre or yarn made from wood pulp, bamboo and eucalyptus stem. The help makes VR of chemical solution, xanthation and dipping into caustic soda. Then it spun from the coagulation bath in spun yarn. VR is a sustainable and non-toxic fibre made by closed-looped process having less environmental hazards.

16.4.11.1 Type of sustainable VR yarns

Lyocell (Fig. 16.45) is manufactured from recyclable bamboo with eco-friendly solvents.

Modal (Fig. 16.46) is made from beech tree pulp and finally forms rayon fibre.

Tencel (Fig. 16.47) is made from sustainably harvested eucalyptus wood pulp using recyclable and earth-friendly solvents.

Figure 16.45 Lyocell.

Figure 16.46 Modal.

Figure 16.47 Tencel.

16.4.12 Tasc

Tasc is a performance fabric manufactured from the primary bamboo viscose. This type of fibre can control moisture and heat and reduce created from bamboo viscose. It prevents humidity and heat, is able to reduce odour and highly UV protective with an ultra-violet protective factor (UPF) = 50+ and at the same time supremely sustainable in nature.

16.4.13 Pinatex

Pinatex (Fig. 16.48) is a leather-like material manufactured from the waste leaves of the pineapple plants. It is breathable and durable fibre and can easily substitute the leather and petroleum-based faux leathers. Pinatex fibre is made from Ananas Anam company in the Philippines, for example, and it is again made by following a closed-looped manufacturing process. This fibre helps the farmers with a notable secondary earning source (Certifications; Unsustainable magazine; Futurelearn; Considerate).

Figure 16.48 Pinatex.

16.5 Conclusion

In this chapter, we have discussed the various organizations and standards to manage sustainability in supply chains in textile, fashion and garment industries. These standards play many pivotal roles in maintaining the environmental, social, product planning and manufacturing process. Many of these initiatives and schemes are well established in the global fashion and textile industries to lift up the poor working and sustainability-related status of the customers and it also already helped many start-up to manage their business sustainability line more efficiently by assisting them in establishing the brand identity and brand positioning (Pandit et al., 2020; Singha et al., 2020).

References

Anandjiwala, R.D., 2007. Textiles for Sustainable Development. Nova Publishers, USA.

Annapoorani, S.G., 2018. Sustainable textile fibers. In: Sustainable Innovations in Textile Fibres. Springer, Singapore, pp. 1—30.

Certifications. https://certifications.controlunion.com/en/industries/textiles/. downloaded on 06.06.2021, at 6.30 pm IST.

Chavan, R.B., 2014. Environmental sustainability through textile recycling. Journal of Textile Science and Engineering 2, 2.

Considerate. https://www.considerate-consumer.com/certified-fashion, downloaded on 06.06.2021, at 6.30 pm IST.

Futurelearn. https://www.futurelearn.com/info/courses/sustainable-fashion/0/steps/13562, downloaded on 06.06.2021, at 6.30 pm IST.

Gullingsrud, A., 2017. Fashion Fibers: Designing for Sustainability. Bloomsbury Publishing USA.

Lee, K.E., 2017. Environmental Sustainability in the textile industry. In: Sustainability in the Textile Industry. Springer, Singapore, pp. 17—55.

Levering, R., Vos, B., 2019. Organizational drivers and barriers to circular supply chain operations. In: Operations Management and Sustainability. Palgrave Macmillan, Cham, pp. 43—66.

Maia, L.C., Alves, A.C., Leao, C.P., 2013. Sustainable work environment with lean production in textile and clothing industry. International Journal of Industrial Engineering and Management 4 (3), 183—190.

Pandit, P., Singha, K., Kumar, L., Shrivastava, S., Yashraj, V., 2020. Business paradigm shifting: opportunities in the 21st century on fashion from recycling and upcycling. In: Recycling from Waste in Fashion and Textiles: A Sustainable and Circular Economic Approach, pp. 151—176.

Periyasamy, A.P., Militky, J., 2020. Sustainability in regenerated textile fibers. In: Sustainability in the Textile and Apparel Industries. Springer, Cham, pp. 63—95.

Poole, A.J., Church, J.S., Huson, M.G., 2009. Environmentally sustainable fibers from regenerated protein. Biomacromolecules 10 (1), 1—8.

Rani, H., Saha, G., 2021. Organizations and standards related to textile and fashion waste management and Sustainability. In: Waste Management in the Fashion and Textile Industries. Woodhead Publishing, pp. 173—196.

Seisl, S., Hengstmann, R., 2021. Manmade cellulosic fibers (MMCF)—a historical introduction and existing solutions to a more sustainable production. In: Sustainable Textile and Fashion Value Chains. Springer, Cham, pp. 3—22.

Singha, K., Pandit, P., Maity, S., Srivasatava, R., Kumar, J., 2020. Sustainable strategies from waste for fashion and textile. In: Recycling from Waste in Fashion and Textiles: A Sustainable and Circular Economic Approach, pp. 199—214.

Sustainability Certifications: Textile Companies Jumping Into the Wagon, Fibre2fashion. https://www.fibre2fashion.com/industry-article/6352/sustainability-certifications-textile-companies-jumping-into-the-wagon, downloaded on 06.06.2021, at 6.30 pm IST.

Tobler-Rohr, M.I., 2011. Handbook of Sustainable Textile Production. Elsevier, USA.

Unsustainablemagazine. https://www.unsustainablemagazine.com/sustainable-fashion-a-list-of-organizations-certifications-terms-and-fabrics/, downloaded on 06.06.2021, at 6.30 pm IST.

Vadicherla, T., Saravanan, D., 2014. Textiles and apparel development using recycled and reclaimed fibers. In: Roadmap to Sustainable Textiles and Clothing. Springer, Singapore, pp. 139–160.

Wong, D.T., Ngai, E.W., 2021. Economic, organizational, and environmental capabilities for business sustainability competence: Findings from case studies in the fashion business. Journal of Business Research 126, 440–471.

Challenges and future directions in sustainable textile materials

Lebo Maduna[1] and Asis Patnaik[2]
[1]Technology Station: Clothing and Textiles, Faculty of Engineering and the Built Environment, Cape Peninsula University of Technology, Cape Town, Western Cape, South Africa; [2]Department of Clothing and Textile Technology, Faculty of Engineering and the Built Environment, Cape Peninsula University of Technology, Cape Town, Western Cape, South Africa

17.1 Introduction

The demand for new clothing and disposal of clothing is increasing, and it is made worse by the fast fashion products which in most cases are not reused or recycled. This creates an enormous amount of waste that predominantly ends up in landfills or incineration facilities. Land space is a limited resource and the incineration process generates toxic chemicals and residues. Consumers must be made aware that disposing of fashion waste has a negative impact on the environment. When the demand for a particular product grows rapidly what is also seen is an increase in the environmental impact of the products as the majority of them end up in landfills. In order to address this, the circular economy concept is being suggested. It promotes reduce, recycle, reuse, repair and remanufacture of goods so that the products or waste are circulated in a closed loop in order to reduce the need for new materials and products and the impact on the environment is reduced.

Traditional linear model is to dispose of products or generate waste. Major barrier in the reduce, recycle, reuse, repair and remanufacture of post-consumer waste is the composition of the waste. Blending causes difficulties in the separation of fibres. Where chemicals are used the mechanical and chemical properties of the recovered fibres are affected. Since the quality of the finish is influenced by fibre properties if damaged fibres are used as a source of new raw materials the properties of the final products will be affected. This compromises the quality of the finished products. However, for fibres which are recovered by the enzymatic treatment, less fibre damage has been reported making them an ideal source of new raw materials to make new products (Dissanayake and Weerasinghe, 2021; Matthes et al., 2021; Moran et al., 2021). This chapter provides an overview of the various factors that have an influence on sustainability and environmental protection. It includes clothing production, consumer behaviour, raw materials (fibres), dyeing, recycling and ecolabelling. In addition, the future direction is also highlighted.

17.2 Clothing production

The global clothing consumption has been increasing rapidly and between 2000 and 2015 it doubled from 50 billion units to 100 billion units. It is being fuelled by fast fashion, online shopping and over-consumption lead to rapid disposal of clothing materials. Washing clothing materials during the home laundry process can release up to 700,000 micro-plastics into water sources and sewage water treatment facilities are not capable of removing them. The environmental impact of the fashion industry is huge as it is the third biggest manufacturing industry behind the automotive and technology industries which occupy the number one and two positions. The industry is estimated to be about $1.3 trillion (Wu and Li, 2019).

Production of clothing has been increasing rapidly due to the advancements in the design, materials and production processes. The product costs that were too high in the past are no longer and the savings are passed to the consumers resulting in products becoming more affordable to the majority of the consumers who previously could not afford them. While these achievements are good for consumers and companies, there is a hidden negative impact on the environment that is associated with these achievements as the textile industry is up there with big polluters when it comes to consumption of water, chemicals and energy. In order to mitigate this, companies are encouraged to adopt cleaner production technologies that improve production efficiency while at the same reducing their environmental impact. Using a combination of various technological innovations in the reduction, recovery, recycling and monitoring of waste and pollutants can improve the production efficiency while at the same time protecting the environment (Dissanayake and Weerasinghe, 2021; Ion et al., 2021; Navone et al., 2020; Shi et al., 2021).

The negative effect of the global production and consumption of products on the environment is not only seen in places or countries where products are being produced as they are seen in faraway places or countries due to the inter-connectedness of the global economy and ecology (Severo et al., 2015).

17.3 Consumer behaviour

Environmental laws and regulations play an important role in reducing pollution but on their own they are limited as other factors such as social norms and attitudes influence the consumer behaviour. Consumers with a limited knowledge about the environment cannot see how their current unsustainable over consumption and disposal of products lead to depletion of resources and damage to the environment and are unlikely to change their behaviour. This is made worse by the culture of fast fashion consumerism. USA, Europe and China are some of the places with the highest disposal of textile waste (Chang, 2020; Shafie et al., 2021). Overconsumption is more prevalent in young consumers than in adults.

Self-appearance also influences people's behaviour as people who compare themselves to others tend to buy more even though they already have more. Change in the consumer behaviour is required to make consumption sustainable in order to reduce the use of raw materials and release of toxic pollutants that damage the environment. It is estimated that about 40% of the damage caused to the environment is due to the consumer purchasing behaviour and if consumers can start to think about sustainability the impact on the environment can be reduced.

Some consumers are aware of the benefits of sustainability; however, they do not align their purchasing with the knowledge they have. Some of the factors that have been reported to be responsible for this behaviour are attitude, spirituality and perceived marketplace purchase. Research and targeted marketing is needed to understand social cultural issues that inhibit sustainable purchasing behaviour in order to find better ways to encourage and reinforce sustainable purchasing behaviour. Individuals who believe their contributions can make a difference in the society and environment need to be encouraged.

Attitudes influence intention, a consumer with a positive attitude towards sustainability issues is more willing to change his/her behaviour and adopt eco-friendly products when purchasing products. Consumers that believe their sustainable purchasing behaviour has an influence on others and organizations will be more likely to maintain their sustainable behaviour. Spirituality influences an individual purchasing behaviour as it is associated with caring for oneself, others and environment. This can lead to sustainable purchasing by spiritual individuals. Age has also an influence as educated young consumers tend to be more willing to learn and adopt sustainable purchasing behaviour (Joshi and Rahman, 2019; Shafie et al., 2021; Yan et al., 2021).

For some consumers, goods are regarded as a societal status symbol and can make them have a feeling of belonging to a certain class. Deprivation of goods can drive this behaviour. Individuals who have been economically deprived tend to be more obsessed with material goods and when they overcome poverty and have the financial resources to purchase goods, there is a need to show their acquired wealth by purchasing more goods (Mainolfi, 2020).

When consumers buy products, they express their concerns about sustainability. Often when consumers purchase products their impatience, impulsivity, and no self-control drive will drive the consumer to consider only their immediate satisfactions with no regard for the future outcomes of those products on the environment. Being aware of the advantages and disadvantages of something does not necessarily mean one will act for the benefit of him/herself. The case in point is an individual who is aware of the benefits of exercising but chooses the immediate benefits of sitting and relaxing (Calderon-Monge et al., 2020; Mahmoodi et al., 2021).

Some consumers are aware of the environmental impact of the products that they want to buy and are not aware of the environmental impact of the manufacturing processes and transportation of the final products to the shops. Environmental issues include the use of raw materials, water and energy, whereas social issues include human rights and opportunities in a sustainable economy that generates profits. In a sustainable environmental space, there are different competing groups that are looking for different issues such as environmental, social and economic issues. In order to achieve sustainability, all the groups should put aside their differences and work together. Often interest groups tend to focus only on their area of interests and ignore the interests of others.

Some groups focus only on the environment issues and ignore the social and economic issues and if sustainability is to be achieved it must address these issues.

Management, workers, consumers and shareholders can work together and come up with better ways of addressing economic and consumption issues that have a negative effect on sustainability. Social consumers know more about societal issues and can be used to persuade their communities to buy sustainable products from businesses that address societal issues. NGOs and online platforms can be used for consumer awareness and promotion of sustainable lifestyles that promote fashion and textiles sustainability (Calderon-Monge et al., 2020; Provin et al., 2021).

17.4 Sustainability approach

The fast-fashion products tend to have a short life cycle, and it has to increase competition. Products need to be made quickly and shipped. It has led to challenges on how to manage waste and whether this is sustainable? Recycling and reuse of clothing products are very low with only about 20% meaning that 80% end up in landfills and incineration facilities. In 2015, it was reported that each person in Europe brought about 13 kg of clothing (Provin et al., 2021; Shafie et al., 2021). A developing country like Bangladesh which has an established textile industry is estimated to generate 2.9 million metric tons of textile waste by the year 2021.

The fashion and textile industry is an intensive user of water-consuming about 79 billion cubic meters annually and it is also one of the biggest polluters due to chemicals used in processing and finishing of products (Niinimaki et al., 2020; Provin et al., 2021). The energy is also required to run production lines. There have been suggestions that the industry needs a rethink and look at the negative effects of the materials it uses. Where possible, it must use new materials which are sustainable and have less impact on the environment (Provin et al., 2021; Tshifularo and Patnaik, 2020).

The current approach of disposing products in a linear manner is not sustainable for the future of the planet and as such circulating waste or products should be adopted. It can be achieved by reduce, reuse, recycle, upcycling and redesigning products to reduce waste, energy and toxic emissions. A concept which incorporates these approaches is called the circular economy model. It emphasizes the need to redesign, reduce, recycle and reuse products, materials or waste that end up in landfills and incineration plants and the energy consumption and toxic emissions such as mercury, lead and nickel will be reduced.

Sometimes it happens that a circular loop is open and no longer closed, such as when waste is not recovered when it ends up in landfills. In 2015, the textile industry was responsible for about 1.2 billion tons of CO_2 emissions, whereas this value is now significantly increased between 1.39 and 3.29 gigatons in 2019 (Schragger, 2021). Fig. 17.1 shows how materials, design, consumption and end of life contribute towards achieving sustainability using a circular loop of a circular economy model (Dissanayake and Weerasinghe, 2021). However, Fig. 17.2 shows the flow of fashion products can be circular and linear.

Challenges and future directions in sustainable textile materials

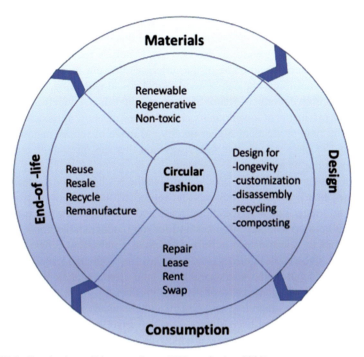

Figure 17.1 Circular loop (Dissanayake and Weerasinghe, 2021).

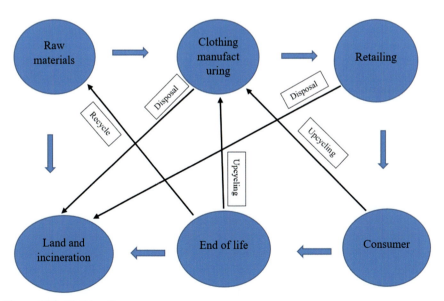

Figure 17.2 Clothing flow.

In a linear flow, the products are disposed of without being used or recycled. Reuse extends the life term of products and reduces waste that ends up in landfills and incineration facilities. Reusing waste to produce raw materials reduces the need for new sources of raw materials and some of the processes that are involved in making the fibres. Recycled fashion waste can be used to make mattress filling, insulation, packaging and automotive products. Reuse of textile products is not widely accepted due to belief that the product is of lower quality and it has led to a new approach called upcycling whereby new products of higher quality are made of reused materials. Reused products tend to have colours that fade away and this can discourage consumers as the colour will not be appealing.

Upcycling requires skills and time to improve the aesthetics and persuasion of the public to accept the product. Recycled and reused waste or products reach a stage where it is no longer viable to recycle or reuse and when such a stage is reached the products can be used as energy sources and for the biodegradable fibres they can be used to produce compost instead of burning them as combustion generates toxic gasses. Redesign involves designing fashion and textile products in such a way that it takes into consideration the sustainability and environmental impact of materials and each production stage and that of the final product.

The use of nontoxic materials, less energy and chemicals is considered during the design. Upcycling has also been suggested which involves forming a circular loop where waste is recycled to make new products. Waste is collected, recycled and processed to make new products which are bought by consumers and when the product is disposed of after use it is recycled again. In this approach, no waste is taken to landfills or incineration facilities. Of the polyester waste that can be recycled, it is possible to use about 72% to make fibres. Fibres made from contaminated waste suffer from breakages and the aesthetics of the products are not good. Eco-labelling is important to make consumers aware that fashion and textile brands comply with environmental standards (Cai and Choi, 2020; Dissanayake and Weerasinghe, 2021; Guo et al., 2021; Herva et al., 2011; Jain et al., 2021; Park and Kim, 2014; Provin et al., 2021; Yousef et al., 2020).

Integration of sustainability into the strategy and at all levels of organization is important. Organizations should begin a process of switching from unsustainable design models to sustainable models such as circular approach design which is sustainable. Management support is important, and there should be clear communication and awareness about the importance of sustainability within the organization and the broader society. There must be a budget for training and upskilling of designers in the knowledge of fibre materials, chemicals, processes and alternatives that are available to them that they can use. Designers tend to focus on the materials neglecting other factors that also have an influence on sustainability. Sustainability costs can be managed by collaborating with other designers, researchers, suppliers, marketers and recyclers (Claxton and Kent, 2020; Islam et al., 2021; Tshifularo and Patnaik, 2020; Cai and Choi, 2020).

Some of the big retailers have made commitments to adopt sustainable supply chain management by sourcing products from suppliers who use sustainable designs, materials and production processes to make products. Eco-labelling a product that was made from organic cotton or recycled materials can get the attention of some of the consumers with regard to its eco-friendliness when they do their purchasing (Cai and Choi, 2020).

17.5 Recycling

After several reuse, a product is damaged beyond recovery and reaches a stage where it can no longer be used. An exposure to heat, oxygen, ultraviolet radiation (UV) and mechanical forces can cause damage. Recycling is recommended which leads to sustainable use of the materials and thereby protects the environment (Tshifularo and Patnaik, 2020). Product must be recycled when it can no longer be used for its intended application. Waste can be used where the requirement of quality is not very high. Recycling is divided into chemical, mechanical and thermal recycling. Chemical breaks a polymer into monomers and oligomers. Chemical recycling is expensive due to the high cost of depolymerization, separation and filtration. The process is not cost-effective as a source of new raw materials when it is compared to the conventional petrochemicals. Biodegradable fibres have minimal impact on the environment when they are disposed of (Feghali et al., 2020; Yousef et al., 2020).

Fashion products like clothing tend to have a short life due to the rapid changing fashion trends and products that are no longer trending are disposed of and this has a negative impact on the environment if products are not recycled and reused. The problem differs from country to country. Some of the biggest contributors to this problem are increasing world population, culture and improving living standards which make it possible for people to buy more. The annual textile waste generated by China, USA and UK is about 26, 15.1 and 1.7 million tonnes, respectively (Yousef et al., 2020). The waste ends up in landfills and incineration plants. In some countries, there are initiatives to collect discarded products and recycle them. Collected clothing items are sorted in special sorting centres and those items that are not damaged are recycled and sold as second-hand clothes. Most end up in third-world countries. Sorting of clothing is still a manual process and automation is not feasible due to the mixture of fibres used to make clothing. Technologies are not yet advanced to identify and sort the blended products. Textile products that are damaged and cannot be sold can be used as a source of energy (Yousef et al., 2020; Shirvanimoghaddam et al., 2020).

17.6 Second-hand clothing

Recycling of second-hand clothing plays an important role in minimizing the negative impact of textile waste on the environment. Trading of second-hand clothing has created a market for clothing that would have ended up in landfills or incineration plants in developed countries. Second-hand clothing if it is not properly managed can create environmental problems. Some of the developing countries that accept these goods do not have the capacity to treat them when they are discarded. It is like the developing countries that accept them serve as a dumping ground of textile waste from developed countries. Second-hand clothing should be properly sorted to remove damaged or unwearable clothing in order to reduce the probability of them being discarded. When they arrive, buyers sort them and those that are damaged and not wearable are discarded and they contaminate water and soil (Yousef et al., 2020).

17.7 Fibres

Natural fibres are sustainable as they are from plant sources that can be replaced provided there is enough land and other resources. The fibres are degraded by microorganisms such as bacteria and fungi as well as some pests when exposed to the soil. Synthetic fibres on the other hand unlike natural fibres are not easily degraded and cause huge environmental burden. Cotton and flax (linen) are the most popular natural plant fibres used to make fashion products, even though there are other plant fibres that can be used such as hemp and jute.

Hemp can be used to make fashion products; however, the fabric suffers from wrinkling and the fibres are rough and rigid. The fibres are treated with NaOH, steam and enzymes to improve their wettability. Jute's main application is found in geotextiles, carpets and packaging sack products even though it can be used to make fashion products. Natural fibres can be blended with each other or synthetic fibres to make a variety of fashion and textile products in order to complement each other fibre properties.

Natural fibres have good moisture absorption which is important for the removal of sweat from the skin surface and make the person feel cool and comfortable. Flax has good strength, and it can be used as a reinforcement material in composites and reduce the environmental impact of the composite product when it reaches its end of life.

Chemicals are used to extract the fibres from the plants which makes the harvesting process not environmentally friendly and switching to alternatives such as enzymes is expensive (Debnath, 2017; Matthes et al., 2021). Man-made cellulosic fibres use a chemical process to extract cellulose and regenerate to make cellulosic fibres from plant sources that contain cellulose, mainly wood. Other alternative feedstocks exist such as leaves, banana stalk, citrus peel, cotton linter, pineapple leaves and sugarcane bagasse which can be used in order to protect the sustainability of trees. Using cotton linter as a feedstock instead of hardwood stems is beneficial to the environment since more cellulose can be extracted from cotton linter (80%–95%) than from hardwood stems (40%–55%). Research is needed to come up with safer technologies that can be used to process feedstock that is not from traditional sources (Matthes et al., 2021).

17.7.1 Blended fibres

Blending of fibres, especially natural and synthetic fibres, should be avoided where possible as it makes it difficult to separate fibres for recycling and reuse purposes (Cai and Choi, 2020). The recovery of fibre materials from fashion and textiles products is considered to be too low. One of the main reasons is that might require that the

Challenges and future directions in sustainable textile materials 393

products be separated based on the fibre type and colour. In order to recover polyester fibres from a blended fabric like polycotton, it might require that the fabric be put in a solution that dissolves the polyester fibres. The polyester solution is then filtered and recovered by cooling it in a CO_2 bath and the undissolved cotton fibres are then washed with water and dried.

For jeans, the colour can be used as an initial step in sorting jeans according to the type of dye that was used. Indigo and sulphur dyes are widely used to impart blue and black colours. Textile products contain toxic chemicals which are used in the processing and untreated waste still contains these chemicals and end up contaminating soil and water. People are also affected by the toxic chemicals which cause various health problems. For treatment, the activated carbon can be used to absorb the leached dyes (Navone et al., 2020; Yousef et al., 2020).

17.8 Dyeing

The processing and finishing of textile products require the use of synthetic chemicals which can cause damage to the environment when untreated effluent is released. Dyes which impart colours to textile products are responsible for the majority of the damage. The focus is to come up with new dyes and dyeing technologies that use less chemicals and water. Spin and supercritical dyeing of fibres uses less water and energy. Waterless dyeing technologies reduce dye waste that can be generated compared to traditional methods.

Natural dyes have a low environmental impact and should be used where possible (Cai and Choi, 2020; Penthala et al., 2022). Natural dyes extracted from plants are seen as a sustainable solution to conventional man-made dyes which are not sustainable and are toxic. The market for natural dyes is expected to grow as countries push sustainability agendas and ban some of the toxic chemicals contained in conventional dyes. Bamboo, wool, cotton fibres can be dyed from extracts from herbal leaves, stems, barks, buds and flowers. When dried red chilli was used to dye bamboo, banana and wool fibres the intensity of the colour was good in wool and followed by banana and with less intensity in bamboo fibres. A similar pattern was observed when dried Munros' dessert-mallow was used. Dried walnut leaves and cloves can also be used for these fibres. This dyeing process is reported to use less water and energy (Thakker and Sun, 2021). Eucalyptus leaves can also be used to dye lyocell fabrics (Cai and Choi, 2020).

Traditional methods of dyeing fibres use a lot of water and chemicals and furthermore because of the non-biodegradability and toxicity the affluent must be treated before it is discharged resulting in increased operational costs. This is not sustainable considering the environmental and health problems and as such alternative waterless methods like supercritical carbon dioxide are being considered as they cause minimal damage compared to traditional methods which use water.

Carbon dioxide is non-flammable and recyclable (Penthala et al., 2022; Saleem et al., 2020). The carbon dioxide can be changed into supercritical liquid by pressurizing it at high pressure and low temperature replacing the need for water. Once the job of cleaning or dyeing is completed, the chamber is depressurized and the liquid turns into gas which can be recovered. Releasing the gas to the environment causes minimal damage compared to other traditional methods. The disadvantage of the liquid carbon dioxide is that it requires high pressures to be maintained during operation and in addition the initial investment cost is high. Supercritical carbon dioxide can also be used for dry cleaning.

Traditional dry-cleaning solvents such as carbon tetrachloride and perchloroethylene pose environmental and health hazards and alternatives such as carbon dioxide and a hydrocarbon called Stoddard can be used which are less harmful. Because carbon dioxide liquid has low viscosity and surface tension, it easily penetrates the fabric products and transports the soil particles that are removed by detergents from the fabrics (Troynikov et al., 2016). Another environmentally friendly dyeing method is spin dyeing which involves dyeing the polymer solution which is going to be used to make fibres. Dyeing pigments which can withstand the higher temperatures and spinning conditions are added. The technique can be used to dye synthetic fibres and is more eco-friendly than traditional methods as it uses less water and chemicals (Liu et al., 2012; Terinte et al., 2014).

17.9 Recycling methods

When a product reaches its end of life, it is recommended that it be recycled which leads to sustainable use of the materials (Tshifularo and Patnaik, 2020). Sustainability of the method is important in minimizing its environmental impact. For a recycling technology to be adopted as a sustainable technology, it must be cost-effective while at the same taking into consideration its own negative impact on the environment. Recycling techniques that can be used are mechanical, thermal, chemical and biological methods. Different techniques have different capabilities and costs associated with them as shown in Table 17.1.

Activated carbon fibres can also be used to treat textile wastewater containing toxic chemicals. Mechanical methods use mechanical parts to cut the waste into smaller pieces which can be used as insulation materials and for reinforcement of concrete or composites. Mechanical methods do not have the capability to separate and process non-homogenous waste. Often the products made from mechanical recycled waste are

Table 17.1 Recycling methods.

Method	Waste separation	Cost
Mechanical	Homogenous waste	Cheap
Thermal	Homogenous waste	Less expensive
Chemical	Non-homogenous waste	More expensive
Biological	Non-homogenous waste	More expensive

of not high quality. Thermal method uses heat to melt the waste and extrudes it to produce new fibres. It is capable of processing homogenous waste just like the mechanical method. It is simple and cost-effective to operate. Chemical methods can be used to digest the fibres. Digested cellulosic fibres such as cotton can be used as feedstock for biofuel production. Biological methods make use of biological processes to degrade fibres. Microorganisms are used to degrade the fibres (Yousef et al., 2020). The cost and energy consumption are the most important factors taken into consideration when one decides which method to recycle the fibres.

17.9.1 Thermal method

Thermoplastic polymers can be melted and used as a source of raw materials to make fibre filaments; however, due to other contaminants, the fibres can be susceptible to breakages. Products tend to have poor dyeability due to the formation of oligomers and the fibres are also susceptible to yellowing due to the intramolecular cross-linking and oxidation. Thermal and hydrolysis change the molecular weight of the fibres. In order to improve the quality of products, a new feedstock is mixed with the recycled polymers.

17.9.2 Chemical method

Chemical recycling breaks down a polymer waste into its monomers or oligomers which are then transformed into finished fibres. The fibres are of better quality than those produced using thermal methods. When looking at the raw materials used to produce the fibres, the cost of the chemical recycling can be higher than that of fibres produced from original petrochemical pellets but when one looks at cost of producing the polymer pellets the cost of petrochemical becomes higher than that of recycled fibres due to the high consumption of energy during the transformation of petrochemicals to polymer pellets.

Further, higher consumption of energy has a bigger impact on the environment. Even though chemical recycling is more expensive than thermal recycling, it is considered to be a more sustainable method due to its ability to produce raw materials that have superior quality. Recycled polyester can be used as a starting raw material for polyurethane synthesis which can be used to make insulation, seating and artificial leather materials. Various chemical processes that can be used to depolymerized recycled polyester fibres are glycolysis, hydrolysis, methanolysis and aminolysis.

The glycolysis method has a lower operating cost, and it uses catalysts such as zinc, lead, cobalt and manganese salts. Glycolysis is widely used on a commercial scale to process recycled polymer waste. The polyester waste is degraded by glycols such as ethylene, diethylene, triethylene or propylene into bis (hydroxyethyl) terephthalate (BHET) and oligomers. The process can be incorporated into existing manufacturing line processes. Catalysts are used to increase the rate of the reaction. Yields up to 90% can be reached in 30 min.

Hydrolysis uses acids such as sulphuric or nitric acid and alkalis such as NaOH at high pressure (1.4−2 MPa) and temperature (200−250°C) to hydrolyse polymer into terephthalic acid (TPA) and ethylene glycol. Alkaline hydrolysis requires less pressure than acid hydrolysis and in addition the final polyester appears colourless which has been attributed to the ability of the process to oxide the dyes. The main disadvantage is the presence of contaminants in the final product.

Methanolysis uses alcohol to degrade waste polyester into dimethyl terephthalate (DMT) and ethylene glycol. This recycling can be used to treat low-quality contaminated recycled waste and the contaminated DMT which is produced is then purified by crystallization and distillation. Low-quality feedstock compensates for its high operating cost.

Aminolysis is not commercially available to be used for recycling. It uses amines such as allylamine, morpholine, hydrazine and polyamines to degrade polyester into terephthalamide (Guo et al., 2021; Park and Kim, 2014).

17.9.3   Biological method

Biological techniques make use of biological organisms and products. Biological degradation of textile waste using enzymes such as cellulase and β-glucosidase is time-consuming and requires that conditions be maintained at a certain temperature and pH. Using the biological process takes a long time and the product yield is not too high. It can take many days and the end product contains a residue. It can take up to 72 h for the enzymes to degrade the waste in order to recover polyester which is not degraded by the enzyme. The temperature is maintained at 45°C and pH at 4.8. Yeasts called Saccharomyces cerevisiae can degrade the cellulosic part of the waste which can be used as a feedstock in the production of ethanol fuel. The biological method can also degrade polyester; however, the cost is too high and the process is complicated to be used on a large industrial scale (Guo et al., 2021; Gholamzad et al., 2014; Hussein et al., 2019; Yousef et al., 2020).

17.9.4   Other techniques

Methods such as energy recovery and activated carbon fibres are not widely used. The energy recovery recycling uses waste fibre materials that cannot be reused or upcycled to make new products. The waste is used to generate heat energy that is used to power the incineration plant. Traditional way is to burn the waste without recovering the energy. Activated carbon fibres can be used to absorb toxic heavy metals contained in clothing waste. Heavy metals are toxic to the environment and humans. They cause health problems such as hypertension, renal and lung damage.

Reducing the presence of the metal in textile waste is important in reducing their negative impact on the environment, plants, animals and human beings. For the treatment of textile wastewater, the fibrous absorbent materials can be used to absorb toxic waste chemicals. The surface area of the sorbent is functionalized to attract the chemicals. Activated carbon fibres have porous structures with large surface area making them good materials for the adsorption of toxic chemicals. The process is influenced by the pH and different adsorbent materials having different optimal pH levels where the maximum sorption can be achieved.

Some of the reused sorbents can maintain their efficiency even after three cycles of sorption-desorption (Bediako et al., 2016; Silva et al., 2018). Fine fibres have high sorption. Textile waste which contains cotton, viscose, linen and acrylic fibre can be used to produce activated carbon fibres. Denims waste can be used to produce activated carbon fibres. One approach to avoiding the contamination is to extract the heavy metals from the fabric before the recovery of fibres. Heavy metals present in fashion and textiles products have been found in the biogas where the feedstock was the source (Silva et al., 2018; Yousef et al., 2020).

17.10 Ecolabel

Ecolabel is an assurance tool that consumers can use to classify whether textile and fashion products comply with environmental standards (Clancy et al., 2015; Herva et al., 2011). Ecolabelling can be used as a sustainability tool that enhances sustainable production and consumption of products. The market share of the ecolabel is still relatively small compared to the conventional non-ecolabel. Sustained growth has only been in the food sector with other sectors either experiencing a decline or no growth. The poor performance has been attributed to various factors such as cost of products, poor marketing and education, political and environmental concerns. Often consumers are not willing to pay more for the ecolabel products.

The mere fact that a product has an eco-labelling does not necessarily mean it is more eco-friendly than non-labelled because some of the important factors considered in certifying whether a product is eco-friendly are its toxicity, resources consumption, biodegradability and packaging. These have been exploited by some manufacturers to use only what they want and exclude the rest. This has caused the public to doubt the claims of ecolabel products (Rex and Baumann, 2007; Saouter et al., 2018).

17.11 Future direction

As more communities, nongovernmental organizations and governments push for more sustainable products and processes, the traditional norms are going to be discarded if companies want to compete and survive. Increasing demand for fibres to make products is going to continue to be a challenge. To meet this demand, the industry will have to come up with new sustainable raw materials, processes and recovery of fibres from waste.

Recycling is going to be increasingly adopted to reduce waste that ends up in landfills. Recycling ensures that sustainability of the raw materials is maintained. In a circular economy, waste or products are circulated in a close loop. The main barrier to recycling is the lack of technologies that can recover the fibres without causing any damage. Fibre blending complicates the recovery as current technologies are not capable of performing such tasks. Biological agents such as enzymes and microorganisms have the ability to separate blended fibres with less damage to the fibres but due to low yield the production cannot be upscale for industrial production. Additional benefit of using biological agents is that both natural and man-made fibres can be degraded depending on the biological agent.

Renewable fibres like plants and animals are going to become more prominent as they are more sustainable and environmentally friendly than non-renewable fossil fuel fibres due to the fact that once they are depleted they can be regenerated again and require less energy. In contrast, the non-renewable fibres made from fossil fuels are finite and consume more energy. The advantage of fossil fuel fibres is that they can be bulk produced quickly which can be cost-effective, whereas the plant and animal fibres require months to harvest.

Dyeing using traditional dyes and methods causes damage and alternative natural dyes and methods can occupy centre stage as they are more sustainable and cause less harm to the environment. Using supercritical carbon dioxide instead of water saves water sources and the liquid can be turned into gas which is recovered or released into the environment causing minimal damage to the environment. Dyes extracted from plant sources show that fibres can be dyed using renewable natural resources. As more plants are identified, the choice of colours is going to increase.

Textile fibres, yarn and fabrics are used in non-clothing products due to their soft feel, lightweight, flexibility and durability. Furthermore, biodegradability and biocompatibility make them good choices for sustainable materials or products. The smart garment and textile products have seen some growth due to unique properties compared to conventional products especially in the fields of sports and health.

Active materials or products are able to react to external stimuli such as heat, voltage or humidity. Active twisted yarns can elongate, contract, bend and rotate depending on the type of stimuli. Voltage can be applied to fabric sandwiched between two electrodes to control fabric thickness by switching it on and off. The benefit of this characteristic is that the fabric's breathability can be controlled to suit the wearer's comfort level (Fu et al., 2022; Wen et al., 2021). Finishing of smart textiles might require the use of water, chemicals and energy to impart performance or aesthetic properties. Alternative methods like gases or plasma are not widely used as the initial capital cost is high and the penetration of the finishing agent might not be high and over-spraying the agent might happen and cause quality problems (Chatha et al., 2019).

Consumer behaviour drives the demand for fashion and textile products and as more consumers become conscious about their purchasing sustainability behaviour and its environmental impact, the choice of products in shops is going to change. Manufacturers that are not into sustainability are going to face declining product sales. Eco-labelling plays an important role in informing the consumer. It can be used to identify products that have been produced in a sustainable manner.

17.12 Conclusion

The challenges facing the textile industry are multifaceted, ranging from raw materials, production processes, recycling to consumer behaviour. Solutions require that all these factors be taken into consideration when new technologies and strategies are adopted to achieve sustainability goals. Traditional ways of only looking at the cost and then selecting the materials and production processes are not going to be the future of the industry. Selection should take into consideration the sustainability and environmental impact of processes and products when they reach end of life. Waste must be recyclable to be used again as a source of raw materials. Consumer behaviour of buying and disposing of products quickly with no regard to the environmental impact is not sustainable as it promotes the rapid depletion of resources and damage to the environment. Even though the cost of new technologies is high, the benefits are high when the cost sustainability and damage to the environment are taken into consideration.

Acknowledgements

The authors gratefully acknowledged various sources for granting permission to reuse figures used in this chapter.

Funding
The author disclosed receipt of the following financial support for the research, authorship and/or publication of this article: This work is based on the research supported in part by the National Research Foundation of South Africa (grant-specific unique reference numbers (UID) 104,840.

References

Bediako, J.K., Wei, W., Yun, Y.-S., 2016. Conversion of waste textile cellulose fibres into heavy metal adsorbents. Journal of Industrial and Engineering Chemistry 43, 6168.

Cai, Y.-J., Choi, T.-M., 2020. A United Nations' Sustainable Development Goals perspective for sustainable textile and apparel supply chain management. Transportation Research Part E 141 (102010).

Calderon-Monge, E., Pastor-Sanz, I., Garcia, F.J., 2020. Analysis of sustainable consumer behavior as a business opportunity. Journal of Business Research 120, 74—81.

Chang, A., 2020. The impact of fast fashion on women. Journal of Integrative Research and Reflection 3, 16—24.

Chatha, S.A., Asgher, M., Asgher, R., Hussain, A.I., Iqbal, Y., Hussain, S.M., Bilal, M., Saleem, F., Iqbal, H.M., 2019. Environmentally responsive and anti-bugs textile finishes—Recent trends, challenges, and future perspectives. Science of the Total Environment 690, 667—682.

Clancy, G., Froling, M., Peters, G., 2015. Ecolabels as drivers of clothing design. Journal of Cleaner Production 99, 345—353.

Claxton, S., Kent, A., 2020. The management of sustainable fashion design strategies: an analysis of the designer's role. Journal of Cleaner Production 268 (122112).

Debnath, S., 2017. Sustainable production of bast fibres. In: Muthu, S.S. (Ed.), Sustainable Fibres and Textiles. Elsevier and Woodhead Publishing, Kidlington, UK, pp. 69—85.

Dissanayake, D.G., Weerasinghe, D., 2021. Towards circular economy in fashion: review of strategies, barriers and enablers. Circular Economy and Sustainability 1–21.

Feghali, E., Tauk, L., Ortiz, P., Vanbroekhoven, K., Eevers, W., 2020. Catalytic chemical recycling of biodegradable polyesters. Polymer Degradation and Stability 179 (109241).

Fu, C., Xia, Z., Hurren, C., Nilghaz, A., Wang, X., 2022. Textiles in soft robots: current progress and future trends. Biosensors and Bioelectronics 196 (113690).

Gholamzad, E., Karimi, K., Masoomi, M., 2014. Effective conversion of waste polyester–cotton textile to ethanol and recovery of polyester by alkaline pretreatment. Chemical Engineering Journal 253, 40–45.

Guo, Z., Eriksson, M., de la Motte, H., Adolfsson, E., 2021. Circular recycling of polyester textile waste using a sustainable catalyst. Journal of Cleaner Production 283 (124579).

Herva, M., Alvarez, A., Roca, E., 2011. Sustainable and safe design of footwear integrating ecological footprint and risk criteria. Journal of Hazardous Materials 192, 1876–1881.

Hussein, Z., Sajjad, W., Khan, T., Wahid, F., 2019. Production of bacterial cellulose from industrial wastes: a review. Cellulose 26, 2895–2911.

Ion, S., Voicea, S., Sora, C., Gheorghita, G., Tudorache, M., 2021. Sequential biocatalytic decomposition of BHET as valuable intermediator of PET recycling strategy. Catalysis Today 366, 177–184.

Islam, M.M., Perry, P., Gill, S., 2021. Mapping environmentally sustainable practices in textiles, apparel and fashion industries: a systematic literature review. Journal of Fashion Marketing and Management 25 (2), 331–353.

Jain, V., Obrien, W., Gloria, T.P., 2021. Improved solutions for shared value creation and maximization from used clothes: streamlined structure of clothing consumption system and a framework of closed loop hybrid business model. Cleaner and Responsible Consumption 3 (100039).

Joshi, Y., Rahman, Z., 2019. Consumers' sustainable purchase behaviour: modeling the impact of psychological factors. Ecol. Econ. 159, 235–243.

Liu, X., Ning, Y., Wang, F., 2012. Processing methods of polyester fibres with deep-coloring. Chemical Fibres International 1–4.

Mahmoodi, J., Patel, M.K., Brosch, T., 2021. Pay now, save later: using insights from behavioural economics to commit consumers to environmental sustainability. Journal of Environmental Psychology 76, 101625.

Mainolfi, G., 2020. Exploring materialistic bandwagon behaviour in online fashion consumption: a survey of Chinese luxury consumers. Journal of Business Research 120, 286–293.

Matthes, A., Beyer, K., Cebulla, H., Arnold, M.G., Schumann, A., 2021. Sustainable Textile and Fashion Value Chains: Drivers, Concepts, Theories and Solutions. Springer.

Moran, C.A., Eichelmann, E., Buggy, C.J., 2021. The challenge of "Depeche Mode" in the fashion industry – does the industry have the capacity to become sustainable through circular economic principles, a scoping review. Sustainable Environment 7 (1), 1975916.

Navone, L., Moffitt, K., Hansen, K.-A., Blinco, J., Payne, A., Speight, R., 2020. Closing the textile loop: enzymatic fibre separation and recycling of wool/polyester fabric blends. Waste Management 102, 149–160.

Niinimaki, K., Peters, G., Dahlbo, H., Perry, P., Rissanen, T., Gwilt, A., 2020. The environmental price of fast fashion. Nature Reviews, Earth and Environment 1, 189–200.

Penthala, R., Oh, H., Park, S.H., Lee, I.Y., Ko, E.H., Son, Y., 2022. Synthesis of novel reactive disperse dyes comprising carbamate and cyanuric chloride groups for dyeing polyamide and cotton fabrics in supercritical carbon dioxide. Dyes Pigments 194 (110003).

Park, S.H., Kim, S.H., 2014. Poly(ethylene terephthalate) recycling for high value added textiles value added textiles. Fashion Textiles 1 (1).

Provin, A.P., de Aguiar Dutra, A.R., Machado, M.M., Cubas, A.L., 2021. New materials for clothing: rethinking possibilities through a sustainability approach − A review. Journal of Cleaner Production 282 (124444).

Rex, M., Baumann, H., 2007. Beyond ecolabels: what green marketing can learn from conventional marketing. Journal of Cleaner Production 15, 567−576.

Saleem, M.A., Pei, L., Saleem, M.F., Shahid, S., Wang, J., 2020. Sustainable dyeing of nylon with disperse dyes in decamethylcyclopentasiloxane waterless dyeing system. Journal of Cleaner Production 276 (123258).

Saouter, E., A, D.S., Pant, R., Sala, S., 2018. Estimating chemical ecotoxicity in EU ecolabel and in EU product environmental footprint. Environment International 118, 44−47.

Schragger, M., 2021. The Climate Impact of Apparel and Textiles. 2021 Progress Report. https://www.sustainablefashionacademy.org/sites/default/files/stica_report_1.0_210210.pdf. (Accessed 15 December 2021).

Severo, E.A., de Guimaraes, J.C., Dorion, E.C., Nodari, C.H., 2015. Cleaner production, environmental sustainability and organizational performance: an empirical study in the Brazilian metal-mechanic industry. Journal of Cleaner Production 96, 118−125.

Shafie, S., Kamis, A., Ramli, M.F., Bedor, S.A., Puad, F.N., 2021. Fashion sustainability: benefits of using sustainable practices in producing sustainable fashion designs. International Business Education Journal 14 (1), 103−111.

Shi, L., Liu, J., Wang, Y., Chiu, A., 2021. Cleaner production progress in developing and transition countries. Journal of Cleaner Production 278, 123763.

Shirvanimoghaddam, K., Motamed, B., Ramakrishna, S., Naebe, M., 2020. Death by waste: fashion and textile circular economy case. Science of the Total Environment 718 (137317).

Silva, T.L., Cazetta, A.L., Souza, P.S., Zhang, T., Asefa, T., Almeida, V.C., 2018. Mesoporous activated carbon fibres synthesized from denim fabric waste: efficient adsorbents for removal of textile dye from aqueous solutions. Journal of Cleaner Production 171, 482−490.

Terinte, N., Manda, B.M., Taylor, J., Schuster, K.C., Patel, M.K., 2014. Environmental assessment of coloured fabrics and opportunities for value creation: spin-dyeing versus conventional dyeing of modal fabrics. Journal of Cleaner Production 72, 127−138.

Thakker, A.M., Sun, D., 2021. Sustainable processing with herbs on bamboo, banana, and merino wool fibres. Journal of Natural Fibres 1−17.

Troynikov, O., Watson, C., Jadhav, A., Nawaz, N., Kettlewell, R., 2016. Towards sustainable and safe apparel cleaning methods: a review. Journal of Environmental Management 182, 252−264.

Tshifularo, C.A., Patnaik, A., 2020. Recycling of plastics into textile raw materials and products (Chapter 13. In: Nayak, R. (Ed.), Sustainable Technologies for Fashion and Textiles. Elsevier and Woodhead Publishing, Kidlington, UK, pp. 311−326.

Wen, F., Xu, B., Gao, Y., Li, M., Fu, H., 2021. Wearable technologies enable high-performance textile supercapacitors with flexible, breathable and wearable characteristics for future energy storage. Energy Storage Materials 37 (94−122).

Wu, X.J., Li, L., 2019. Sustainability initiatives in the fashion industry. In: Beltramo, R., Romani, A., Cantore, P. (Eds.), Fashion Industry - an Itinerary between Feelings and Technology. Intech Open Limited, London, UK.

Yan, R., Diddi, S., Bloodhard, B., 2021. Predicting clothing disposal: the moderating roles of clothing sustainability knowledge and self-enhancement values. Cleaner and Responsible Consumption 3 (100029).

Yousef, S., Tatariants, M., Tichonovas, M., Kliucininkas, L., Lukosiute, S., Yan, L., 2020. Sustainable green technology for recovery of cotton fibres and polyester from textile waste. Journal of Cleaner Production 254 (120078).

Life cycle analysis of textiles and associated carbon emissions

Yamini Jhanji

Fashion and Apparel Engineering Department, The Technological Institute of Textile and Sciences, Bhiwani, Haryana, India

18.1 Introduction to life cycle assessment (LCA)

The textile and fashion industry is ecologically one of the most environmentally polluting industries owing to unsustainable characteristics of the life cycle of textile and clothing. The excessive consumption of toxic and hazardous chemicals, dependence on non-renewable, fossil-based energy sources, wasteful utilization of land and water resources and high fuel consumption for transportation and generation of solid and gaseous waste are all accounted for adverse environmental effects. It thus becomes imperative to assess the environmental impact of textiles by the implementation of assessment tools, which can provide a holistic and comprehensive viewpoint of products and processes involved in the textile supply chain on the environment. The environmental impact of textiles can be assessed by a number of assessment tools such as life cycle assessment (LCA), Higg material sustainability index (MSI), and made by fibre benchmark. LCA, a promising tool for environment improvement takes into account all stages of a product's life cycle for precise and conclusive assessments. The use phase is a very crucial part of LCA analysis since the fibre ranking tool compares fibres on basis of an assessment of clothing's production stage. Nevertheless, the functionality and manner in which consumers use and handle their clothing are affected by fibre content (Kalliala and Nousiainen, 2000).

The non-inclusion of functional, aesthetic and performance attributes of garments during the use phase will lead to an inaccurate comparison of garments owing to the negligence of consumer handling and care during use and life span. Comparison of clothing of different fibres without taking into their account their usage will equate to disposable and durable clothing. If lifetime and usage of clothing are not taken into consideration, plastic and man-made cellulosic fibre clothes will turn out to be suitable candidates owing to less engagement of economic and environmental input in contrast to natural fibre fabric and clothing, which will exhibit high environmental cost at the material production stage.

The best assessment of the overall impact of product, process or services on the environment can be accomplished by LCA. LCA outshines the other two assessment tools, which primarily exclude the use phase in the life cycle of the product and mainly focus on the production aspects. LCA, on the other hand, is a technique implemented for the assessment of environmental impacts of products, processes and services that involves the quantification and evaluation of consumed resources and environmental emissions at varying stages of the product life cycle from resource extraction, production of materials, product components and product and eventually to the usage, reuse, recycling and final disposal of the product. Thus, LCA attempts to include impacts from raw materials to the disposal phase of product (cradle to grave) or raw material to point of sale (cradle to gate) (Laitala et al., 2018). The successful implementation of LCA in the textile arena can also be attributed to the inherent characteristics of the textile and clothing supply chain being the longest, most intricate, dispersed and globally stretched system with less transparent manufacturing processes. Each phase in the life of the textile product may impact social and environmental factors and thus assessment of the product over its entire life cycle by a robust and holistic tool like LCA is crucial for considering the environmental impacts of the production process.

LCA is considered to be a powerful tool in environmental management capable of assessing the environment-friendly attributes of the product. LCA is defined as a method to analyse and determine the environmental impact along the product chain of systems. LCA encompasses the chemical conversion of material during the manufacturing process, material formulation, structure and removal of material leading to an increase of output over input, and assembling of material leading to a decrease of output over inputs.

LCA (as per ISO 14040) is the technique for assessment of environmental aspects and potential impact of the product via:

- Compilation of an inventory of inputs and outputs required for product development.
- Evaluation of potential environmental impacts as a result of selected inputs and outputs.
- Inventory analysis interpretation and impact assessment phases in accordance with the objectives of the study being undertaken.

As per SETAC (Society of Environmental Toxicology and Chemistry), LCA is an evaluation process for environmental burdens in the context of the product, process or activities by identification and quantification of energy and materials utilized and release of waste to the environment for assessment of the impact of energy and material used and released to the environment and for identification and evaluation of opportunities targeting at environmental improvements. The assessment takes into consideration the product's entire life cycle, process, raw material procurement, manufacturing, transportation and distribution, maintenance, recycling and eventually the disposal (Fig. 18.1).

Qualitative and quantitative methods are adopted for conducting LCA. The qualitative method involves drawing explicit conclusions from the life cycle. The quantitative method involves the mathematical processing of data describing a product's life cycle for evaluation of environmental impact. The preceding discussion will enable the readers to understand that LCA serves as an effective tool for the assessment of environmental impacts of products, processes and services and environmental emissions at varying stages of the product life cycle. Before discussing in detail the tools, phases, merits and demerits of LCA, it is crucial to understand environmental impact of textiles. The following section shall discuss the environmental impact and carbon emissions associated with wasteful, unsustainable textile practices and fast fashion.

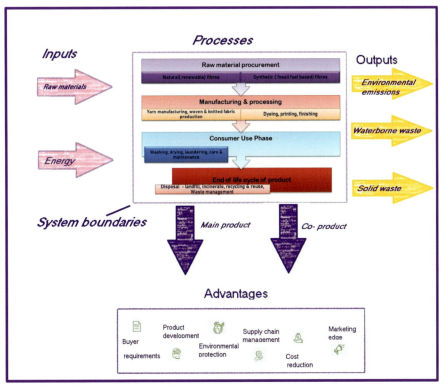

Figure 18.1 Overview of life cycle assessment.

18.2 Environmental impact, carbon emissions & the ardent need of LCA

The textile and clothing supply chain is one of the longest, most intricate, dispersed and globally stretched systems with less transparent manufacturing processes. Each phase in the life of a textile product may impact social and environmental factors and positive attributes at some points of the supply chain may have negative repercussion at a later stage. It thus becomes crucial to identify the impact of each phase of the textile supply chain on the environment. The raw materials, manufacturing process, location of manufacturing unit, packaging, distribution, consumers consumption and disposal patterns of textile end products can all influence the environment and contribute to carbon emissions.

The textile and garment industry owing to its energy-intensive processes results in 6%–8% of total global carbon emissions. Wet processing stages namely dyeing and finishing use energy for steam generation to heat water, for fabric drying. The carbon intensity of energy sources like coal, and natural gases used for processes result in high emission intensity during textile manufacturing. A significant cost is also associated with energy consumption in the manufacturing phase. Fig. 18.2 highlights the environmental impact of textiles at different phases of the product life cycle.

Apparel and footwear industry together account for approximately 8.1% of global climate impacts with apparel leading to 6.7% of the global climate change. The indicators of pollution impact include climate change, freshwater withdrawal, resources, ecosystem, quality and human health. The life cycle phases were the main drivers of global pollution impact for all the above-stated indicators including fibre production, yarn preparation, dyeing and finishing while the distribution and disposal phase contributed least to global pollution impact (Intl.com; Pacific Report, 2021).

18.2.1 Environmental impact of different textile raw materials

The raw material procurement for textile applications in accordance with sustainable principles and for minimizing environmental impact is still a challenging task for manufacturers and other stakeholders since some fibres may be user-friendly and obtained from renewable energy sources but may still contribute to environmental impact during their cultivation owing to the massive requirement of water, pesticides and fertilizers etc. Similarly, some fibres may demand less natural resources during their cultivation but may be fossil fuel-based. Bamboo for instance is natural renewable material utilized for rayon production however rayon processing is a chemically intensive process. Likewise, cotton holds the lion's share of the market as far as textile and

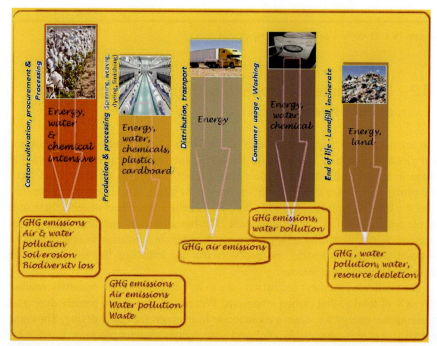

Figure 18.2 Environmental impact at each phase of textile product's lifecycle.

apparel application is concerned. Although being a natural fibre, cotton is not a very conducive environment owing to the huge demand for fertilizer, pesticides during its production, and environmental emissions as a result of cotton seed growing and irrigation processes. Furthermore, extensive utilization of insecticides leads to eco toxicity. The massive usage of fertilizers and agro-chemicals during agricultural production makes the cotton fabric the highest contributor to environmental impact compared to other fibre types. Also, a large amount of water is used during cotton cultivation. Additionally, pre-treatment process like scouring carried out on cotton is energy-intensive processes. Different fibres vary in their functionality and environmental impact. The washing pattern of clothing varies in accordance with fibres, which comprise the clothing. Wool unlike cotton demands lesser quantities of chemicals and energy. The production of polyester is an energy-intensive process and contributes to energy-related indicators like climate change and ionizing radiation. The higher environmental impact of synthetic fibres in contrast to renewable-based fibres can be attributed to the depletion of fossil resources. Synthetic fabrics require more frequent washes as they become dirty faster. Moreover, synthetic fabrics result in microplastic release during use and at the end of the life cycle of clothing by fragmentation of micro and nano-sized plastic particles (Laitala et al., 2018).

Table 18.1 compares the climate impact of various textile fibres. It can be observed that silk fibre results in maximum climate impact followed by wool with climate impact lying in the range of $1.7-36.2$ KgCO$_2$ equivalent per Kg fibres.

Composting of waste in the case of silk fibres is accounted for high climate impact. However, the impact can be reduced by waste incineration. The climate-intensive attribute of wool is accounted to methane emissions from sheep farming. The climate impact of cotton fibre lies in the range of $0.5-$four KgCO$_2$ equivalent per Kg fibres. However, the climatic impact of organic cotton is less than conventional cotton owing to the reduced utilization of artificial fertilizers required for the production of the former (Muthu; Sandin et al.). Flax exhibits the lowest carbon footprint followed by hemp in contrast to other fibres. Table 18.2 shows the cradle-to-gate global warming potential (GWP) of different fibres. It can be observed that polyester tops the list with a maximum contribution to GWP while lyocell exhibits the lowest GWP.

The environmental burden of varied fibres namely cotton, polyester, nylon, acryl and elastane were studied and it was inferred that apart from base material, yarn

Table 18.1 Environmental impact of different fibres.

Fibre type	Climate impact (Kg CO$_2$ equivalent per kg fibre)
Wool	$1.7-36.2$
Silk	$52.5-80.9$
Cotton	$0.5-4$
Flax	$0-0.8$
Hemp	$0-3.6$
Polyester	$1.7-4.5$

Table 18.2 Global warming potential of different fibres.

Fibre type	Cradle to factory gate GWP (100 years) (tCO$_2$ equivalent/fibre
Polyester	4.1
Cotton	2
Lyocell	0.05
PLA	0.62

thickness has a bearing on the environmental burden. The cradle-to-grave analysis demonstrated acryl and polyester exhibited the least environmental input followed by elastane, nylon and cotton (Rosa and Grammatikos, 2019; Velden et al., 2014).

Bayda et al. (2015) compared the LCA of eco t-shirts produced using originally cultivated cotton and conventional t-shirts. They concluded that eco t-shirts resulted in lower environmental impact due to the elimination of nitrogen and phosphorous-containing chemical-based fertilizers during their production. The researchers emphasized on utilization of sustainable raw materials in all life cycle stages of cotton textile products and the ardent need to focus on consumer behaviour and sustainable practices in the usage phase of the product.

Kalliala and Nousiainen (2000) studied the environmental impact of hotel textiles by employing LCA methodology. LCA is useful in providing information pertaining to quantities of energy and resources consumed and emissions associated with production systems. The inventory calculations suggested that cotton cultivation consumed 40% less energy than that required for the production of polyester fibres. The researchers concluded that higher durability and lower laundering energy requirements of 50/50 CO/PES sheets in hotels resulted in their lower environmental impact compared to sheets composed of 100% cotton fibre.

Life cycle analysis of hemp associated with the environmental impact of cultivation, harvesting, pressing of oil, decortication, steam pressure digestion and textile production suggested that 17% of climate change, 36% of acidification and 10% of energy usage could be accounted for cultivation and harvesting of hemp crop.

Werf (2004) in his studies on life cycle analysis of hemp fibre production found that hemp and flex were low input and low impact crops in comparison to potato and sugar beet crops.

Velden et al. (2014) studied the environmental burden of textiles composed of raw materials such as cotton, polyester, nylon, acryl and elastane to identify the raw materials and life cycle stage that have the strongest environmental impact.

LCA of cotton shopping bags suggested that the environmental impact of cotton bags was higher in the raw material and production stage, thus necessitating the quest for alternative sustainable options like organic and recycled cotton (Muthu).

The cradle to grave LCA analysis of the bathrobe in accordance with waste potential, energy expenditure and environmental impact of each phase of the bathrobe's life cycle was methodologically carried out. The study suggested against the preconceived notion of cotton being environmentally friendly due to environmental impact at each stage from growing through manufacturing to consumer usage and recovery stage of bathrobes composed of 100% cotton (Tekstiller et al., 2009).

Environmental impact of manmade cellulosic fibres, namely, visocse, modal, and tencel (Lenzing viscose Asia, Lenzing viscose Austria, Lenzing Modal, Tencel Austria, Tencel Austria, 2012) was assessed using LCA, and the results were compared with cotton (as a benchmark), novel bio-based fibre (PLA) and synthetic fibres like polyester, polypropylene. The environmental indicator for the assessment included resource and impact indicators. Resource indicators included renewable and non-renewable energy usage — non-renewable and cumulative energy demand, water, energy and land usage. The impact indicators included global warming potential 100a, abiotic depletion and depletion of ozone layer, fresh water aquatic and terrestrial eco-toxicity, human toxicity, photochemical oxidant formation, acidification and eutrophication (Shen and Patel, 2010).

Cradle to factory gate approach was deployed for LCA however, the global warming potential and energy profile of man-made cellulosic fibres were assessed by the inclusion of waste incineration with energy recovery. Lenzing Viscose Austria and Lenzing Modal were found to exhibit the least environmental impact (except for land and water utilization) in contrast to Lenzing Viscose Asia. Among all fibres under consideration, Tencel Austria 2012, Lenzing Viscose Austria, Lenzing Modal and Tencel Austria outshined all other fibres in terms of reduced non-renewable energy usage, and global warming potential, land and water utilization and toxicity impacts. The LCA analysis and comparison revealed cotton contributed the most to environmental impacts as far as land and water usage, eutrophication and eco-toxicity impacts were concerned.

It is an established fact from several studies that recycled fibres in contrast to virgin fibres have lower environmental impacts as far as fibre, textile and garment production is concerned. Nevertheless, the magnitude by which recycled fibres lower the impact needs to be assessed in a quantitative manner. Accordingly, several apparel brands like H & M are working towards collection and recycling of textiles and garments to obtain spinnable fibres in contrast to the traditional practice of sourcing virgin raw materials.

The environmental impact of denim fabric produced by utilizing mechanically recycled cotton fibre in place of virgin cotton fibre and installation of combined heat and power plant instead of grid energy as means of energy source during fabric production was assessed using LCA approach. Cotton cultivation resulted in the most environmental impact termed as a hot spot in denim fabric production accounting for 53% and energy consumption during the spinning process accounting for 16% of environmental impact. LCA results suggested the highest environmental impact improvement in terms of water use by 98%, eutrophication potential by 90%, acidification potential by 74%, cumulative energy demand by 63% and global warming potential by 54% by using 100% recycled cotton and CHP plant use.

The researchers drew some concrete conclusions based on LCA and suggested that apart from raw material, yarn thickness also served as a crucial factor to determine environmental burden. It was proposed that acryl and polyester-based textiles exhibited the least environmental impact followed by elastane, nylon and cotton (Saxce et al., 2012).

18.2.2 Environmental impact of manufacturing processes

The fibre, yarn, fabric and garment production processes along with textile chemical processing like dyeing, printing and finishing have been globally recognised to be significantly affecting the environment. Fibre production offers the highest impact on fresh water withdrawal and ecosystem quality owing to cotton cultivation. The energy-intensive processes and reliance on fossil fuel during yarn preparation, dyeing and finishes contributed to high environmental impact (Intl.com; Pacific Report, 2021).

The four focal points in the supply chain where environmental impacts are concentrated involve:

- Weaving, dyeing and finishing process.
- Energy consumption in textile manufacturing and garment assembling.
- Textile waste in cutting and sewing room operation (Fig. 18.3).

The impact of the manufacturing process, usage and consumers' discarding pattern on the environment was studied by LCA. Manufacturing conditions were observed to be the greatest influencers on the environmental assessment of bedsheets since textile product quality is determined by manufacturing conditions, it is crucial to assign a specific lifetime to the corresponding manufacturing process (Qian et al., 2021).

Velden et al. (2014) conducted LCA benchmarking study on textiles and concluded that the knitting fabrication process was better in terms of environmental impacts compared to weaving.

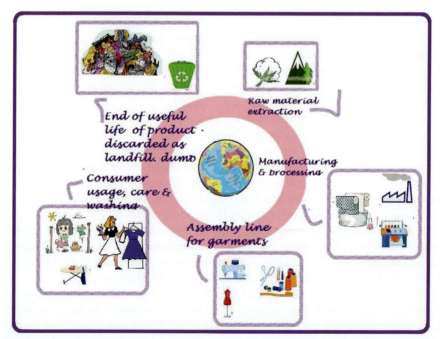

Figure 18.3 Life cycle assessment from raw material extraction till the end of the product's useful life.

Rosa and Grammatikos (2019) performed cradle-to-grave analysis on textiles composed of various fibres and concluded that the spinning and weaving process posed a higher environmental impact compared to knitting and the environmental burden of spinning, weaving and knitting was found to be a function of one/yarn size.

Qian et al. (2021) compared the carbon footprint of textiles composed of virgin and recycled polyester. The carbon footprint of virgin polyester textiles was 119.59 Kg CO_2/100 kg while the carbon footprint of waste polyester recycling was approximately 10 times that of the former. LCA polygon method employed to evaluate carbon footprint suggested that irrespective of the polyester source i.e., whether virgin or recycled, the production process of polyester fabric resulted in high environmental impact.

Alay et al. (2016) developed a sustainable, eco-friendly and anti-bacterial knitted structure consisting of 80% polylactic acid, 15% lyocell and 5% chitosan. The cloth designed from this improved knit structure demanded less washing cycle and ironing at a lower temperature during consumer use phase in contrast to cotton clothing. Anti-bacterial activity was achieved in structure without the usage of any toxic antibacterial agents like silver, zinc, copper, cobalt etc. Further, the improved fabric showed slower degradation when compared to cotton control fabric.

18.2.3 Environmental impact of transportation, usage & disposal phase

The environmental impact of transportation is associated with modes of transportation for supplying raw materials from vendors or suppliers' sites to manufacturing units, location of manufacturing units and shipment of finished goods from warehouse to outlets. Likewise, the impact of the use phase on the environment depends on a myriad of factors like consumer behaviour, the geographical zone in which the product is being used and that zone's climatic conditions (Zamani et al., 2017; Muttu, 2015; Bloomberg; Choudhury, 2014).

The overall environmental impact of garments is affected by post-production phase namely cut and sew, consumer usage and end life of clothing.

The impact on account of consumer usage and disposal can be minimized by modification of laundering behaviour by consumers by making a transition from machine drying to line drying, cold water washing and utilizing energy-efficient washing machines. Energy utilization during the consumer usage phase is considered to be a crucial factor of environmental impact. Energy-saving and reduced carbon emissions amounting to a 10% reduction of CO_2 equivalent emission can be achieved by the corresponding reduction in washing machine energy to 40% by employing an efficient washing machine with full loading capacity.

The t-shirt subjected to the lowest laundry cycles led to lower impact in contrast to woven pants that needed a higher number of washes over their lifetimes and thus contributing to greater impact.

The transition toward fast fashion leads to the garment's shorter practical service life. The replacement of the conventional model of ownership-based consumption with collaborative consumption (involving maintaining clothing libraries) can prolong the practical service life of the garment thus lowering the environmental impact. Environmental performance of clothing libraries (stocking clothing ensembles like jeans, t-shirts, and dresses) was explored by LCA thereby comparing merits and demerits to conventional business models. The implementation of clothing libraries was associated with an enhanced service life of clothing. However, the benefits of reduced production achieved through clothing libraries are offset by customer transportation. Therefore, logistics need to be taken into consideration during the implementation of collaborative consumption business models.

LCA methodology was used to assess the environmental impact of men's shirts by Netherland-based retailers to identify the phase of the shirt's life cycle responsible for maximum pollution and selection of the most eco-friendly fibre option for shirt making.

The four phases of a shirt's life cycle were assessed for environmental impact:

- Manufacturing (Selection of raw material, cultivation conditions of selected fibres, spinning, weaving, dyeing and processing).
- Transportation.
- Consumer usage and care (washing, drying, ironing).
- Disposal at the end of the shirts life cycle (reuse, recycling, composting, incineration).

Transportation for distribution of merchandise to retail outlets and consumer use phase were the two phases in the shirt's life cycle that contributed the most to environmental impact (Bevilacqua et al., 2011).

Bevilacqua et al. (2011) applied a LCA methodology for the determination of the carbon footprint of a wool sweater. System boundaries were defined as part of LCA initiating from sheep breeding to the production phase, distribution to retail outlets globally and eventually user phase (that included washing and final disposal). Functional units considered for the study consisted of fibre composition, colour and year in which the collection was launched. The major source of greenhouse effect in the woollen sweater life cycle was the electric and thermal energy consumption mode of transportation owing to the delocalized nature of the production phase.

LCA methodology was deployed to assess the impact of the textile supply chain of cotton, wool and polyester apparel on the environment (Moazzem et al., 2018; Zhang and Chen, 2019). The assessment involved the consideration of climate change (related to greenhouse gas emissions to the environment) as the environmental impact category. The researchers concluded that a major contributor to climate change associated with cotton and polyester apparel was the usage stage of the life cycle of apparel composed of these fibres while a different trend was observed with wool apparel with the production process of wool apparel contributing to environmental impact as against use phase.

The environmental impact of three cotton garments namely t-shirt, woven casual pants and knit casual collared shirt was assessed by LCA carried out in three primary phases:

- Artificial production (Bale preparation separating leave, seed from fibre by ginning).
- Textile processing (yarn to fabric and fabric to cut and sew garment).
- Usage (Consumer usage & disposal at end of clothing's life cycle.

The consideration of the entire cotton life cycle suggested that irrespective of fibre type, the greatest contributors to environmental impacts were textile manufacturing and consumer usage phase, which can be credited to garment laundering, high electricity consumption in fibre processing and energy-intensive processing conditioning, heating, dyeing, finishing etc. The overall environmental impact was also due to the contribution of agricultural production however the contribution of the former was less compared to textile manufacturing and consumer usage (Qian et al., 2021).

Socially responsible consumption associated with disposal methods preferred by consumers at the end of the product's life cycle has a great bearing on environmental impact. Three modes of clothing disposal were identified by females in Scotland and America: selling online on platforms like eBay, second-hand shopping and a donation to charities and gifting to siblings and peer group (Muthu).

The environmental impact associated with fibre type, manufacturing processes, transportation, consumer usage and disposal phase along with their assessment by LCA has been discussed in the previous section in detail. The carbon emissions associated with textiles shall be covered in the following section.

18.3 Carbon footprint, classification & related parameters

A carbon footprint measures the number of greenhouse gases (GHG) produced by burning fossil fuels for energy production like the generation of electricity, heating and transportation. CO_2, CH_4, and N_2O are the gases naturally present in the atmosphere with their concentration increasing continuously owing to human activities. Chlorofluorocarbons also impact green house effect however these do not occur naturally in the environment and human activities are solely responsible for their creation.

Carbon footprint denotes the GHG emission evaluation as a result of human activities.

Carbon footprints can be classified into primary and secondary footprints based on direct and indirect emissions of greenhouse gases:

- Combustion of fossil fuels results in direct emission of GHG and the primary carbon footprint. Transportation and domestic energy consumption also contribute to the primary carbon footprint.
- Secondary carbon footprint - indirect GHG emission during the entire life cycle of products is primarily responsible for the secondary carbon footprint. Usage of clothing, recreation and leisure goods results in secondary GHG emissions.

The major sources of GHG emissions include residential, industrial, and commercial sectors along with transportation, agricultural resources and waste management.

Carbon footprint is expressed in terms of Global warming Potential (GWP), which represents the relative effect of climate change over a prescribed time period like 20 years (GWP20) or 100 years (GWP100). The measure of heat trapped by GHG in the atmosphere contributing to global warming and the comparison of the amount of heat trapped by a fixed mass of gas to the amount of heat trapped by a similar mass of CO_2 is analysed by GWP.

Carbon footprint is calculated in terms of GWP as per Eq. (18.1):

$$\text{Climate change} = \sum_i GWP_{ai} \, X \, m_i \tag{18.1}$$

GWP_{ai} — GWP for a substance i integrated over a particular number of years.
m_i (Kg) — quantity of emitted substance (i).
Climate change — Kg or tons CO_2 equiv.

18.3.1 Carbon footprints along with various phases of textile supply chain

There has been an alarming rise in carbon footprints and greenhouse gas emissions (GHG emissions) owing to ever-increasing consumption of food, clothing and electricity and thus rapid and continued growth of related industries. Carbon emission across the value chain is affected by:

- Varying fibre types.
- Textile & garment production.
- Consumer use phase.
- Distribution & end of life.

The high carbon footprint contributed by the textile industry is attributed to its dependence on hard coal and natural gas as means of electricity and heat production. High energy demands and dependence on fossil fuels as energy sources are primarily responsible for a high share of carbon emissions. Variation of raw materials (natural versus synthetic fibres), the carbon intensity of energy sources in manufacturing units, the overall volume of production length and intensity of use phase are the governing factors influencing the carbon emissions. Moreover, the defiance of sustainable principles during material procurement, production, consumer use and handling phase of clothing and reckless disposal of clothing at end of life cycle accounts for the largest share of emissions and hence climate change with raw material production, garment assembly and distribution contributing to comparatively much smaller shares of emissions. The contribution of the consumer use phase to carbon emission is difficult and challenging to access precisely owing to lack of information pertaining to consumers' behaviour as far as washing frequency, washing and drying method, detergent type and washing temperature is concerned.

18.3.1.1 Carbon footprints associated with fibre types

The fibre constituents of the clothing influence carbon footprints and GHG emissions. Leather, silk and wool are the biggest contributors to emissions by Kg of material.

The crucial consideration related to carbon footprint in the case of synthetic fibres is the fossil fuel origin of synthetic fibres. A high amount of energy requirement and thus

CO_2 emission results as a result of oil extraction from earth and production of polymers, unlike the natural fibres.

Among synthetic fibres, nylon requires the highest amount of energy in its production followed by acrylic and polyester. The inability of synthetic fibres to decompose and release heavy metals, and additives in soil and ground water when discarded as landfills further add to environmental adversities.

Polypropylene and acrylic however are reported to contribute least to carbon emissions. The highest environmental impact of natural fibres is attributed to large volumes of water and toxic, hazardous pesticides required during their cultivation. Although, synthetic fibres are credited to lower carbon emissions and environmental impact compared to their natural counterpart but the main cause of concern when working with synthetic fibres is the release of micro-plastic, non-renewable source of origin and release of micro-plastic during consumer use and disposal phase of clothing.

The GHG emission of jute fibre yarn in different phases namely cultivation, retting stage, manufacturing and disposal were studied and it was concluded that the cultivation and retting phase of jute yarn resulted in negative GHG emissions implying that the plantation process of jute acted as carbon absorber.

18.3.1.2 Carbon footprints associated with manufacturing, garment assembling, usage & disposal phase

The manufacturing process, textile chemical processing, garment assembling, consumer usage and disposal are all responsible for carbon footprints.

The style and usage of the garment also impact the emission associated with different constituent fibres. High-end, durable garments are generally composed of wool, silk and leather while synthetic fibres are generally preferred for fad styles and clothing that follow fast fashion. The inception and widespread acceptance of fast fashion have further resulted in increased climate change owing to the production of fast, cheap and low-quality apparel completely ignoring the incorporation of sustainable principles in the textile and fashion supply chain.

The production of raw materials accounts for significant GHG emissions while the fabric and garment production process and product disposal phase are considered to be the least GHG intensive processes. However, the usage phase of cotton garments contributes the maximum to CO_2 emissions.

A comparison of cotton and acrylic garments suggested that while the manufacturing and use phase of acrylic jackets resulted in a lower carbon footprint compared to cotton counterparts nevertheless, the disposal of acrylic garments results in considerably higher CO_2 emissions compared to cotton garments.

The relative GHG emissions produced during the manufacturing process of linen shirt was high in China due to dependence on coal-fired power plants for electricity generation in contrast to France as nuclear power was the main energy source there thus leading to lower GHG emissions.

The major source of CO_2 emissions for the production of pashmina shawls comprising of viscose rayon fibre includes production of rayon staple fibre, transportation of raw material and fuel to the site of production, transportation of finished product from production site to warehouse.

Zhang and Chen (2019) studied the men's shirt sewing assembly line for its carbon emission and concluded fabric is the main culprit of carbon emissions. Likewise, studies on the environmental impact of cotton t-shirt production suggested that the dyeing process contributed the largest with 35% to carbon emissions while garment assembly contributed 32% to the carbon emission.

The use phase of linen shirts demands higher energy owing to the high energy requirements for ironing the linen shirt. Cultivation of raw material (flax) led to a low amount of energy utilization and GHG emissions.

Marks and Spencer in their studies observed that washing and drying during the consumer use phase were the most energy-intensive processes and hence the greatest source of carbon emission.

Levi Strauss and co. also observed consumer phase contributed the most to carbon emissions compared to cotton cultivation and fabric production phases. Distribution and end-of-life of products also contribute to carbon emissions with transportation of merchandise from a retail outlet by a consumer, the biggest source of carbon emission in contrast to all other transportation included in the value chain.

Different cotton garment styles were also compared in terms of their carbon emissions and interestingly, it was inferred that cotton long shirts resulted in a lower carbon footprint during raw material, manufacturing and disposal stages while significantly higher carbon footprint in the distribution phase when compared to sweat jacket also composed of cotton. A comparison of different natural and synthetic fibres for their energy consumption and CO_2 emissions suggested that except for hemp, wool fibre exhibited lower energy consumption and carbon footprint compared to nylon, acrylic, polyester, polypropylene and cotton etc.

Although the GHG emission during the cultivation and retting phase of jute resulted in negative GHG emissions disposal of jute end products in landfills contributed to the GHG effect owing to CH_4 emissions however the impact could be considerably reduced by disposing of the jute products through incineration (Zhang and Chen, 2019; Fidan et al., 2021; Rana et al., 2015).

18.3.2 Effective strategies to mitigate environmental impact & carbon footprint

The environmental impact, carbon emissions and various factors contributing to GHG emissions were discussed in the previous section. This section will discuss the effective strategies to mitigate the environmental impact and carbon footprint.

A myriad of effective strategies have been adopted to reduce environmental impact and carbon footprint as far as raw material procurement, technological up-gradation in manufacturing, textile chemical processing, usage and disposal phase of textile end products are concerned.

- The optimum solution for reducing the high environmental impact in the textile industry is shifting towards organic cotton cultivation since the amount of water and energy consumption is lower compared to traditional cultivation. Natural fibres are biodegradable and possess ability to sequester carbon from the atmosphere thus leading to a lower carbon footprint compared to synthetic fibres. Therefore, the increased usage of natural and organic fibres can assist in bringing down carbon emissions. Furthermore, innovative fibres like Dupont's Sorona is renewable and composed of the agricultural product thereby exhibiting lower carbon emissions compared to petrochemical-based synthetic fibres. Other natural, sustainable fibres such as jute, and hemp are also gaining momentum to subvert the deleterious environmental impact.

- The usage of environmentally friendly technologies in energy utilization and switching over to recyclable raw materials can be an effective measure toward sustainable approaches in the textile supply chain (Fidan et al., 2021; Rana et al., 2015).
- Innovations in technologies, the transformation of business models, endorsing energy efficiency and switching to renewable sources of energy like wind, solar and hydro power can go a long way in ensuring reduced carbon emissions.
- Energy consumption, climate change and other environmental impacts at the yarn manufacturing stage can be taken care of by usage of open-end spun yarns instead of ring-spun yarns.
- Carbon footprints can be significantly reduced in textile processing by switching over to low-energy and water-intensive processes and products. Reduced water consumption during pretreatment, dyeing, washing and finishing can be achieved by employing low and ultralow liquor ratio machines. Less water requirement results in reduced energy requirement thereby lowering the effluent load. The introduction of continuous dyeing processes like thermosol and new processing techniques such as waterless dyeing, CO_2 dyeing, foam dyeing, finishing and coasting can significantly reduce the water and energy utilization. The processing time and energy cost can further be reduced by adopting one-step process of textile dyeing and finishing.
- Selection of textiles and finishes with lower environmental and social impact, substituting dyed fabrics with naturally coloured cotton or waterless dyeing are some measures to minimize the environmental impacts.
- Life cycle impact of cotton-based end products like shopping bags can be reduced by increasing functional limits and reuse capacities of the bags.
- Carbon footprints in the textile sector can further be reduced by the reuse and recycling of textiles. Released discarded, disposed of clothes into the marketplace through second-hand shops, a donation to charitable trusts etc can be effective ways of reusing textiles (Rana et al., 2015; Payne, 2011). The economic and environmental benefits as a result of clothing reuse have been highlighted by M & S. A considerable amount of energy could be saved by the distribution and processing of second-hand clothing. Environmental impacts can be considerably brought down by extending the garment's life span and encouraging low-impact transportation throughout the supply chain. Marks and Spencer plan to eradicate its carbon footprint in the next decade by focusing on rapid de-carbonization, inculcating a better understanding of carbon-related risks to retailers, and staff and launching incentive programmes encouraging sustainable fibre procurement and customers to donate discarded garments rather than throwing them away.
- Implementation of a holistic assessment methodology such as LCA for evaluation of environmental impacts owing to complex and multi-tiered approach of textile supply chain. The environmental impact of the entire life cycle of a product or service can be analysed in a systematic and scientific manner by the implementation of LCA (Payne, 2011; Koszewska, 2015). LCA measures green improvement, contributes to waste minimization, for estimation of CO_2 and GHG emissions, and investigates energy and water flow in the process (Bevilacqua et al., 2011).

18.3.3 Tools for environmental impact determination

The extensively used tools for environmental impact and greenhouse emissions associated with the textile supply chain include:

- Higg Material Sustainability Index (Higg MSI)
- Made by Fibre Benchmark
- Greenhouse gas protocol accounting
- LCA

The first two assessment tools primarily focus on sustainable aspects of raw materials i.e., fibre type overlooking the use phase altogether.

Higg MSI by Sustainable Apparel Coalition is a custom tool specifically developed for textile industries enabling common methodologies and procedures for LCA.

Made by fibre benchmark is similar to Higg MSI however it just includes the raw materials prior to the spinning phase and excludes fabrication and processing phases. The data from production is used by tools for the evaluation of the environmental impact of textiles composed of varied fibre types.

The tools assess the impact of fibre type on the environment in the use phase since use, durability, care and handling by consumers depend on fibre constituents of clothing. However, both the tools exclude the use phase in the life cycle of the product with a prime focus on merely the production aspects thereby omitting the environmental impact of the consumer use phase like emissions due to washing and release of micro-plastics by synthetic fibres (Muttu, 2015; Bloomberg). The care and usage of clothing by consumers are based on the fibres constituents of the clothing. The one-sided focus on material production ignoring the use phase, further excludes the environmental impact associated with the usage of a product like the importance of the product's life span, quality and functional attributes thereby leading to equating short-lived disposable end products with the durable clothing. The consumers are thus left in predicament as far as reduction in clothing's environmental burden is concerned owing to the comparison of dissimilar garments.

Greenhouse gas protocol is deployed by textile industries for measurement related to corporate level emissions across value chain and operations.

Calculation of greenhouse gas emissions or carbon footprint is one of the dimensions of LCA with assessment capabilities in myriad of concern areas like ozone depletion, eutrophication and impact on human health (Bevilacqua et al., 2011; Moazzem et al., 2018). LCA measures carbon emissions as a result of energy utilization and other environmental impacts of the product and process stage. The impacts are subsequently scaled up for a comprehensive understanding of the effect of production and product consumption on the environment.

LCA is a standardized framework methodology to quantify the impact of products, processes and services on the environment. The environmental performance of products in terms of energy expenditure, greenhouse gas emissions, water footprints and pollutants can be quantified by employing LCA tools and techniques. The assessment method further assists in the optimization of eco-efficiency of production processes and supply chain thereby enabling judicious decisions related to material selection and procurement. The versatility and holistic approach of LCA with the inclusion of each phase of the product life cycle makes it a standardized and most preferred assessment tool in comparison to the other three assessment tools discussed above.

18.4   LCA framework methodology

The assessment tools for the determination of the environmental impact and greenhouse emissions associated with the textile supply chain along with their relative merits and demerits were discussed in the previous section. The following sections of the chapter will attempt to present readers with a comprehensive framework of LCA along with its phases, merits, demerits and application areas.

18.4.1   Social & environmental labels on basis of LCA

The textile and clothing supply chain is one of the longest, most intricate, dispersed and globally stretched systems with less transparent manufacturing processes. Each phase in the life of a textile product may impact social and environmental factors and positive attributes at some points of the supply chain may have negative comportment at a later stage. The assessment of a product over its entire life cycle is thus crucial for consideration of both social and environmental impacts of the production process. Social and environmental labels correspond to approaches for furnishing information about social and environmental aspects of the production process. The labels provide useful information to consumers regarding constituents of products they buy in terms of environmental and social responsibility standards (Koszewska, 2015; Payne, 2015). The social and environmental labelling scheme enables the availability and visibility of sustainable products, thereby convincing consumers to buy products that contribute to social and environmental benefits.

Social and environmental labels are divided into three groups on basis of LCA:

- Single attribute labels such as organic content standard (OCS) or Oeko Tex Standard 100 comprises particularly one stage of the product's life cycle generally extraction of raw material or product usage stage.
- Multi-attribute labels — considers each stage of the product life cycle offering a more fragmented version of the product life cycle.
- Life cycle-oriented labels are targeted towards assessment of the entire life cycle initiating from the design phase.

18.4.2   Different phases of a product's life cycle

Life cycle-oriented labels are most relevant in the context of LCA with assessment being performed at different phases of product life cycle such as:

- Designing Phase
- Production & processing

- Distribution
- Consumer Usage
- End of life

Designing Phase - The assessment of processes and design decisions to achieve sustainability by designers is possible by employing underlying thinking of life cycle analysis. Each stage in the life of a garment right from fibre and textile through to consumer use and eventually disposal, reuse and disassembling for recycling of fires can be considered by the designer by employing LCA. LCA serves as an effective tool for designers to plan the environmental impact of their designs at both input (impact of raw material in pre-production) and output (emission and waste generation during production, usage and disposal). Thus designers can plan the impact of garments in the design stage.

The production and processing phase - is regarded as the most efficient phase owing to land and fertilizer requirements during cultivation and impact on the environment like eutrophication, agricultural land occupancy and transformation of natural land. Waste water generation and its contamination, water, energy expenditure and raw material characteristics are crucial in the production and processing phases. Impact of energy and water consumption resulting in depletion of fossil fuel, climate change, ozone depletion etc are detrimental consequences occurring along the textile supply chain.

Distribution phase - pertains to transportation of raw materials to factory site and dispatch of finished goods from warehouse to retail outlet. The mode of transportation and location between factory site and outlet can influence carbon emissions owing to fuel consumption and thus LCA at the distribution phase is crucial to assess the environmental impact (Fig. 18.4).

Consumer usage phase - Toxicity indicators associated with human beings and water ecosystems are a result of detergent and energy utilization during the washing process. The use phase owing to high water consumption during washing becomes more important than the production and processing phase.

End-of-life phase - The disposal treatments like incineration, landfill and recycling processes are part of the end-of-life phase of a product and have a smaller environmental impact compared to all other phases of the product's life cycle (Zhang and Chen, 2019; Fidan et al., 2021; Rana et al., 2015; Payne, 2011).

18.4.3 LCA classification & prerequisites for the LCA model

LCA is capable of measurement, analysis and aggregation of environmental impacts in association with energy and raw material consumption, emissions and other factors like materials inputs, waste generated from raw material procurement to final product distribution and disposal. LCA attempts to include impacts from raw materials to disposal phase of product (cradle to grave) or raw material to point of sale (cradle to gate) in case the consumers are not traceable. The different phases of the product's life cycle were discussed in the previous section. This section will highlight the classification of LCA on the basis of the phases of the product's life cycle under consideration and prerequisites for the LCA model.

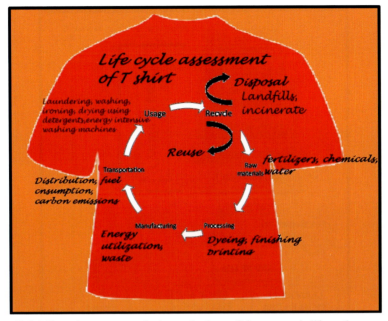

Figure 18.4 Life cycle assessment during each phase of the garment's life cycle.

LCA can be classified into the cradle to gate and cradle to grave on the basis of phases of product's life cycle under consideration.

Cradle-to-gate assessment excludes the use and disposal phase of the product and involves a partial product life cycle from resource extraction or raw material procurement (cradle) to factory gate (till the end product is shipped to consumers). Cradle-to-gate assessments serve as the basis of business-to-business EDPs (Environmental Product Declarations).

Cradle-to-gate LCA is related to quantifying emissions till the product is ready for dispatch from the factory while cradle-to-grave LCA includes the complete life cycle of the product to its disposal and reuse, recycling etc. Cradle-to-gate LCA for cotton clothing for instance shall include assessment related to energy utilization, water consumption, emissions during cotton cultivation, fibre production, manufacturing and garment assembling (Fig. 18.5).

18.4.3.1 Prerequisites for the LCA model

The process of LCA demands defining system boundary and functional unit. The former entails the parts or components included in the system while the latter relates to for what entity (e.g., shirt or trouser) or unit environmental impact is quantitatively measured. The functional unit defined by service delivered is the product's measuring unit and signifies the attainment of quantified performance within a prescribed time frame.

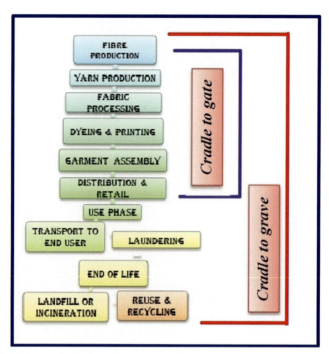

Figure 18.5 Cradle to gate and cradle to grave LCA.

As the functional unit is selected, the next step involves the development of a life cycle inventory (LCI) to describe the inflow to outflow from nature with respect to the functional unit.

A simplified LCA model for implementation in textile industries can be designed by considering energy and chemical utilization excluding transportation. The process begins with the selection of fibre type. A product comprising blended fibres requires a separate model for each fibre in the blend with due consideration of fibre proportion in the blend. Identification of unit operations namely spinning, weaving, knitting, processing, cutting and sewing room operations for garments are crucial. The selection of fibre type is followed by assigning default values for energy, chemical, detergent usage, disposal, incinerate, data for fibre, fabric yield and cost for each unit operation.

Default values, source of data and conditions in which data is valid are found in the database. The elementary and money flow data for different unit operations are listed below:

- Resource utilization of crude oil, coal, natural gas, water, fossil fuel, water, arable forest and another land.
- Air emission of CH_4, SOx, NOx, NH_3, CO_2 and particulate matter.
- Water discharge with BOD, COD, total phosphorus and Nitrogen.
- Cost.

Further, the model includes inventory data aggregation, characterizing global warming, acidification, eco-efficiency and photo oxidant creation potential.

Transportation becomes essential to be included in the model in case assessment of non-efficient mode of transporting garments from warehouse to outlets is followed. Packaging materials may be included if a significant amount of packaging materials are used for packaging final products. Waste water treatment plants and data for chemicals can also be included in the model.

Identification, assessment and measurement of material inputs and energy requirement outputs of useable products and wastage emissions are crucial at each phase of the product life cycle. Optimal improvement measures are identified followed by an estimation of eco-efficiency (Payne, 2011; Koszewska,; Payne, 2015).

18.4.4 Phases of Life cycle assessment (LCA)

LCA also referred to as environmental LCA, enables objective and subjective evaluation of product resource requirements and in turn its environmental impact during each phase of the product life cycle (Saxce et al., 2012).

The prime attribute of LCA is the addressal of environmental aspects and potential impacts on the environment in terms of resource utilization and emissions to the environment throughout the entire life cycle of the product spanning from acquisition of raw material to production, usage, recycling and disposal. The product development phase and its associated activities namely the selection of materials and processes can majorly impact the environment.

The LCA framework comprises four phases as enlisted below:

- Goal and scope definition
- Life cycle Inventory analysis (LCI)
- Life cycle Impact assessment (LCIA)
- Life cycle Interpretation

Goal and scope definition comprises the first phase of LCA. This phase defines the study and the central idea for conducting it. The goal and scope should be defined at the outset of the study before the data collection stage. Although this phase is associated with the simple introduction of LCA but it is an important phase of LCA since it enables the determination of the exact approach to be followed while implementing LCA (Fig. 18.6).

Life cycle inventory analysis (LCI), the second phase of LCA is a comprehensive analysis that takes into account the inventory of energy and materials required at each stage in the life cycle of the product in relation to the system under consideration. It involves the listing of utilized resources (environmental inputs) and emissions (environmental outputs) that are associated with the product and results in a comprehensive understanding of environmental significance. The phase involves data collection essential for fulfilling the defined goals. LCI.

Life cycle impact assessment phase (LCIA), the third phase of LCA provide additional information for assessment of product systems. The phase comprises

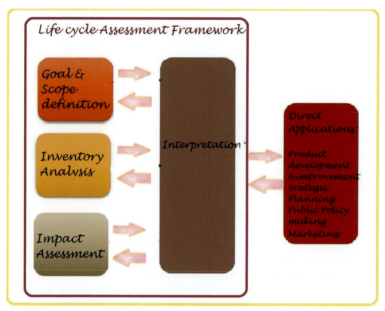

Figure 18.6 Phases of life cycle assessment.

characterization factors (a linear factor) for deriving the quantitative relationship between inflow and outflow to determine their environmental impacts (Measuring,; Reducing, 2021).

Life cycle interpretation, the final phase of LCA involves summarization and discussion of results of LCI thereby drawing out conclusions, recommendations and decision making as per the defined goal and scope in phase 1. Interpretation includes direct applications, product development and improvement, strategic planning, public policy making and marketing.

The phases of LCA have been discussed in this section. The following sections of the chapter will emphasize on relative merits and demerits of LCA in the context of textile industries.

18.4.5 Merits associated with LCA implementation in textile supply chain

The life cycle of textile and fashion products follows a complex and multi-tiered approach and has a massive environmental impact. It thus becomes crucial to implement a versatile and robust assessment methodology such as LCA for evaluation of environmental impacts at various phases of the textile product's life span. LCA has proved to be a lucrative and promising tool offering several benefits for the textile and fashion industries. The following section shall discuss some of the merits associated with LCA implementation in the textile supply chain.

Life cycle analysis of textiles and associated carbon emissions 425

- LCA implementation enables analysis of the environmental impact of the entire life cycle of a product or service in a systematic and scientific manner.
- LCA enables holistic examination of environmental impact in accordance with the utilization of resources, raw materials, its usage and disposal at end of life.
- LCA serves as an effective tool for quantitative assessment of the environmental advantages of recycling fibres. Consideration of energy, water and chemical impacts of a product system from cradle (raw material stage) through to production, distribution, consumer usage and disposal forms the basis of LCA. Apart from its quantitative aspects, LCA is also a lucrative qualitative framework for tracing material flow through complex systems like interconnected fibre, textile and apparel supply chains (Payne, 2011; Koszewska, 2015).
- Life cycle analysis is an effective approach for successful and precise interpretation of the whole cycle of chosen products, processes and services. LCA is regarded as a decision support tool that includes the compilation and evaluation of inventory data pertaining to the products and services (Bayda et al., 2015).
- Furnishing useful information pertaining to strategic planning, priority setting, designing and redesigning products or processes to decision-makers and stakeholders in industries.
- Improvement of product's environment performance at various stages of life cycle by identification of opportunities for improvement.
- LCA methodology serves as a guiding tool for companies to evaluate and communicate the impact of their processes and products on the environment. Moreover, it offers several benefits to manufacturing units and stakeholders by addressing eco-efficiency. Carbon footprint analysis is considered to be a crucial criterion for suppliers' evaluation from an environmental viewpoint thereby endorsing means and measures for improvement of the supply chain and realization of green products via eco-design (Moazzem et al., 2018).
- Identification and selection of key indicators of environment performance (such as measurement techniques)
- LCA measures green improvement, contributes to waste minimization, for estimation of CO_2 and greenhouse gas emissions and investigates energy and water flow in the process (Kalliala and Nousiainen, 2000).
- Marketing via adopting several strategies such as the implementation of eco-labelling scheme, making environmental chain.

18.4.5.1 Open & closed loop recycling for LCA — A boon for designers & brands

The innumerous merits of LCA implementation in the textile and fashion supply chain have already been discussed in the previous section. Apart from systematic and scientific analysis of the environmental impact of the entire life cycle of a product, tracing material flow through complex systems like interconnected fibre, textile and apparel supply chain, measurement of green improvement and contribution to waste minimization, LCA also serves as a boon for designers and brands by providing innovative avenues such as open and closed-loop recycling to deal with disposal related environmental impact.

The assessment of processes and design decisions to achieve sustainability by designers is possible by employing underlying thinking of life cycle analysis. Each stage in the life of a garment right from fibre and textile through to consumer use and eventually disposal, reuse and disassembling for recycling of fires can be considered by the designer by employing LCA. LCA serves as an effective tool for designers to plan the environmental impact of their designs at both input (impact of raw material in pre-production) and output (emission and waste generation during production, usage and disposal). Thus designers can plan the impact of garments in the design stage.

The raw material subjected to recycling can undergo either open-loop or closed-loop recycling. *Open-loop recycling* also referred to as down-cycling involves reclaiming the fibre for usage in an altogether different product thereby entering into a new product life cycle after the end of the useful life of the garment. The recycling process is referred to as down-cycling as the new product developed from reclaimed fibres is of less economic value as compared to the virgin garment being recycled. The collection and recycling of fibres for new products from pre-consumer textile offcuts from apparel units is an example of down-cycling or open-loop recycling. OLR results in the temporary diversion of waste from landfills or incineration.

Open-loop recycling includes:

(i) Pre-consumer textile waste like off-cuts from the cutting room in apparel industries.
(ii) Post-consumer textile waste (whole garment).
(iii) Post-consumer PET bottles being manufactured into recycled PET fibres.

The amount of solid waste ending up in landfills can be considerably reduced via waste from one product system serving as feedstock for another with open-loop recycling.

LCA on open-loop recycling of polyester bottles has suggested utilization of recycled polyester bottles in fibre for apparel applications shows energy benefits.

LCA conducted for examination of the mechanical and chemical recycling of polyester suggested environmental benefits offered by recycled polyester in contrast to virgin polyester in terms of energy consumption, greenhouse gas potential, eutrophication and acidification.

Closed-loop recycling involves reprocessing of collected textiles into new fibres to be used for new garments re-entering the same production system that was used for virgin textiles.

The carbon footprint of products was also observed to reduce by recycling pre-consumer textile waste or process waste in LCA analysis (Payne, 2015; Wendin, 2016).

Cradle-to-cradle design and manufacturing bypassing the grave stage enables reusing the valuable fibres via closed-loop manufacturing methods. Fibre choice for fashion designers can be critical and trade-off between environmental impact and sustainable aspects. Polyester, petro chemical-based fibre is obtained from non-renewable sources but its processing utilizes a fraction of the water required for organic cotton cultivation. Bamboo is renewable and sustainable fibre however fibre processing is energy-intensive and polluting process.

Implementation of cradle to cradle approach demands watchful sourcing and selection of all garment elements, thus imposing limits on feasible design. At present, the cost incurred (in terms of energy) in disassembling garments, and reprocessing to fibre and textile is higher compared to energy expenditure for the production of virgin polyester (Koszewska). However, brands are keen to adopt the approach to address environmental issues.

Patagonia, dealing in winter sportswear has adopted closed-loop recycling of synthetic textiles. The common thread program aims at recycling polyester polar fleece jackets, repolymerizing the cloth for extrusion to the new fibre of virgin quality. Likewise, H & M has begun using recyclable polyester chiffon.

18.4.6 Application areas of life cycle analysis (LCA)

The preceding section elaborately discussed the several advantages of LCA in the textile and fashion supply chain. The holistic assessment approach of LCA over the entire lifecycle of the product enables the widespread acceptability and application of LCA. The applications of LCA are not just restricted to the textile and garment sector but several other industrial establishments have successfully implemented and benefitted from LCA as well:

- Consumer goods
- Livelihood
- Heavy industry
- Transportation
- Energy production
- Process Industry
- Infrastructure

The textile and fashion industries have been successfully implementing LCA in their supply chain for quantification of environmental impacts of textile and garment sectors. The implementation of LCA can serve as an effective tool for highlighting the impact of a garment's life cycle on the environment and the selection of low-impact materials and processes in the entire supply chain.

- Further LCA can be effective for drawing comparisons between products in terms of their environmental impacts.
- For understanding emissions in the textile sector at varying scales e.g., emissions in the production process
- To have an overview of the entire global apparel system spanning from fibre production to yarn and fabric preparation, fabric processing, garments, distribution and end of product's life cycle (transportation and laundry in use phase).
- Streamlined LCA for determination of energy footprints of the garment, enabling industries to accomplish the assessment of life cycle energy consumption.

18.4.7 Demerits & challenges associated with life cycle assessment

LCA has been successfully implemented by several commercial establishments owing to numerous benefits it offers in terms of environmental impact assessment as discussed in previous sections. However, there are certain demerits and challenges associated with LCA, which need to be addressed to explore the full potential of this assessment tool.

428   Sustainable Fibres for Fashion and Textile Manufacturing

- Environmental LCA considers an input-output model to make an inventory of the flow of raw materials, energy and emissions overlooking the procedures and protocols followed within the industry with the production process being considered as a black box. The non-inclusion of social criteria like wages, working hours and discrimination generally encountered during the production process in industries makes the black box model and irrelevant in the context of social impacts (Payne, 2015; Wendin, 2016; Bianchi and Birtwistle, 2010).
- LCA serves as a tool rather than a system.
- LCA cannot be classified as a local risk assessment tool due to its inability to address localized aspects.
- LCA follows a steady rate approach rather than a dynamic one.
- Technical assumptions and value choices considered during LCA do not have a scientific basis for assumption (Bianchi and Birtwistle, 2010).
- Does not take into consideration market mechanisms or technological development associated with secondary effects.
- Processes in economy and environment are regarded as linear by LCA.
- Complex and resource-intensive assessment.
- Demands huge investment of time and money.
- Assessment is not possible in-house by textile and garment industries limiting the adoption of LCA in the textile industry.
- Validity of LCA approaches for incremental changes in product and specific geopolitical regions seems abstruse.
- Long-term planning using LCA is difficult.
- LCA is inadequate if product evaluation demands social and economic aspects since LCA focuses on environmental aspects without paying any heed to social and economic factors.
- Use phase analysis takes into consideration the consumer behaviour, which is dynamic across the globe. Thus the inclusion of the use phase owing to variable behaviourism of consumers in LCA may present marked uncertainties thus making the assessment questionable. Most LCA models emphasize on cradle-to-gate impact of clothing and do not take into consideration the environmental impact of the end of life or use phase.
- Challenges in analysis of environmental benefits of recycling textiles by employing LCA methodology include setting boundaries or cut-off points for LCA and ways of allocation of energy savings from recycling between different production systems (Bianchi and Birtwistle, 2010, LCA).

18.5   Conclusion

The textile and fashion industry is ecologically one of the most environmentally polluting, energy and chemical-intensive industries owing to unsustainable characteristics of the life cycle of textile and clothing. Further, the inception of fast fashion to provide cheap and affordable clothing and consumers' negligence in usage and disposal of end products result in deleterious environmental impacts and carbon emissions. The intricate, dispersed and globally stretched system with less transparent manufacturing processes necessitate watchful monitoring and assessment of each phase of the textile supply chain and its associated environmental impact. The assessment of sustainable textile products demands an integrated and holistic approach since protocol followed in one phase of a product's life cycle can directly or indirectly affect other phases and the overall performance of the environment. Several assessment tools are claimed effective for environmental impact however most of the tools disregard the usage phase of textile products, which significantly impacts the environment. LCA is a standardized framework methodology to quantify the impact of products, processes and services on the environment. The environmental performance of products in terms of energy expenditure, greenhouse gas emissions, water footprints and pollutants can be quantified by employing LCA tools and techniques.

The textile and garment industry has been an ardent follower of LCA methodology for evaluation of myriad of aspects and life cycle stages of end product right from fibre cultivation to yarn spinning, fabrication techniques until fabric processing, garment making and disposal at the end of the useful life of the product.

The textile industry is undoubtedly one of the largest consumer-intensive sectors however recycling and reclaiming practices adopted by the industry are not considered adequate. Thus open and closed-loop recycling by LCA enables effective recycling techniques for textiles.

The impact of technology and factory settings along with policy changes, consumers' preferences and buying behaviour on the environment can also be evaluated by LCA implementation.

The major barrier to compliance with a sustainable approach in the textile and fashion industry and thus the implementation of LCA is the short life span of textile end products. The manufacturing of a durable end product demands additional processes and different materials thereby leading to enhanced energy consumption and more waste generation. Consequently, increased lifetime does not tend to be so advantageous owing to the negative effect of additional processes on the product's global environmental impact. Moreover, the textile and fashion industry is a dynamic industry however LCA follows a steady rate approach rather than a dynamic one and does not take into consideration market mechanisms or technological development associated with secondary effects. The consumer behaviour that forms the basis of use phase analysis is dynamic across the globe thereby presenting marked uncertainties and inappropriate LCA owing to variable behaviourism of consumers.

The challenges associated with LCA need to be judiciously addressed to explore the full potential of this assessment tool. The limitation presented by LCA have been addressed by the introduction of other simplified tools, however, the tools are in the development stage and reliability of data obtained from the tool is questionable (as per standards of LCA).

References

Alay, E., Duran, K., Korlu, A., 2016. A sample work on green manufacturing in textile industry. Sustainable Chemistry & Pharmacy 3, 39–46.

Bayda, G., Ciliz, N., Mammadov, A., 2015. Life cycle assessment of cotton textile products in Turkey. Resources, Conservation and Recycling 104, 213–223.

Bevilacqua, M., Ciarapica, F.E., Giacchetta, G., Marchetli, B., 2011. Carbon footprint analysis in the textile supply chain. International Journal of Sustainable Engineering 4 (1).

Bianchi, C., Birtwistle, G., 2010. Sell, give away or donate, an exploratory study of fashion clothing disposal behaviour in two countries. The International Review of Retail, Distribution & Consumer 20 (3).

Bloomberg. bloomberg.com/news/articles.

Choudhury, A.R., 2014. Environmental impacts of the textile industry and its assessment through life cycle assessment. In: Roadmap to Sustainable Textiles and Clothing, Textile Sciences & Clothing Technology. Springer Science +Business Media, Singapore. https://doi.org/10.1007/978-981-787-110-7-1.

Fidan, F.S., Aydogan, E.K., Uzal, N., 2021. An integrated life cycle assessment approach for denim fabric production using recycled cotton fibres and combined heat and power plant. Journal of Cleaner Production 287.

Kalliala, E., Nousiainen, 2000. Life cycle assessment environmental profile of cotton and polyester fabrics. Autex Research Journal 1 (1), 8—20.

Koszewska, 2015. Life cycle assessment and the environmental and social labels in the textile and clothing industry. In: Handbook of Life Cycle Assessment (LCA) of Textiles & Clothing. https://doi.org/10.1016/B978-0-08-1001691.0015-0.

Laitala, K., Klepp, I.G., Henry, B., 2018. Does use matter? Comparison of environmental impacts of clothing based on fibre type. Sustainability 10 (7), 2524.

LCA Update of Cotton Fibre and Fabric Life Cycle Inventory Resource. cottoninnc.com/LCA/2016-LCA-Full-Report.

Measuring Fashion, Environmental Impact of the Global Apparel and Footwear Industries Study Quantis — Intl.com/wpcontent/.

Moazzem, S., Daver, F., Crossin, E., Wang, L., 2018. Assessing environmental impact of textile supply chain using life cycle assessment methodology. Journal of The Textile Institute 109 (12), 1574—1585.

Muthu, S., LCA of cotton shopping bags. In: Handbook of Lifecycle Assessment (LCA) of Textiles and Clothing.

Muttu, S.S., 2015. Environmental Impacts of the Use Phase of the Clothing Life Cycle, Textiles & Clothing. Woodhead Publishing Series in Textiles, pp. 93—102.

Payne, A., 2011. The Life Cycle of the Fashion Garment and the Role of Australian Mass Market Designers.

Payne, A., 2015. Open and closed loop recycling of textile and apparel products. In: Handbook of Lifecycle Assessment (LCA) of Textile & Clothing, pp. 103—123.

Qian, W., Ji, X., Xu, P., 2021. Textile Research Journal. https://doi.org/10.1177/0040517521100.

Rana, S., Pichandi, S., Moorthy, S., Bhattacharyya, A., 2015. Carbon footprint of textile and clothing products. In: Handbook of Sustainable Apparel Production.

Reducing the Footprint? How to Assess Carbon Emissions in the Garment Sector in Asia/LO Asia — Pacific Report, 2021, ISBN 978-922-0347-34-8.

Rosa, A.D., Grammatikos, S.A., 2019. Comparative life cycle assessment of cotton and other natural fibres for textile applications. Fibers 7, 101. https://doi.org/10.3390/fib7120101.

Sandin, G., Roos, S., Johansson, M. Environmental Impact of Textile Fibres — What We Know and what We Don't Know the Fibre Bible Part 2, Mistra Future Fashion Report, vol. 219: 03.

Saxce, M.D., Pesnel, S., Perwulez, A., 2012. LCA of bedsheets — some relevant parameters for lifetime assessment. Journal of Cleaner Production 37, 221—228.

Shen, L., Patel, M.K., 2010. Life cycle assessment of man made cellulose fibres. Lenzinger Berichte 88, 1–59.

Tekstiller, P., Cevre, V., Degerlendirmese, Y.D., 2009. Cotton textiles and the environment. In: Proceedings, Tckskiller.

Velden, N., Patel, M., Vogtlander, 2014. LCA bechmarking study o textiles made of cotton, polyester, nylon, acryl or elastene. International Journal of Life Cycle Assessment 19, 331–356.

Wendin, M., 2016. LCA on Recycling Cotton, Technical Report. https://doi.org/10.13140/RG2.2.222598.57927.

Werf, H., 2004. Life cycle analysis of field production of fibre hemp, the effect of production practices on environmental impacts. Euphytica 140, 13–23.

Zamani, B., Sandin, G., Peters, G.M., 2017. Life cycle assesment of clothing libraries: can collaborative consumption reduce the environmental impact of fast fashion. Journal of Cleaner Production 162, 1368–1375.

Zhang, Chen, 2019. Carbon emission evaluation based on multi-objective balance of sewing assembly line in apparel industry. Energies 12 (14). https://doi.org/10.3390/en12142783.

Index

'*Note:* Page numbers followed by "f" indicate figures and "t" indicate tables.'

A

A/O process. *See* Anaerobic-aerobic process (A/O process)
Abrasion resistance, 189
Absorbable organic halides (AOX), 166
ACBC. *See* Anything can be changed (ACBC)
Accessories, 131
 industry, 129
Acetic acid, 263–264
Acidic amino acids, 207
Acids, 262
 effect of, 253
 dyes, 263–264
Acremonium species, 166
Acrylic fibres, 6, 14–15
Actinase, 174
Acyl chlorides, 251
Additives, 238
Adenosine Triphosphate (ATP), 150
Agricultural land occupation (ALO), 171
Agricultural wastes, 135, 273
Ahimsa silk, 170
Air-jet texturing process, 296
AIRMEM, 296
Alexandra K (Italian clothing brand), 155
Alkali, 262
 deacetylation process, 42
Alkaline hydrolysis, 395–396
Alkalis, effect of, 253
ALO. *See* Agricultural land occupation (ALO)
Alpaca (*Vicugna pacos*), 187–188
 abrasion resistance, 189
 durability, 191
 fibre, 188–191
 flame resistance, 190
 prickle factor, 190
 resistance to compression, 189

shine, 190
simple preparatory process, 189–190
 low allergenic, 190
 odour resistance, 190
strength, 189
warmth, 188–189
water resistance, 189
wrinkle and shrink resistance, 190
Alpha (industrial hemp), 81
Alternaria alternate, 166
Amber hued vegan leather, 145–146
American forests, 355, 355f
Amino acids, 195
Aminolysis, 396
Ammonium polyphosphate (APP), 202
Amni Soul Eco material, 329
Anaerobic-aerobic process (A/O process), 264
Angora
 fibres, 188
 rabbit, 187
 wool, 186–188
Animal fibres, 28–29, 182
Animal leather
 for fashion ensembles—nonconformity of sustainable principles, 130–133
 leather apparels and accessories, 131f
 mushroom leather outshines, 138–139
 properties and benefits of mushroom leather over, 137–139
Animal protein fibres, 182–203
 alpaca fibre, 188–191
 angora wool, 186–188
 casein fibre, 198–203
 cashmere wool, 186
 chicken feather fibre, 195–198
 microscopic views of wool, silk and degummed, 183t
 sustainable silk fibre, 191–195
 sustainable wool fibres, 184–185

Animal sources, 42
Antheraea assamensis. See Muga
  (*Antheraea assamensis*)
Antheraea mylitta. See Tasar (*Antheraea mylitta*)
Anti-microbial efficacy of orange fibre, 282–283, 283t
Anything can be changed (ACBC), 154
AOS. *See* Australian Organic Standard (AOS)
APP. *See* Ammonium polyphosphate (APP)
Apparel industry, 129
Apple, 148
Aquatic lotus (*Nelumbo nucifera*), 95
Aquatic macrophytes, 123–124
Aramid, 287
Ardil, 216
Arsenic (As), 15
ATP. *See* Adenosine Triphosphate (ATP)
Aulive (Indian brand), 143
Australian Merino wool, 164–165
Australian Organic Standard (AOS), 65
Autoclaving methods, 135–136
Azadirachta indica. See Neem (*Azadirachta indica*)
Azo dyes, 131

B
Baby wear, 215
Bamboo, 406–407
  bamboo-based fabrics, 230
  fibre, 39–40, 229–232, 341
    advantages, 234
    blend with silk and, 202
    manufacturing process of regenerated, 231–232
    research on, 232–234
  products, 229–230
  sustainability aspects of natural fabrics and knitting from, 78–89
  trees, 229
  viscose, 229–230, 232
Bast fibres, 79, 88, 338, 345–346
  chemical composition of plant fibres, 85t
  mechanical and physical properties of natural fibres, 86t
  properties of, 85
Batch-wise polymerization process, 250
Bathrobe's life cycle, 408–409

BCI. *See* Better Cotton Initiative (BCI)
BCSS. *See* Better Cotton Standard System (BCSS)
Better Cotton Initiative (BCI), 66–68, 365
  Bt cotton, 68–69
    production *vs.* organic cotton production, 68–69
Better Cotton Standard System (BCSS), 66
Bio materials, 133–134
Bio recycling methods, 247
Bio-logical cotton. *See* Organic cotton
Bio-nylon, 328–329
Biochemical extraction, 99
Biocompatibility, 43
Biocomposites, 297–298, 336–337
Biodegradability, 4, 6
Biodegradable fibres, 390
Biodegradable material, 275–276
Biodegradable synthetic nylon, 329
Biodegradable wastes, composites derived from
  aspects of biodegradability of natural cellulose-based fibres, 337–341
  biocomposites, 336–337
  design innovations, 343–344
  designing out waste with circularity approach, 335
  material thinking, 335
  natural fibres as reinforcement for composites materials, 342
  opportunities and challenges around natural fibres reinforced polymers, 342–343
  streamlining waste, 344
  textile composites, 336
  textile waste, 333–335
Biogenic methane, 164
Biological enzymes, 27–28
Biological extraction, 98
Biological methods, 35–36, 394–396
Biological micro-organisms, 337
Biomass waste, 297–298
Biomaterials, 166, 172–173
  biomaterial-based anti-moth treatments, 166
Biomedical applications of silk, 173
Bioplastic, 153
Bioskin, 150–151
Biovoltines (BV), 169

Bis(2-hydroxyethyl) terephthalate (BHET), 35, 250, 395
Bisphenol A (BPA), 311
Bisphenol S (BPS), 311
Blending
  of casein fibres, 202–203
    blend with cotton and cashmere, 202
    blend with silk and bamboo fibres, 202
    blend with wool and cashmere, 202
  of groundnut fibre, 217
  programs, 57
  of soybean fibre, 212–213
Blue water footprint (BWF), 171
Bluesign standard, 362
Bolt thread, 134, 143
Bombyx mori, 29, 168
BPA. *See* Bisphenol A (BPA)
BPS. *See* Bisphenol S (BPS)
Braiding process, 166–167, 301
Brands promoting recycled nylon, 328–329
  bio-nylon, 328–329
  biodegradable synthetic nylon, 329
  ECONYL, 328
  EcoRib, 328
Brodo (Indonesian footwear brand), 146
BSCI. *See* Business Social Compliance (BSCI)
Bt cotton, 68–69
  production *vs.* organic cotton production, 68–69
Business Social Compliance (BSCI), 371–372
Business supply chain, 355
Butane-1,4-diol, 251
Butanediol, 251–252
BV. *See* Biovoltines (BV)
BWF. *See* Blue water footprint (BWF)

C
C2C certification. *See* Cradle-to-cradle certification (C2C certification)
Cactus, 148
Cadmium (Cd), 15
CAGM. *See* Centre for Advancement of Garment Making (CAGM)
CAGR. *See* Compound Annual Growth Rate (CAGR)

Calotropis gigantea. *See* Milkweed (*Calotropis gigantea*)
Camel, 187
Canopy, 359
Caprolactam, 257–259
  recovery of, 265
Caprolactam fibre, 247, 257–265
  application of sustainable, 262–263
  caprolactam or nylon fibre, 257–265
  characteristics of nylon, 261
  chemical property, 262
    alkali and acids, 262
    electrical and flammability, 262
    light and biological impact, 262
    organic solvent and bleaches, 262
  dyeing and finishing, 263–264
  LCA, 264
  manufacturing process, 257–261
  physical property, 261–262
    heat effect, 262
    tenacity, density and moisture regain %, 261
  polyester fibre, 248–257
  properties, 261–262
  recovery of caprolactam, 265
  recycling, 264–265
Carbohydrates, 80
Carbon dioxide (CO_2), 3–4, 88, 393–394
Carbon dioxide equivalent (CO_2-e), 164, 411
Carbon disulphide (CS_2), 14, 40
Carbon emissions
  ardent need of LCA, 405–413
  environmental impact
    different textile raw materials, 406–409
    manufacturing processes, 410–411
    transportation, usage & disposal phase, 411–413
Carbon footprint (CF), 162–163, 413–419
  effective strategies to mitigate environmental impact & carbon footprint, 416–417
  with phases of textile supply chain, 414–416
    associated with fibre types, 414–415
    associated with manufacturing, garment assembling, usage & disposal phase, 415–416
  of textiles, 411

Carbon footprint (CF) (*Continued*)
    tools for environmental impact
        determination, 418–419
Carbon gas, 247
Carbon Neutral standard (CN standard), 360
Carbonizing process, 161
Carboxylase, 174
Carding, 161
Carotenoids, 198
Casein fibre, 41–42, 182, 198–203
    blending of, 202–203
    care of, 203
    drying, 203
        dry cleaning, 203
        ironing, 203
    manufacture of, 199–200
    merits-demerits of, 199
    properties of, 201–202
    uses of, 203
        properties of SPF milk protein fibres,
            204t
    washing, 203
    wet processing of, 200–201
Cashmere, 182, 188
    blend with cotton and, 202
    blend with wool and, 202
    wool, 186
CB. *See* Crossbreed (CB)
CCS. *See* Content Claim Standard (CCS)
CE. *See* Circular Economy (CE)
CED. *See* Cumulative Energy Demand
    (CED)
CELC. *See* European Confederation of Flax
    and Hemp (CELC)
Cellulose
    cellulose-based composites, 343
    extraction, 40
Cellulosic fibres, 35, 41, 78–80, 167
    recycling, 35
Cellulosic polymers, 337
Cellulosic waste, 35
Cement, 78
Centre for Advancement of Garment
    Making (CAGM), 357
Certification, 64, 355
CF. *See* Carbon footprint (CF)
CFF. *See* Chicken feather fibres (CFF)
CHD. *See* Coronary heart disease (CHD)
Chemical analysis of lotus fibre, 101–102

Chemical composition
    of lotus fibre, 101
    of macrophyte and wetland plant fibres,
        121–123
Chemical extraction, 99
Chemical fibres, 7
Chemical methods, 35, 394–396
Chemical oxidation treatment, 255–256
Chemical process(ing), 16, 40
    of groundnut fibre, 217–218
        bleaching, 218
        dyeing, 218
        scouring, 217–218
    of polyester, 314
    of SPF, 213–214
        desizing, 213
        dyeing, 214
        scouring and bleaching, 213–214
Chemical techniques, 27–28
Chemicals, 35–37
Chicken feather fibres (CFF), 195–198, 221
Chiffon, 12
Child labour, 57
Chip preparation, 40
Chitin fibre, 42
Chitosan fibre, 42
Chlorine, 12
Chromate, 131
Chrome tanned leathers, 132
Chromium (Cr), 15, 138
    chromium-based tanning agents, 132
Chymotrypsin, 174
Circular Economy (CE), 335
    concept, 27–28
    model, 388
Citrus peels, 276
Clean clothes campaign, 356
Cleaning process, 61, 316
Climate change effect on wool, 165
Climate Neutral certification, 360
Clipping, 42
Clipping method, 187
Closed tray system, 135–136
Closed-loop process, 140–142, 237
Clothing, 386
    business, 309
    from linen and hemp, 82–85
CN standard. *See* Carbon Neutral standard
    (CN standard)

Index

CO. *See* Common Objective (CO)
Coarse wool, 160
Cobalt, 131
Cocona fabric, 379
Cocona fibre, 288
Coconut (*Cocos nucifera* L.), 119−120
 fibre, 341
Cocoons, 170
Cocos nucifera. See Coir (*Cocos nucifera*)
Coffee bean, 288
 skins, 289
Coffee botanicas, 288−291
 coffee oil derived from SPG, 290t
 coffee waste, 289f
 extracted coffee oil parameters from SPG,
  291t
Coffee fabric manufacturing
 fabric manufacturing, 301
 fibre production, 299−300
 sustainability of, 299−303
 wet processing, 303
 yarn manufacturing, 300
Coffee fibres, 293−297, 299−300
 coffee botanicas, 288−291
 in footwear, 296−297
 products of, 293−296
  AIRMEM, 296
  coffee yarns, 294−295
  eco^{2}sy, 296
  P4DRY, 295
  S. Café AIRNEST, 295
  S. Café mylithe, 296
  sefia fabric, 295−296
 rPET, 291−293
 sustainability of coffee fabric
  manufacturing, 299−303
 sustainable products from coffee waste,
  297−299
Coffee grounds, 16
Coffee husks, 289
 composite from, 297−298
Coffee oil, 290−291
Coffee silver skin (CS), 288−289
Coffee sneakers from RENS, 297
Coffee waste, 289
 composite from coffee husk, 297−298
 extract of spent coffee ground as natural
  dye, 299
 sustainable products from, 297−299

Coffee yarns, 294−295
Coir (*Cocos nucifera*), 110
 fibres, 119−120
Colour fastness, 254
Commercial organic cotton, 58
Commercial product, application of lotus
  fibre for, 103−104
Commercial-scale fibre production, 11
Commercialized polyester, 247−248
Common Objective (CO), 356
Composites, 173
 materials, 297−298
  from coffee husk, 297−298
  natural fibres as reinforcement for,
   342
Compost, 39
Composting methods, 135−136
Compound Annual Growth Rate (CAGR),
  78, 87, 310
Compression process, 137
Conceptualization stage, 3−4
β-conglycinin, 207
Consumers, 385
 behaviour, 28, 42−43, 386−388
 sustainability, 42−43
Contaminated fibres, 27
Content Claim Standard (CCS), 377−378
Continuous polymerization (CP), 251
Control Union certifications (CU
  certifications), 367
Conventional agriculture, 67−68
Conventional chlorination process, 166
Conventional cotton, 67−68. *See also*
  Organic cotton
 comparison of organic cotton with, 55−56
  comparison between conventional and
   organic cotton, 55t
Conventional feedstock, 35
Conventional nylon, 321
Conventional petrochemical fibres, 29
Conventional process of polyester fibre,
  250−252
Conventional wet milling, 150
Conventional wool, 185
Convolutions, 52−53
Copper (Cu), 81, 131
'Corn + Cotton Initiative',
  154−155
Corn eco-leather, 150−151

Corn fibre, 148–156. *See also* Mushroom fibre
  application areas of corn, 153–154
  challenges associated with corn fibre production and usage, 155–156
  corn leather, 150–153
  corn plastic, 153
  prime choice of sustainable fashion brands, 154–155
  production process of, 150
  salient features of, 149
Corn leather, 148, 150–154
Corn plastic, 153
Corn protein fibre, 218–219
  properties of sustainable protein fibres, 220t
Corn starch, 40
Corn-based products, 154
Corona methods, 84
Coronary heart disease (CHD), 14
Cotton, 3–7, 9–10, 28–30, 38–41, 51, 66, 70, 160–161, 236–237
  blend with cashmere and, 202
  fabric, 299
  ginning process, 338
  impacts of cotton farming, 30–31
Cotton fibres, 10, 28, 52–53, 287, 338
  comparison of lotus fibre with, 102–103
  cotton fiber structure, 52–53
Cotton Made in Africa (CMiA), 362–363
CP. *See* Continuous polymerization (CP)
Crabs, 42
Cradle to grave LCA analysis, 408–409
Cradle-to-cradle certification (C2C certification), 363–364
Cradle-to-gate assessment, 421
Crop residue management, 60
Cross-linked lyocell, 240
Crossbreed (CB), 169
Crude sorting method, 316
Crystallinity, 101–102
CS. *See* Coffee silver skin (CS)
CU certifications. *See* Control Union certifications (CU certifications)
Cumulative Energy Demand (CED), 171, 242
Cuprammonium rayon (Cupro), 379
Curaua, sustainability aspects of natural fabrics and knitting from, 78–89
1,4-cyclohexane-dimethanol, 249

Cysteine, 195
Cystine, 207

D
Damaged fibres, 34
Datura stramonium L., 166
Defoliants, 36–37
Degraded polymer, 174
Degummed silk, 170
Degumming, 170
  methods, 84
  of bast fibrous plants, 80
Denaturation, 208
Dendritic polymers, 84
Dew retting, 39, 79–80
Differential scanning calorimetry (DSC), 273–274
Dimethyl formamide (DMF), 14–15
Dimethyl terephthalate (DMT), 396
Disinfectants, 12
Disposal phase, environmental impact of transportation, usage &, 411–413
DMF. *See* Dimethyl formamide (DMF)
DMT. *See* Dimethyl terephthalate (DMT)
Domestic silk, 183
Dry cleaning, 203
Dry fabrics, 336
Drying, 203
DSC. *See* Differential scanning calorimetry (DSC)
Dupont's Sorona, 416–417
Dyeing, 393–394
  process of polyester fibre, 254–255
  process of textile fabric, 165
Dyes, 44

E
Earth, recycling nylon sweet for, 325
Eco leather, 133–134
Eco-efficiency, 160
Eco-friendly soybean baby clothing, 215
Eco^{2}sy, 296
Ecofriendly fibre, 229
Ecolabel(ling), 365, 390, 397
Ecological fibre, 232
Ecological textiles, 57
Econyl, 261, 326, 328
  fibre, 379
EcoRib, 328

Index

Ecovative Design, 143
EcoVero fibre, 380
EG. *See* Ethylene glycol (EG)
Eichhornia crassipes. See Water hyacinths
    (*Eichhornia crassipes*)
Electro-spun fibre, 182
End-of-life (EOL), 4
Energy consumption in MJ, 299—300
Energy-efficient buildings, 167
Environment, 386
    detrimental impact of textile and fashion
        supply on, 129—133
    polluting process, 138
Environmental effect, 247
Environmental hazards, 133
Environmental impacts
    carbon emissions & ardent need of,
        405—413
    of textile fibre production, 9—16,
        28—42
Environmentally friendly cotton.
        See Organic cotton
Environmentally preferred leather, 133—134
Enzymatic methods, 84
Enzymatic retting, 39
EOL. *See* End-of-life (EOL)
Epipremnum aureum. See Money plant
        (*Epipremnum aureum*)
Epoxy resin, 297—298
Eri (*Philosamia ricini*), 183
Escaro (Indian brand), 143
Esprit, 57
Ethanol, 35
Ethylene glycol (EG), 35, 249
Ethylene polyester, 249
Ethylene-Vinyl Acetate (EVA), 297
European Confederation of Flax and Hemp
        (CELC), 75
European Union (EU)
    Eco Label, 365
    regulations, 64
        of organic cotton European Union
            Regulation 2092/91,
            65
Eutrophication, 123—124
EVA. *See* Ethylene-Vinyl Acetate (EVA)
Everlane, 320
Expandable graphite, 221
Extra-long staple fibres, 52

F

Fabrics, 34—35, 75—78, 83, 173, 283
    making, 170
    manufacturing process, 301
    materials, 378—382
        Cocona, 379
        CUPRO, 379
        econyl, 379
        EcoVero, 380
        Filium, 380
        Luxe, 380
        Pinatex, 382
        Refibra, 378
        Repreve, 378
        Supima, 378
        Tasc, 382
        type of sustainable VR yarns, 381
        viscose, 381
        ZQ Merino wool, 378
    weaving, 170
Fabscrap non-profit recycling program,
        360—361
Facultative wetland (FACW), 110
Fair Labor Association (FLA), 372
Fair Trade movement, 366
Fair Wear Foundation, 372
Farm Development Program of Organic
        Exchange, 51
Fashion
    detrimental impact of textile and fashion
        supply on environment, 129—133
    industry, 27, 129, 386, 403, 405—406
    natural fibres, yarns, fabrics and knitting
        for, 75—78
    products, 27, 391
    revolution, 358
Fashion Forum Fellowship 500, 358
Fashion Transparency Index, 359
Fashionable shops, 57
Fast fashion, 129—130
FE. *See* Freshwater eutrophication (FE)
Feather fibres, 196—197
Fermentation, 150
Fertilizers, 36—37
Fibrefill, 249
    stuffing, 249
Fibres, 16—21, 39—40, 95, 173, 343
    drawing process, 100
    extraction method, 273—274, 277—278

440 Index

Fibres (*Continued*)
    manual extraction process of fibrous, 277f
    fibre-based composites, 16–21
    fibre-reinforced composite materials, 333
    formation, 210
    from lotus peduncle, 96
    from lotus rhizome, 98
    moisture, 99
    morphology, 121
      and properties, 279
    production process, 299–300
Fibroin, 170
Fibrous matrixes, 278–279
Filament fibres, 242
Filium, 380
Film, 173
Fine wool, 166–167
Finishing
    of caprolactam or nylon fibre, 263–264
    process of polyester fibre, 254–255
FITR. *See* Fourier transform infrared spectroscopy (FITR)
FLA. *See* Fair Labor Association (FLA)
Flame resistance, 190
Flame retardant treatments, 84
Flavobacterium sp., 264
Flavonoids, 273, 282–283
Flax (*Linum usitatissimum* L.), 3–4, 6–7, 28, 39, 41, 76, 79–80
    application of, 85–89
    fibres, 29, 339
    plant, 29
    sustainability aspects of natural fabrics and knitting from, 78–89
Food biomass, 293
Footwear
    coffee fibre in, 296–297
    coffee sneakers from RENS, 297
    Xpressole Panto, 297
Foreign Trade Association (FTA), 371–372
Forest, 120
Formaldehyde, 12
    resins, 131
Fossil fuel, 165
Fourier transform infrared spectroscopy (FITR), 102, 273–274, 281–282
Freshwater eutrophication (FE), 171
Fructose, 35

Functionalization methods, 84
Fungi, 135
Fungus, 42
Fusarium species, 166
Future footwear, 145
Fyrol PNX, 221

G
Ganoderma, 135
    G. lucidum, 135
Garment industries, 355
GDP. *See* Gross Domestic Product (GDP)
Genetically modified (GM), 58
    natural fibres, 80
    plants, 58
    seeds, 39
Georgette, 12
GHG emissions. *See* Greenhouse gas emissions (GHG emissions)
Gilling process, 161
Ginning process, 61
Global fibre market, 7–8, 77–78
Global industrial hemp market, 87
Global market of textile fibres, 30, 31t
Global Organic Textile Standard (GOTS), 368–369
Global Recycle Standard (GRS), 367
Global textile fibre production, 7
Global warming potential (GWP), 407, 413–414
Globular proteins, 208
Glucose, 35
Glycine max. See Soybean (*Glycine max*)
Glycinin, 207
Glycolysis method, 395
Goat, 187
Golden fibre. *See* Jute fibre
Golden pothos, 120
Gossypium, 52
GOTS. *See* Global Organic Textile Standard (GOTS)
GPF. *See* Groundnut protein fibre (GPF)
Gram-negative *Klebsiella pneumonia*, 299
Gram-positive *Staphylococcus aureus*, 299
Green composites, 219–221, 336–337
Green cotton. *See* Organic cotton
Greenhouse effect, 412

Index

Greenhouse gas emissions (GHG emissions), 11, 88, 160, 287–288, 337, 413, 418
Greige fabrics, 303
Gross Domestic Product (GDP), 333
Groundnut fibres, 215–218
  blending of groundnut fibre, 217
  chemical processing of groundnut fibre, 217–218
  production of, 216–217
    extraction of oil, 216
    extraction of protein, 216
    fibre formation, 217
    preparation of spinning solution, 216
    after treatments, 217
  properties of, 217
Groundnut protein fibre (GPF), 216
  uses of, 218
GRS. *See* Global Recycle Standard (GRS)
GWP. *See* Global warming potential (GWP)

H
Halogen-free flame retardant additives, 221
Harvesting, 61
Hazardous pollutants, 131
Heat techniques, 27–28
Hemp (*Cannabis sativa*), 4, 28, 38–40, 80–82, 91, 148, 392
  application of, 85–89
  clothing from, 82–85
  fibres, 338–339
  sustainability aspects of natural fabrics and knitting from, 78–89
Hennes & Mauritz (H&M), 57
Herbicides, 36–37
Hermes (clothing and accessory brand), 143
Hesperidin, 273–274
Higg index, 160
Higg MSI, 418
Himalayan poplar (*Populus ciliata*), 120
Home laundry process, 386
Hot-air sterilization, 12
Household goods, 229
Hugo Boss, 155
Hurd core, 338–339
Hydrogel, 173
Hydrogen peroxide bleaching, 213
Hydrolysis-alkalization process, 229–230
Hydroxyl groups (OH groups), 102

I
IKEA, 320–321
Imperial Chemical Industries (ICI), 248
In vitro cultivation of mushroom leather, 140
Incineration, 33
Indian brands, 143
Industrial hemp, 80–81
Ingeo fibre. *See* Corn fibre
Insects, 42
Integrated pest control (IPC), 37–38
Intercultural operations, 60
International Coffee Organization (ICO), 288
International Federation of Organic Agricultural Movements (IFOAM), 37, 54, 64–65
International Labour Organization conventions (ILO conventions), 370
International Oeko-Tex association, 369
Internationale Verband der Naturtextil wirtschaft (IVN), 376–377
Ironing, 203
Irradiation, 99
Irrigation, 60
ISO Certification, 356

J
Japanese Agricultural Standard (JAS), 65
Jute, 3–4
  fibre, 339

K
Kapok fibre, 341
Kenaf, 28
  fibre, 338
Keratin, 167–168, 173, 184, 195
  fibres, 183
Knit(ting), 83, 91
  fabrics, 214, 301
  for fashion, 75–78
  from flax, hemp, ramie, curaua, bamboo, pineapple fibres, sustainability aspects of natural fabrics and, 78–89
  technique, 301

L
Lama, 187
Lamb's wool, 188

Lead (Pb), 81
Leather
  apparel, 132
  manufacturing process, 131, 138
Lenpur fibre, 42
Life cycle assessment (LCA), 160, 403–404, 408
  application areas, 427
  of caprolactam fibres, 264
  carbon footprint, classification & related parameters, 413–419
  environmental impact, carbon emissions & ardent need of, 405–413
  framework methodology, 419–428
    application areas of life cycle analysis LCA, 427
    demerits & challenges associated with LCA, 427–428
    different phases of a product's life cycle, 419–420
    LCA classification, 420–423
    merits associated with LCA implementation in textile supply chain, 424–427
    phases of life cycle assessment LCA, 423–424
    prerequisites for LCA model, 420–423
    social & environmental labels on basis of LCA, 419
  of Indian silk, 171
  of silk, 170–172
  study of Brazilian silk, 172
  of sustainable fibres, 255–256
    effluent treatment, 255–256
  of wool, 162–165
    energy and CO_2 emission of wool and polyester fibre, 163t
Life cycle impact assessment phase (LCIA), 423–424
Life cycle interpretation, 424
Life cycle inventory (LCI), 421, 423
Lignocellulosic coir fibre, 119–120
Lime, 12
Limonene, 282–283
Limonoids, 273–274
Linen, 29, 79, 91
  clothing from, 82–85
  fibres, 339

Linum usitatissimum L. *See* Flax (*Linum usitatissimum* L.)
Liquid ammonia methods, 84
Lithium bromide (LiBr), 170
Long staple fibres, 52
Lotus (*Nelumbo nucifera*), 110–111, 121
  cultivation, 96
    classification of lotus plant based on DNA markers, 96t
    product development of lotus fibre, 97f
  farmers, 95
  peduncles, 99
  plant, 95
  propagation, 96
  robe, 103–104
Lotus fibre, 111
  application of lotus fibre for commercial product, 103–104
    lotus textile merchandizing around world, 104t
  biochemical extraction, 99
  biological extraction, 98
  chemical analysis of lotus fibre, 101–102
    chemical composition of lotus fibre, 101
    chemical structure, 102
    crystal structure and crystallinity, 101–102
  chemical extraction, 99
  comparison of lotus fibre with cotton fibre, 102–103
  drawing, 96–99
  fabric, 103–104
  fibre drawing from lotus, 97f
  fibre physical properties, 99–100
    fibre moisture, 99
    physico-mechanical properties, 100
  irradiation, 99
  lotus cultivation, 96
  lotus fibre production process, 97t
  lotus inspired design culture, 105–106
  manual drawing of lotus fibres, 98
  mechanical extraction, 98
  production process, 96
  retting, 99
  steam explosion, 99
Lotus inspired design culture, 105–106
Louis Vuiton (fashion luxury brand), 154
Low chemical cotton (LCC), 16, 37–38
Low water cotton, 16, 37–38

Index 443

Lululemon, 143–144
Luxe cotton, 380
LUXTRA London (Australian luxury brand), 155
Lycra, 236–237
Lyocell, 6, 234–242
  fabric, 295–296
  fibres, 41, 236
  manufacturing process of lyocell, 236–242, 236t
  origin of, 236
Lysine, 195

M
Macrophyte fibres, 121
  chemical composition, 121–123
  classification of wetland plants and, 110–120
    coir, 119–120
    lotus fibre, 111
    milkweed, 115
    money plant, 120
    munja, 115–118
    poplar, 120
    reed, 119
    rice, 119
    typha, 114
    water hyacinth, 111–113
    water lily, 113
    wild cane, 118–119
  fibre morphology, 121
  physicomechanical properties, 121
Made-By, 356
Man-made cellulosic fibre clothes, 403
Mannans, 290
Manufacturing process, 14–15
  of caprolactam or nylon fibre, 257–261
    caprolactam, 257–259
    ECONYL®, 261
    nylon 6, 259–260
    nylon 6,6, 260–261
  of lyocell, 236–242
    chemical properties, 242
    comfort properties, 241–242
    cross-linked lyocell, 240
    dissolving cellulose, 237
    drying and finishing, 238
    durability properties, 240–241, 241t
    environmental impact of lyocell, 242

fibrillation, 240
filtering, 238
lyocell cross-sectional view, 238, 239t
lyocell longitudinal view, 239
moisture related properties, 241
pore structure, 239–240
properties of lyocell, 240
spinning, 238
standard lyocell, 240
thermal comfort properties, 241
types of lyocell, 240
uses of lyocell, 242
washing, 238
of nylon, 323–325
  recycle and utilize nylons or tights, 324
  recycle nylon carpet, 318–319
  recycle or utilize nylon fabric, 324
  recycled nylon, 323
  recycling nylon sweet for earth, 325
  recycling of nylon baggage, 324
of polyester, 313–317
  chemical, 314
  mechanical, 313–314
  methods of textile recycling, 313
  recycled polyester yarn process, 316–317
of polyester fibre, 250–252
  conventional process, 250–252
of regenerated bamboo fibre, 231–232
  ageing, 231
  dissolving, 231
  preparation, 231
  pressing, 231
  shredding, 231
  spinning, 232
  steeping, 231
  sulfurization, 231
  xanthation, 231
Material sustainability index (MSI), 403
MBF. *See* Multi-bore hollow fibre (MBF)
MBR. *See* Membrane bioreactor (MBR)
Mechanical extraction, 98
Mechanical process, 16
  of polyester, 313–314
Medium wool, 166–167
Melanins, 197
Melt-spinning method, 247
Melting method, 317
Membrane bioreactor (MBR), 264

Mercury, 131
Metal oxides, 221
Metal-organic Frameworks (MOF), 84, 88
Methane (CH_4), 11, 40
  emissions, 407
Methanolysis, 396
Methionine, 195
Methylmorpholine-N-oxide hydrolyse
  cellulosic fibres, 35
MHET. *See* Mono(2-hydroxyethyl)
  terephthalic acid (MHET)
Micro irrigation, 37–38
Micro-organisms, 254
Micro-plastics, 30, 319
Microencapsulation, 275–276
Microfibers, 310
Milk, 184
  fibre, 202
  protein fibre, 41–42
Milkweed (*Calotropis gigantea*), 110, 115
  fibres, 115
    chemical constituents of macrophyte and
      wetland plant fibres, 117t
    physico-mechanical properties of
      macrophyte and wetland plant fibres,
      116t
  seed fibres, 121
MOF. *See* Metal-organic Frameworks
  (MOF)
Mohair, 28, 182, 186
Moisture regains of polyester fibre, 253
Molecular weights (MW), 207
Money plant (*Epipremnum aureum*), 110, 120
  fibres, 120
Mono(2-hydroxyethyl) terephthalic acid
  (MHET), 35
Mori silk, 168
Moriculture, 168–169
Morus alba, 168–169
Moulds, 42, 134
MSI. *See* Material sustainability index (MSI)
Muga (*Antheraea assamensis*), 183
Mulberry
  plantation, 168
  silk, 168, 183
Multi-bore hollow fibre (MBF), 105–106
Multivoltine (MV), 169
Munja (*Saccharum munja*), 110
  fibre, 115–118

Mushroom fibre. *See also* Corn fibre
  detrimental impact of textile and fashion
    supply on environment, 129–133
    animal leather for fashion
      ensembles—nonconformity of
      sustainable principles, 130–133
    fast fashion, 129–130
  eco leather/environmentally preferred
    leather, 133–134
  mycelium and mushroom leather, 134–148
Mushroom leather, 133–148
  challenges for mass adoption of mushroom
    leather, 147–148
  making process, 138–140
  preferred choice of sustainable fashion
    brands, 142–147
  production of, 135–137
  properties and benefits of mushroom leather
    over animal leather, 137–139
    comparison of mushroom leather with
      animal leather, 138t
    mushroom leather outshines animal
      leather, 138–139
  raw materials for, 134–135
  sustainable and beneficial attributes of
    mushroom, 140–142
Mushrooms, 134, 140
MuSkin, 134
MV. *See* Multivoltine (MV)
MW. *See* Molecular weights (MW)
Mycelium, 134–148
  mycelium-based biocomposites,
    141–142
  patterns, 145
MYCL. *See* Mycotech Lab (MYCL)
Myco Works, 135
Mycotech Lab (MYCL), 146
MycoTEX, 145
MycoWorks, 145–146
Mylo, 134, 143–144

N
N-methylmorpholine N-oxide (NMMO),
  235
Nanotechnology, 275–276
National obligatory standards for organic
  cotton and organic cotton certifiers,
  64–66
  AOS, 65

Index

European Union Regulation 2092/91, regulation of organic cotton, 65
IFOAM, 64—65
JAS, 65
organic exchange, 66
Native Shoes, 155
Natural cellulose-based fibres
aspects of biodegradability of, 337—341
bamboo, 341
coconut fibre, 341
cotton, 338
flax, 339
hemp, 338—339
jute, 339
kapok, 341
kenaf, 338
nettle, 340
pineapple leaf, 340
ramie, 339—340
sisal, 341
Natural colours, 172—173
Natural dyeing, 166
Natural dyes, 393
extract of spent coffee ground as, 299
Natural fabrics and knitting from flax, hemp, ramie, curaua, bamboo, pineapple fibres, sustainability aspects of, 78—89
Natural fibres, 4—7, 28—29, 44, 75—79, 84—85, 210, 236—237, 273, 297—298, 333, 342, 392
natural fibre-based composites, 345—346
natural fibre-reinforced polymer matrix composites, 333
opportunities and challenges, 342—343
as reinforcement for composites materials, 342
Natural fibrous plants, 77
Natural manure, 39
Natural plant fibres, 299—300
Natural protein fibres, 206
Natural rubber, 297—298
Natural textiles
fibres, 3
recycling of natural textiles as sustainable solution, 89—90
Naturtextil, 376—377
Neem (*Azadirachta indica*), 166
Negative carbon footprints, 142—143

Nelumbo nucifera. *See* Aquatic lotus (*Nelumbo nucifera*); Lotus (*Nelumbo nucifera*)
Nettle fibre, 340
Nettle plants, 340
NIKE, 321
NMMO. *See* N-methylmorpholine N-oxide (NMMO)
Non-biodegradable synthetic polymer fibres, 44
Non-biodegradable wastes, 342—343
Non-hazardous waste production, 15
Non-Renewable Energy Use (NREU), 242
Nonwoven technique, 301
NOP. *See* US National Organic Standards (NOP)
NREU. *See* Non-Renewable Energy Use (NREU)
NTCWV. *See* United States' National Technical Committeefor Wetland Vegetation (NTCWV)
Nylon, 4, 6—7, 13—14, 287, 310, 321—329. *See also* Polyester
advantages, 326
applications, 326
conventional nylon, 321
disadvantages, 327
manufacturing process, 323—325
properties, 325
recycled nylon, 321—322
recycling of nylon baggage, 324
sources, 322
sustainability, 327—329
brands promoting recycled nylon, 328—329
Nylon 6, 259—260
Nylon 6,6, 260—261
Nylon fibres, 13—14, 247—248, 257—265
application of sustainable caprolactam fibres, 262—263
characteristics of nylon, 261
chemical property, 262
alkali and acids, 262
electrical and flammability, 262
light and biological impact, 262
organic solvent and bleaches, 262
dyeing and finishing, 263—264
LCA of caprolactam fibres, 264
manufacturing process, 257—261

Nylon fibres (*Continued*)
  physical property, 261–262
    heat effect, 262
    tenacity, density and moisture regain -, 261
  properties, 261–262
  recovery of caprolactam, 265
  recycling, 264–265
Nylon Waste Recycling Market report, 310

O
OCS. *See* Organic Content Standard (OCS)
OCS blended standard, 376
ODOP. *See* One District One Product (ODOP)
OE. *See* Organic Exchange (OE)
Oeko-Tex association, 369–371
  Oeko-Tex standard 1000, 370
  SA8000 certified, 370–371
Oil extraction
  of groundnut fibres, 216
  of sustainable soybean fibre, 209
One District One Product (ODOP), 115–118
One% for the Planet, 357
Optical microscopy, 279
Orange fibre, 30, 273
  anti-microbial efficacy of, 282–283
  benefits of textiles made of orange peel extracts, 283–284
  burning behaviour of, 280
  chemical composition of, 279–280
  fibre extraction method, 277–278
  fibre morphology and properties, 279
  FTIR spectroscopy, 281–282
  moisture absorbency behaviour of, 280, 281f
    burning behaviour of, 280t
    solubility of orange peel fibre, 281t
  orange fruit, 274
  orange peel waste as textile raw material, 274–276
  preparation of film from orange peel extracts, 278–279
  solubility behaviour of, 280
  structure and chemical composition of orange peel, 276–277
  thermal characterization of, 282
Orange fruit, 274

Orange peel extracts
  benefits of textiles made of, 283–284
    textiles made of orange peel fibres, 284f
  fibres, 279
  preparation of film from, 278–279, 278f
Orange peels, 16, 277
  structure and chemical composition of, 276–277
    chemical composition of citrus peel, 276t
    structure of flavonoid, 277f
  waste as textile raw material, 274–276
    Amazon rainforest, 275f
    orange peel waste into textiles, 275f–276f
Organic agriculture, 54, 67
Organic Content Standard (OCS), 373, 419
  OCS-100, 375–376
Organic cotton, 10, 16, 51, 54–66
  application of organic cotton, 67–68
  arguments against expanding organic cotton, 63
    processing, 63
    production, 63
    retail, 63
  arguments in favour of expanding organic cotton, 62–63
    processing, 62
    production, 62
    retail, 62–63
  BCI, 66–68
  comparison of organic cotton with conventional cotton, 55–56
  cotton fibre, 52–53
  cultivation, 36–37, 51
  current status of, 57–58
  farming, 51
  fibre, 36–37
    average production of various natural fibres per hectare of land, 37t
  future prospects of organic cotton market, 62–64
  growing organic cotton, 59–61
  history of organic cotton production, 56–57
  logo, 66
  national obligatory standards, 64–66
    for organic cotton and organic cotton certifiers, 64–66
  opportunities for organic cotton, 63–64

Index 447

processing, 64
production, 63
retail, 64
production, 56
history of, 56–57
systems, 55–56, 58
standard production practices, 58–59
sustainable organic cotton, 61–62
Organic Exchange (OE), 66, 372
Organic fabrics, 70
Organic Farming, 65
Organic fibre, 184
Organic ingredients, 39
Organic silk, 194
Organic solvent and bleaches, 262
Organic spun silk, 194–195
Organic textile fibres, 185
Organic wastes, 31, 273
Organic wool, 38, 184–185
Organizations, 355
Organza, 12
Oryza sativa. See Rice (*Oryza sativa*)
Osmotic degumming method, 80
Oxygen (O$_2$), 4, 88
Oyster mushrooms, 134

P
P4DRY, 295
Pad-dry-cure technique, 299
Paio (Indian brand), 143
Pasteurization methods, 135–136
Patagonia, 320
PBA. *See* Polyhydroxybutyrate (PBA)
PBAT. *See* Polybutylene adipate
terephthalate (PBAT)
PBT. *See* Polybutylene terephthalate (PBT)
PCDT. *See* Poly-1,4-cyclohexylene-
dimethylene terephthalate (PCDT)
PE. *See* Polyethylene (PE)
Peace silk, 170
Peel waste, 274
Pelletized plastic, 317
Penicillium, 166
Pest control, 60
Pesticides, 36–37
PET. *See* Polyethylene terephthalate (PET);
Polythene terephthalate (PET)
PETA standard, 370
Petrochemical-derived fibres, 273

Petroleum, 342
PHA. *See* Polyhydroxyalkanoates (PHA)
Phellinus ellipsoideus, 134, 144
Philosamia ricini. See Eri (*Philosamia
ricini*)
Phosphoric acid, 35
Photoreforming, 90
Phragmites australis. See Reed plants
(*Phragmites australis*)
Pinatex material, 382
Pineapple, 4, 148
leaf fibres, 340
sustainability aspects of natural fabrics and
knitting from pineapple fibres, 78–89
PLA. *See* Polylactic acid (PLA)
Plant by-product straw, 119
Plant fibres, 6, 28–29, 78–79, 249–250,
342
Plant peduncles, 99
Plant-based natural fibres, 4
Plant-based polyester, 249–250
Plant-based textile fibres, 337
Plasma methods, 84
Plasma treatment, 36
Plastic bottle, 313–314
Plastic waste, 288
Poly-1,4-cyclohexylene-dimethylene
terephthalate (PCDT), 249
Poly(methyl methacrylate), 255–256
Polyamides, 7, 287
Polybutylene adipate terephthalate (PBAT),
298
Polybutylene terephthalate (PBT), 251
Polyester, 4, 6–7, 13, 16, 38–40, 236–237,
248, 254, 292–293, 310–321. *See
also* Nylon
advantages, 318–319
keeping plastics from attending to
lowland and ocean, 319
RPET, 319
applications, 317–318
disadvantages, 319–320
fabrics, 13
filament, 249, 294
manufacturing process, 313–317
polyester-cellulosic blends, 255
polyester-polyurethane, 255–256
properties, 317
recycled polyester, 311–313

Index

Polyester (*Continued*)
  recycling, 35—36, 287—288
  sources, 313
  sustainability, 320—321
    brands promoting RPET, 320—321
  virgin polyester, 310—311
  waste, 35
Polyester fibres, 3, 28, 247—257, 287—288, 291—292, 294
  applications of sustainable PET fibres, 254
  characteristics of, 253
  chemical properties, 253—254
    effect of alkalis, 253
    colour fastness, 254
    effects of acids, 253
    effects of solvents, 254
    sunlight and micro-organisms, 254
  classification, 248
  dyeing and finishing, 254—255
  impact on environment, 256
  LCA of sustainable fibres, 255—256
  manufacturing process, 250—252
  physical properties, 253
    heat effect, 253
    moisture regains %, specific gravity and tenacity, 253
  properties, 253—254
  recycling, 256—257
  types, 249—250
    ethylene polyester, 249
    fibrefill, 249
    PCDT polyester, 249
    plant-based polyester, 249—250
Polyethylene (PE), 287
Polyethylene terephthalate (PET), 13, 146—147, 248, 291—292, 310—312
Polyhydroxyalkanoates (PHA), 80
Polyhydroxybutyrate (PBA), 80
Polylactic acid (PLA), 4, 6, 40, 150, 336—337
  fibre, 40—41
Polylactide. *See* Polylactic acid (PLA)
Polymer, 251
Polymerization, 4—6, 15
Polyoxymetalates (POM), 84
Polypropylene (PP), 287, 297—298
Polysaccharides, 288
  in SCGs, 290
Polystyrene, 297—298
Polythene terephthalate (PET), 314

Polytrimethylene terephthalate (PTT), 251—252
Polyurethane (PU), 295
Polyvinyl alcohol (PVA), 206
POM. *See* Polyoxymetalates (POM)
Poplar (*Populus*), 110
  colour values of aquatic macrophyte and wetland plant fibres, 122t
  fibres, 120
Populus ciliata. See Himalayan poplar (*Populus ciliata*)
Porphyrins, 198
Post-consumer waste, 313
  material, 292
PP. *See* Polypropylene (PP)
Pre-consumer waste, 313
Primary polymer ethylene, 249—250
Principle of care, 54
Principle of ecology, 54
Principle of fairness, 54
Principle of health, 54
Product life cycle, 404
Production process of mushroom, 140—141
Proline, 195
Propane-1,3-diol (PDO), 251—252
Protein extraction
  of groundnut fibres, 216
  of sustainable soybean fibre, 209—210
Protein fibres, 167—168
Proteolytic enzymes, 174
Pseudomonas sp., 264
PTT. *See* Polytrimethylene terephthalate (PTT)
PU. *See* Polyurethane (PU)
Pupa, 169
PVA. *See* Polyvinyl alcohol (PVA)

Q
Qmilk, 30
Qualitative methods, 404
Quantitative methods, 404

R
r-PET. *See* Recycled PET (r-PET)
Radiation treatment, 36
Rainfall, 39
Rainfed cotton, 16
Ramie
  fibre, 339—340

Index 449

sustainability aspects of natural fabrics and knitting from, 78–89
Raw materials, 247–248
of mushroom leather, 137–138
Rayon, 4, 14, 274–275, 406–407
RDS. *See* Responsible Down Standard (RDS)
Recover products, 35
Recovered fibres, 35
Recron Green Gold, 321
Recycle, reuse and repair (3R), 33–34
Recycle and utilize nylons or tights, 324
Recycle nylon carpet, 318–319
Recycle or utilize nylon fabric, 324
Recycle products, 35
Recycled Claim Standard (RCS-100), 375
Recycled fashion waste, 390
Recycled fibres
nylon, 321–329
polyester, 310–321
textile recycling, 310
Recycled materials, 249
Recycled nylon, 321–323, 326
Recycled PET (r-PET), 249–250, 287–288, 291–293, 311–312
brands promoting RPET, 320–321
everlane, 320
IKEA, 320–321
NIKE, 321
patagonia, 320
recron green gold, 321
different types of fiber, 291f
different types of polyester fiber, 292f
limitations of, 292–293
material and energy, 293t
virgin polyester, 319
Recycled polyester, 249, 311–313, 319, 395
challenge, 309–310
yarn process, 316–317
cleaning, 316
crude sorting method, 316
melting, 317
shredding, 316–317
sorting method, 316
Recycled polypropylene, 297–298
Recycled technology, 309
Recycling, 32–33
of caprolactam or nylon fibre, 264–265
methods, 391, 394–397
biological methods, 396

chemical methods, 395–396
thermal method, 395
of natural textiles as sustainable solution, 89–90
of nylon baggage, 324
orange peel waste, 284
of plastics, 16
process, 256–257, 310, 326, 420
sustainability by, 33–36
Reduction bleaching, 214
Reebok, 154–155
Reed fibres, 119
Reed plants (*Phragmites australis*), 110, 119
Reeling, 168, 170
Refibra cotton, 378
Regenerated bamboo fibre, 232
manufacturing process of, 231–232
Regenerated cellulosic fibres, 30
Regenerated fibres, 6, 28
Regenerated protein fibres, 206, 215
Reishi mushroom, 135
Reisi, 135
Renewability, 4
Renewable corn starch polymer, 153
Renewable Energy Use (REU), 242
Renewable fibres, 28, 398, 406–407
Renewable resources, 148
Renewable sources, 40
Rens, coffee sneakers from, 297
Repreve fibre, 378
Resin systems, 333
Resistance to compression, 189
Responsible Down Standard (RDS), 374–375
Responsible Wool Standard (RWS), 373–374
Retting, 79–80, 99
physical characteristics of lotus fibre reported in researches, 100t
REU. *See* Renewable Energy Use (REU)
Rhizomes, 96
Rice (*Oryza sativa*), 110, 119
fibres, 119
husk, 119
Ring spinning technique, 300
Riparian plants, 110
Roving, 83, 91
Rutin, 273–274
RWS. *See* Responsible Wool Standard (RWS)

S

S café spun yarn spun, 294
S. Café AIRNEST, 295
S. Café mylithe, 296
Saccharum bengalense fibres, 118–119
Saccharum munja. See Munja (*Saccharum munja*)
Saccharum munja grass (SMG), 115–118, 123–124
Saccharum spontaneum. See Wild cane (*Saccharum spontaneum*)
SAI aspects. *See* Social accountability international aspects (SAI aspects)
Saris (traditional attire worn by woman in India), 173
Sawdust, 135
Scaffold, 173
Scanning electron microscopy (SEM), 100, 115, 207–208, 273–274
SCG. *See* Spent coffee grounds (SCG)
Scoured wool fibre, 161
SeaCell fibre, 42
Seaweed, 42
Second-hand clothing, 391–392
Secreted fibres, 183
Sedex, 368
Sefia fabric, 295–296
Self-standing silk fibroin thin films, 173
SEM. *See* Scanning electron microscopy (SEM)
Semi-cellulosic chemicals, 99
Semi-products, 83, 91
Semi-rigid composites of polyurethane foams (SRPUF), 221
Semi-worsted yarns, 161–162
Sericin, 170
Sericulture, 168–170
  life cycle of mulberry silk, 169f
SETAC. *See* Society of Environmental Toxicology and Chemistry (SETAC)
SFTMS. *See* Sustainable Fair Trade Management System (SFTMS)
Sheep, 160, 187
  wool, 167
Short-staple coarse wool, 166–167
Shredding method, 316–317
Shrimps, 42

Shrinkage, 166
Silane-treated coffee husk fibre (T-CH), 298
Silk, 4, 12–13, 28–29, 160, 168–174, 182, 206, 210
  blend with bamboo fibres and, 202
  degumming process, 171
  fabrics, 12
  fibroin, 173
  LCA of, 170–172
  production, 168–170
    moriculture, 168–169
    reeling, degumming, and fabric making, 170
    sericulture, 169–170
  satin, 12
  sustainability in silk processing, 172–173
  utilization of silk as sustainable material, 173–174
Silkworm, 29, 170
  breeding, 193
Silver vines, 120
SimPro software, 171
Singtex, 293–294
Sisal, 4
  fibre, 341
Sludge volume index (SVI), 264
SMG. *See* Saccharum munja grass (SMG)
Sneature (sock shoe), 145
Social accountability international aspects (SAI aspects), 370–371
Society of Environmental Toxicology and Chemistry (SETAC), 404
Sodium bicarbonate, 161
Sodium chloride, 131
Sodium hydroxide (NaOH), 10, 14
Soft fibers, 79
Soil, 60
  soil-borne diseases, 338–339
Solvents, effect of, 254
Sorting method, 316
Sound organic cotton production packages, 58–59
Sources
  nylon, 322
  polyester, 313
Soy protein isolate (SPI), 215
Soy silk, 205

Soybean (*Glycine max*), 205–206
 fibre, 182, 205–206, 209, 368–369
 proteins, 207–208
 soybean/cashmere blended fabric, 212
 soybean/cotton blended fabric, 212
 soybean/lycra blended fabric, 212
 soybean/silk blended fabric, 212
 soybean/synthetic blended fabric, 212
 soybean/wool blended fabric, 212
Soybean protein fibres (SPF), 41, 203
Spandex, 38–40
Specific gravity of polyester fibre, 253
Spent coffee grounds (SCG), 288–290, 295
 extract of spent coffee ground, 299
SPF. *See* Soybean protein fibres (SPF)
SPI. *See* Soy protein isolate (SPI)
Spin-draw process, 251
Spinning
 preparation of spinning solution, 210
 process, 35–36, 214, 411
Sponge, 173
Sprinkler irrigation, 37–38
SRPUF. *See* Semi-rigid composites of
 polyurethane foams (SRPUF)
Stachybotrys species, 166
Staple fibre, 242
Steam explosion, 39, 99
Steaming, 12
Steel, 78
Steeping process, 103–104
Stella McCartney (clothing and accessory
 brand), 143
Stretching process, 100
Substrate dampening process, 135–136
Sucrose, 35
Sugar beet, 40
Sugarcane, 40
Sulfur dioxide (SO$_2$), 15
Sulphate, 131
Sulphide, 131
Sulphur-containing amino acids, 207
Sun drying, 12
Sunlight, 254
Supima cotton, 378
Surfactants, 238
Sustainability, 4, 159, 181, 233, 249, 387
 approach, 388–391
 aspects of natural fabrics and knitting from
 flax, hemp, ramie, curaua, bamboo,
 pineapple fibres, 78–89

 application of flax and hemp, 85–89
 clothing from linen and hemp, 82–85
 properties of bast fibres, 85
 assessment studies, 170–171
 in building, 167
 of coffee fabric manufacturing, 299–303
 nylon, 327–329
 polyester, 320–321
 by recycling, 33–36
 in silk processing, 172–173
 textile fibres, 28–42
Sustainable approach, 129–130
Sustainable caprolactam fibres, application
 of, 262–263
Sustainable designing, 43–44
Sustainable development, 181
Sustainable Development Goals, 335
Sustainable Fair Trade Management System
 (SFTMS), 365–366
Sustainable fashion brands, 154–155
 mushroom leather—preferred choice of,
 142–147
Sustainable fibres, 338–339
 fair labour schemes and initiatives, 371–378
 business social compliance, 371–372
 content claim standard, 377–378
 fair labor association, 372
 fair wear foundation, 372
 naturtextil, 376–377
 OCS blended—organic content standard,
 376
 organic content standard, 375–376
 textile exchange, 372–375
 for fashion and textiles
 consumer behaviour and sustainability,
 42–43
 sustainable designing, 43–44
 textile fibres-environmental impacts and
 sustainability, 28–42
 key sustainability organizations and
 certifications, 355–371
 American forests, 355
 Better Cotton Initiative, 365
 bluesign, 362
 canopy, 359
 carbon neutral, 360
 centre for advancement of garment
 making, 357
 clean clothes campaign, 356
 climate neutral, 360

452

Index

Sustainable fibres (*Continued*)
  CMiA, 362—363
  cradle to cradle, 363—364
  EU eco label, 365
  fabscrap, 360—361
  Fair Trade, 366
  Fairtrade Certified Cotton, 366—367
  Fashion Forum Fellowship 500, 358
  Fashion Revolution, 358
  Fashion Transparency Index, 359
  global organic textile standard, 368—369
  global recycle standard, 367
  Good on You, 359
  ISO Certification, 356
  made-by, 356
  oeko-tex, 369—371
  one% for the planet, 357
  PETA standard, 370
  sedex, 368
  SGS, 368
  sustainable fair trade management
    system, 365—366
  upcycling certification, 361
  LCA of, 255—256
  sustainable textile fibres and fabric
    materials, 378—382
Sustainable natural fibres, 36—42
Sustainable organic cotton, 61—62
Sustainable PET fibres, applications of, 254
Sustainable polyester
  caprolactam or nylon fibre, 257—265
  polyester fibre, 248—257
Sustainable principles, animal leather for
    fashion ensembles—nonconformity
    of, 130—133
Sustainable protein fibres
  animal protein fibres, 182—203
  green composites, 219—221
  vegetable protein fibres, 203—219
Sustainable resolution projects (Su. Re
    projects), 115
Sustainable silk fibre, 191—195
  organic silk, 194
  organic spun silk, 194—195
Sustainable soybean fibre, 204—215
  after-treatments, 210
  baby wear, 215
  biomedical, 215
  blending of soybean fibre, 212—213

  chemical processing of SPF, 213—214
  extraction of oil, 209
  extraction of protein, 209—210
  fibre formation, 210
  knit fabric, 214
  preparation of spinning solution, 210
  production of soybean fibre, 209
  properties of soybean fibre, 210—212
  uses of SPF, 214
  woven fabric, 215
  yarn, 214
Sustainable textile fibres, 378—382
Sustainable textile materials
  clothing, 386
  consumer behaviour, 386—388
  direction, 397—398
  dyeing, 393—394
  ecolabel, 397
  fibres, 392—393
    blended fibres, 392—393
  recycling, 391
  recycling methods, 394—397, 394t
  second-hand clothing, 391—392
  sustainability approach, 388—391
Sustainable wool fibres, 184—185
  organic wool, 185
Sustainable wool processing, recent
    development in, 166
SVI. *See* Sludge volume index (SVI)
Sylvania, 145—146
Synthetic fabric, 30
Synthetic fibres, 3—7, 236—237, 247, 309,
    333—334, 342—343
Synthetic petroleum-based fibre, 256
Synthetic textiles, 310
Synthetic-based polymers, 337

T
Take-make-dispose model, 335
Tanning process, 132
Tasar (Antheraea mylitta), 183, 194
Tasc fabric, 382
Tawing process, 132
TE. *See* Textile exchange (TE)
Teijin's recycling process,
    287—288
Tenacity of polyester fibre, 253
Tencel (Lyocell fibre brand name), 6,
    234—235

Tencel A100, 240
Tencel LF, 240
Terephthalic acid (TPA), 35, 250, 395—396
Terylene, 3
Tetra hydrocanabinol (THC), 80—81
Textile business, 309
Textile composites, 333, 335—336
Textile exchange (TE), 367, 372—375
  Organic Content Standard, 373
  Recycled Claim Standard, 375
  Responsible Down Standard, 374—375
  Responsible Wool Standard, 373—374
Textile fabrics, 287—288
Textile fibres, 3—9, 28—42, 287
  environmental impacts, 30—33
    fibre water consumption, 32f
    impacts of cotton farming, 30—31
    of textile fibre production, 9—16
  global market of textile fibres, 30
  sources, 28—30
  summary of impacts of textile fibre
    production, 15—16
  environmental impacts of textile fibres in
    decreasing order, 16t
  summary of environmental impacts of
    major textile fibres, 17t—20t
  sustainability by recycling, 33—36
    cellulosic fibres recycling, 35
    intertwined sustainability challenges, 34f
    polyester recycling, 35—36
  sustainable natural fibres, 36—42
    bamboo fibre, 39—40
    chitin and chitosan fibre, 42
    flax, 39
    hemp, 38—39
    Lenpur fibre, 42
    low water and low chemical cotton,
      37—38
    lyocell fibre, 41
    milk protein or casein fibre,
      41—42
    organic cotton, 36—37
    organic wool, 38
    polylactic acid fibre,
      40—41
    seacell fibre, 42
    soybean protein fibre, 41
    wild silk, 38
  types, 4—7

environmental factors involved in
  production of cotton and polyester
  fibres, 6t
natural and synthetic textile fibres, 5t
usage, 7—9
wool reuse and sustainability, 36
Textile industry, 27, 129, 287, 355, 388, 403,
  405—406
Textile manufacturers, 291—292
Textile processing, 166
Textile products, 27, 309
Textile raw materials
  environmental impact, 406—409
    of different fibres, 407t
    GWP of different fibres, 408t
  orange peel waste as, 274—276
Textile recycling, 310
  methods of, 313
Textile supply chain
  boon for designers & brands, 425—427
  merits associated with LCA implementation
    in, 424—427
  open & closed loop recycling for LCA,
    425—427
Textile system, 160
Textile waste, 247, 333—335
  type of, 334t
TGA. See Thermogravimetry analysis
  (TGA)
THC. See Tetra hydrocanabinol (THC)
Thermal method, 394—395
Thermal recycling method, 35
Thermogravimetry analysis (TGA),
  273—274
Thermoplastic polymers, 395
Thermosol, 254
Three-dimensional linked fabrics (3D linked
  fabrics), 84
Timber crop, 120
Tissue engineering, 160
TLF. See Typha leaf fibres (TLF)
Toxic emissions, 388
TPA. See Terephthalic acid (TPA)
Traditional fibres
  environmental impacts of textile fibre
    production, 9—16
    acrylic, 14—15
    cotton, 9—10
    nylon, 13—14

454   Index

Traditional fibres (*Continued*)
   polyester, 13
   rayon, 14
   silk, 12–13
   summary of impacts of textile fibre
     production, 15–16
   wool, 11–12
   textile fibres, 4–9
   textile raw materials for environmental
     sustainability, 21t
Traditional footwear making process, 154
Traditional leather brands, 143
Traditional linear model, 385
Transported waste, 274
Treatment methods, 135–136
Trimethylene glycol, 251–252
Tropical climates, 89
TSF. *See* Typha seed fibres (TSF)
Tubes, 173
Tufting technique, 301
Turkey feather barbs, 197
Typha (*Typha domingensis*), 114
   fibres, 114
Typha leaf fibres (TLF), 114
Typha seed fibres (TSF), 114

U

Ultimate tensile strength (UTS), 221
Ultra-violet radiation (UV radiation), 230,
   294, 379, 391
Ultrasound methods, 84
Ultrasound-assisted dyeing, 173
Ultrasound-assisted scouring of wool,
   166
Ultraviolet protection factor (UPF),
   294
United Nations conventions (UN
   conventions), 370
United Nations Sustainable Development
   Goals (UNSDG), 372–373
United States (US), 339
United States' National Technical
   Committeefor Wetland Vegetation
   (NTCWV), 110
Upcycling certification guarantees,
   360–361
US Department of Agriculture (USDA),
   37
US National Organic Standards (NOP), 64

V

Vegetable fibres, 78–79
Vegetable protein fibres, 182, 203–219
   corn protein fibre, 218–219
   groundnut fibres, 215–218
   sustainable soybean fibre, 204–215
Veja (French sustainable footwear brand), 154
Vesicular arbuscular mycorrhizae (VAM),
   172–173
Veterinary medicines, 185
Vicara, 219
Vicugna pacos. See Alpaca (*Vicugna pacos*)
Virgin polyester, 310–311
   RPET simply nearly as good as, 319
Viscose, 30
   fibre, 14
Viscose rayon (VR), 368–369

W

w-aminocaproic acid, 257–258
Warm-cool feeling, 294–295
Washing, 203
Wastes, 16, 27–28, 344–345
   textiles, 343
Wastewaters, 255–256
Water hyacinths (*Eichhornia crassipes*),
   110–113
   fibres, 111–113
Water lily (Nymphaeaceae), 110, 113
   fibres, 113
Water purity, 123–124
Water resistance, 189
Water retting, 79–80
Weaving technique, 301, 411
Weed control, 60
Weeding, 60
Wet finishing processes, 334–335
Wet milling process, 150
Wet processing, 303
   of casein fibres, 200–201
     bleaching, 200
     carbonizing, 201
     crease resistant finishing, 201
     desizing, 200
     drying, 201
     dyeing, 200
     finishing, 201
     scouring, 200
     softening, 201

Index

Wet-spinning process, 206
Wetland plant fibres
  application of macrophytes in effluent
    treatment, 123—124
  chemical composition, 121—123
  classification of macrophyte and, 110—120
    aquatic/wetland plants according to US
      NTCWV definition, 110t
    coir, 119—120
    lotus fibre, 111
    macrophytes potential and general
      characteristics, 112t
    milkweed, 115
    money plant, 120
    munja, 115—118
    poplar, 120
    reed, 119
    rice, 119
    typha, 114
    water hyacinth, 111—113
    water lily, 113
    wild cane, 118—119
  fibre morphology, 121
  physicomechanical properties, 121
While wild silk, 183
White pine, 42
Wild cane (*Saccharum spontaneum*), 110,
    118—119
  fibres, 118—119
Wild silk, 38
Wood, 30, 78
Woody core, 338—339
Wool, 3—4, 7, 11—12, 28—29, 160—168,
    182—183, 206, 210, 236—237
  blend with cashmere and, 202
  climate change effect on, 165
  fibres, 11, 29, 36, 161
  LCA of, 162—165
  production, 161—162

prospective applications of wool as
    sustainable choice, 166—168
recent development in sustainable wool
    processing, 166
reuse, 36
scouring, 161
sustainability, 36
Wool-producing animal breeds, 187
Woolen yarn, 161—162
World fibre production, 76
Worsted yarn, 161—162
Woven fabrics, 215, 301

X

X-ray diffraction (XRD), 101—103
X-ray photoelectron spectroscopy (XPS),
    101
Xanthation process, 14, 40
Xpressole Panto, 297

Y

Yarns, 75—78, 83, 91, 214
  manufacturing process, 300
Yeasts, 42, 134
Yoga accessories, 143—144
Yoga bags, 143—144
Yoga mats, 143—144

Z

Zein, 219
Zero Discharge of Hazardous Waste
    (ZDHC), 362
Zero Grado Espace, 144
Zinc (Zn), 15, 81
ZQ Merino wool, 378
Zvnder (German-based mushroom leather
    accessory company), 146